I0069850

CATASTROPHIC INCIDENT
SEARCH AND RESCUE ADDENDUM

LAND SEARCH AND RESCUE MANUAL

Volume I: Land Search and Rescue Addendum

National Search and Rescue Committee

Volume II: Catastrophic Incident Search and Rescue Addendum

National Search and Rescue Committee

Volume III: Incident Command System Field Operations Guide for Search and Rescue

Robert J. Koester

Volume IV: Management Field Operations Guide for Search and Rescue

Robert J. Koester

Volume V: Tactical Field Operations Guide for Search and Rescue

Robert J. Koester

Volume VI: Endangered & Vulnerable Adults and Children

Robert J. Koester

CATASTROPHIC INCIDENT SEARCH AND RESCUE ADDENDUM

to the National Search and Rescue Supplement
to the International Aeronautical and Maritime
Search and Rescue Manual

Version 3.0 Illustrated

National Search and Rescue Committee
June 2012

Department of Homeland Security
Department of Interior
Department of Commerce
Department of Defense
Department of Transportation
National Aeronautics and Space Administration
Federal Communications Commission

dbS

dbS Productions
Charlottesville, Virginia

Copyright © 2013 dbS Productions LLC

No claim is made on original US. Government works.

All other Rights Reserved. No part of this book may be reproduced in any form or by electronic means including storage and retrieval systems without permission or license in writing from the publisher except by a reviewer who may quote brief passages in a review. For information contact dbS Productions.

All photographs and illustrations in this work are from the U.S. Government. Use of the NSARC logo and logos of other federal and non-governmental agencies involved in SAR are used for educational and informational purposes, as incorporated within the body of original US government work and are not intended to state or imply any endorsement of this publication or publisher. Use of NSARC and USCG logos is used with written permission.

Cover photo: Photograph by Bob McMillan/FEMA Photo. Jake Pelk of California Task Force 3 Urban Search and Rescue coordinates air and boat assets during a rescue mission in the flooded University area of New Orleans. September 16, 2005.

dbS

Published by dbS Productions LLC
P.O. Box 94
Charlottesville, Virginia USA 22902-0094
+1.434.293.5502 / 800.745.1581
www.dbs-sar.com

Library of Congress Cataloging-in-Publication Data

Catastrophic incident search and rescue addendum to the National search and rescue supplement to the International aeronautical and maritime search and rescue manual / National Search and Rescue Committee. -- Version 3.0.
      p. cm. -- (Catastrophic incident search and rescue addendum ; v. 2)
   Includes index.
   ISBN 978-1-879471-45-0 (trade paper : alk. paper) -- ISBN 978-1-879471-46-7 (e-book) -- ISBN 978-1-879471-42-9 (series, bound : alk. paper) 1. Search and rescue operations--Handbooks, manuals, etc. I. National Search and Rescue Committee (U.S.) II. IAMSAR manual.
   TL553.8.L362 2012
   363.12'381--dc23
                          2012027424

Printed in the United States
10 9 8 7 6 5 4 3 2 1

Department of Defense   Department of Homeland Security
Department of Interior   Department of Transportation
Department of Commerce   Federal Communications Commission
National Aeronautics and Space Administration

National Search and Rescue Committee

Letter of Promulgation

Catastrophic Incident Search and Rescue Addendum (Version 3.0)
To the
National Search and Rescue Supplement

Washington, D.C.

JUN 2 1 2012

This third revision of the *Catastrophic Incident Search and Rescue (CISAR) Addendum to the National Search and Rescue Supplement to the International Aeronautical and Maritime Search and Rescue (IAMSAR) Manual* continues the National Search and Rescue Committee's (NSARC) work in developing guidance for Federal SAR responders in the conduct of disaster response search and rescue (SAR) operations.

MSARC revised the CISAR Addendum to address the continuing process of improving national preparedness and SAR response to catastrophic incidents in support of States, Tribes, Territories, and local authorities. Version 3.0 provides additional information and lessons learned from other disasters that have occurred since Version 2.0 was promulgated.

This version addresses several changes to existing Sections, as well as incorporating several new Sections and a new Part on specific natural disasters that may impact CISAR responders. In order to better organize the information, the CISAR Addendum (Version 3.0) has been reformatted into five Parts: CISAR Organization, CISAR Management, and CISAR Supplemental Considerations, Natural Disasters, and Chemical, Biological, and CBRNE Incidents.

Mass rescue operations are complex, confusing and extremely challenging for multiagency CISAR responders. Providing effective and up to date guidance to those that save lives will continue to be a primary purpose of NSARC.

On behalf of the National Search and Rescue Committee,

Mary E. Landry

Mary E. Landry
Director of Incident Management & Preparedness Policy
United States Coast Guard
Chair, National Search and Rescue Committee

This page intentionally left blank.

Foreword

Disasters will continue to challenge Federal, State, Tribal, Territorial, and local Search and Rescue (SAR) responders in the planning and execution of large-scale mass rescue operations. The National Search and Rescue Committee (NSARC) realized that even though the Catastrophic Incident Search and Rescue (CISAR) Addendum (Version 2.0) was last updated in 2009, significant additional changes were required to ensure Federal SAR responders remained current with new information and guidance in the conduct of interagency mass rescue operations.

The NSARC directed the NSARC Correspondence Work Group to research and revise the current CISAR Addendum. With significant additional information and a complete reorganization of the existing Sections, the Correspondence Work Group decided that instead of updating Version 2.0, the CISAR Addendum would be reissued as Version 3.0.

Significant changes to Version 3.0 include:

- The addition of a new Part on specific natural disasters (earthquakes, flooding, hurricanes, tornadoes, and tsunamis);

- A complete reorganization and updating of the existing material;

- New Sections on Tribes and Territories have been added;

- A detailed glossary and index have been added;

- The States Section, as well as the Sections on Aircraft Management and Temporary Flights Restrictions have been revised and updated;

- Significant improvements were incorporated in the Section on Emergency Support Function (ESF) #9 – Search and Rescue, as well as on the responsibilities of the ESF #9 Primary Agency and overall Primary Agency; and the

- ESF #9 – Search and Rescue Annex to the National Response Framework has been added as Appendix A.

As in the previous versions of the CISAR Addendum, this Version 3.0 attempts to provide baseline guidance and information that can improve the effectiveness of the unified command in the conduct of the CISAR, and provide important guidance for the CISAR responder.

Correspondence Work Group
National Search and Rescue Committee

This page intentionally left blank.

Table of Contents

List of Figures

List of Tables

Glossary

(Note: Not all of the following terms are used in the CISAR Addendum. Additional NIMS and other important terms used in the response to a disaster have also been included.)

Air Burst: A nuclear weapon explosion that is high enough in the air to keep the fireball from touching the ground. Because the fireball does not reach the ground and does not pick up any surface material, the radioactivity in the fallout from an air burst is relatively insignificant compared with a surface burst.

Aircraft Coordinator (ACO): A person who coordinates the involvement of multiple aircraft in SAR operations.

ALARA (As low as reasonably achievable): A process to control or manage radiation exposure to individuals and releases of radioactive material to the environment so that doses are as low as social, technical, economic, practical, and public welfare considerations permit.

Area Command: An organization established to oversee the management of multiple incidents that are each being handled by a separate Incident Command System organization or to oversee the management of a very large or evolving incident that has multiple incident management teams engaged. An agency administrator/executive or other public official with jurisdictional responsibility for the incident usually makes the decision to establish an Area Command. An Area Command is activated only if necessary, depending on the complexity of the incident and incident management span-of-control considerations.

Area of Responsibility (AOR): Geographical area associated with a command within which the commander has the authority to plan and conduct operations – in addition to geographic delineation, an area of responsibility may be relative to subject, mission, or other factors.

Authority Having Jurisdiction (AHJ): The elected or appointed official(s) who have jurisdiction over an area, or over certain contemplated or ongoing actions. This may include having the authority to request resources. In some circumstances, there may be more than one Authority Having Jurisdiction. See Jurisdiction.

Biological Attack: The deliberate release of germs or other biological substances that can make you sick.

Biological Incident: An event in which a biological agent harms or threatens to harm humans, livestock, or agricultural or economic assets.

Bioterrorism: The deliberate release of viruses, bacteria, or other germs (agents) used to cause illness or death in people, animals, or plants. These agents are typically found in nature, but it is possible that they could be changed to increase their ability to cause disease, make them resistant to current medicines, or to increase their ability to be spread into the environment. Biological agents can be spread through the air, through water, or in food. Terrorists may use biological agents because they can be extremely difficult to detect and do not cause illness for several hours to several days. Some bioterrorism agents, like the smallpox virus, can be spread from person to person and some, like anthrax, cannot.

Blast Effects: The impacts caused by the shock wave of energy through air that is created by detonation of a nuclear device. The blast wave is a pulse of air in which the pressure increases sharply at the front, accompanied by winds.

Blister Agent: Also known as a vesicant, a blister agent is a chemical warfare agent which produces local irritation and damage to the skin and mucous membranes, pain and injury to the eyes, reddening and blistering of the skin, and when inhaled, damage to the respiratory tract.

Blood Agent: A chemical warfare agent that is inhaled and absorbed into the blood. The blood carries the agent to all body tissues where it interferes with the tissue oxygenation process. The brain is especially effected. The effect on the brain leads to cessation of respiration followed by cardiovascular collapse.

Branch: The organizational level having functional or geographical responsibility for major aspects of incident and accident operations. A branch is organizationally situated between the section and the division or group in the Operations Section, and between the section and units in the Logistics Section. Branches are identified by the use of Roman numerals or by functional area.

Catastrophic Incident: Any natural or manmade incident, including terrorism that results in extraordinary levels of mass casualties, damage, or disruption severely affecting the population, infrastructure, environment, economy, national morale, and/or government functions.

Catastrophic Incident Search and Rescue (CISAR): Civil SAR operations carried out as all or part of the response to an emergency or disaster declared by the President, under provisions of the NRF and ESF #9 Annex

Chemical, Biological, Radiological, Nuclear (CBRN) Emergency Response Force Package (CERFP): A National Guard unit. The CERFPs provide a regional response capability comprised of existing traditional National Guard units task organized to respond to Weapons of Mass Destruction (WMD) attacks. The CERFPs are capable of performing search and extraction, casualty/patient decontamination, mass medical triage, and treatment at a CBRNE incident.

Chemical, Biological, Radiological, Nuclear, and High Yield Explosives (CBRNE): An emergency resulting from the deliberate or unintentional release of nuclear, biological, radiological, toxic, or poisonous chemical materials, or the detonation of a high yield explosive.

Chemical Attack: The deliberate release of a toxic gas, liquid or solid that can poison people and the environment.

Chief: The Incident Command System title for individuals responsible for management of functional Sections: Operations, Planning, Logistics, Finance/Administration, and Intelligence/Investigations (if established as a separate Section).

Coastal Flooding: Flooding that occurs along the Great Lakes, the Atlantic and

Pacific Oceans, and the Gulf of Mexico.

Command Staff: In an incident or accident management organization, the Command Staff consists of the Incident Command and the special staff positions of Public Information Officer, Safety Officer, Liaison Officer, and other positions as required, e.g., legal and medical advisor who report directly to the Incident Commander. They may have an assistant or assistants, as needed.

Common Operating Picture : An overview of an incident by all relevant parties that provides incident information enabling the Incident Commander/Unified Command and any supporting agencies and organizations to make effective, consistent, and timely decisions.

Communications Unit: An organizational unit in the Logistics Section responsible for providing communication services at an incident or an Emergency Operations Center (EOC). A Communications Unit may also be a facility (e.g., a trailer or mobile van) used to support an Incident Communications Center.

Confined Space: A space that has limited openings for entry and exit and has poor natural ventilation.

Contamination (general): A release of hazardous material from its source to people, animals, the environment or equipment.

Contamination (radioactive): 1) The deposition of unwanted radioactive material on the surfaces of structures, areas, objects, or people where it may be external or internal. 2) Contamination means that radioactive materials in the form of gases, liquids, or solids are released into the environment and contaminate people externally, internally, or both. An external surface of the body, such as the skin, can become contaminated, and if radioactive materials get inside the body through the lungs, gut, or wounds, the contaminant can become deposited internally.

Continuity of Government: A coordinated effort within the Federal Government's executive branch to ensure that National Essential Functions continue to be performed during a catastrophic emergency (as defined in National Security Presidential Directive 51/Homeland Security Presidential Directive 20).

Continuity of Operations: An effort within individual organizations to ensure that Primary Mission Essential Functions continue to be performed during a wide range of emergencies.

Critical Infrastructure: Assets, systems, and networks, whether physical or virtual, so vital to the United States that the incapacitation or destruction of such assets, systems, or networks would have a debilitating impact on security, national economic security, national public health or safety, or any combination of those matters.

Decontamination: The process of making any person, object, or area safe within acceptable limits by absorbing, making harmless, or removing contaminated material clinging to or around it.

Decontamination Station: A building or location suitably equipped and organized where personnel and material are cleansed of radiological and other hazardous or toxic contaminants.

Defense Coordinating Officer (DCO): Individual who serves as the Department of Defense (DoD)'s single point of contact at the Joint Field Office (JFO) for requesting DoD assistance. With few exceptions, requests for Defense Support of Civil Authorities (DSCA) requests originating at the JFO are coordinated with and processed through the DCO. The DCO may have a Defense Coordinating Element consisting of a staff and military liaison officers to facilitate coordination and support to activated Emergency Support Functions.

Defense Support of Civil Authorities (DSCA): Support provided by U.S. military forces (Regular, Reserve, and National Guard), DOD civilians, contract personnel, agency, and component assets, in response to requests for assistance from civilian Federal, State, and local authorities for domestic emergencies, designated law enforcement support, and other domestic activities.

Director: The Incident Command System title for individuals responsible for supervision of a Branch.

Dirty Bomb: The use of common explosives to spread radioactive materials over a targeted area. Also known as a radiation attack, a dirty bomb is not a nuclear blast, but rather an explosion with localized radioactive contamination.

Division: The organizational level having responsibility for operations within a defined geographic area. Divisions are established when the number of resources exceeds the manageable span of control of the Section Chief. See Group.

Earthquake: A sudden slip on a fault, and the resulting ground shaking and radiated seismic energy caused by the slip, or by volcanic or magmatic activity, or other sudden stress changes in the earth.

Electromagnetic Pulse (EMP): A sharp pulse of radiofrequency (long wavelength) electromagnetic radiation produced when an explosion occurs near the earth's surface or at high altitudes. The intense electric and magnetic fields can damage unprotected electronics and electronic equipment over a large area.

Emergency: Any incident, whether natural or manmade, that requires responsive action to protect life or property. Under the Robert T. Stafford Disaster Relief and Emergency Assistance Act, an emergency means any occasion or instance for which, in the determination of the President, Federal assistance is needed to supplement State and local efforts and capabilities to save lives and to protect property and public health and safety, or to lessen or avert the threat of a catastrophe in any part of the United States.

Emergency Management Assistance Compact (EMAC): A congressionally ratified organization that provides form and structure to interstate mutual aid. Through EMAC, a disaster-affected State can request and receive assistance from other member States quickly and efficiently, resolving two key issues up front: liability and reimbursement.

Emergency Operations Center (EOC): The physical location at which the coordination of information and resources to support incident management (on-scene operations) activities normally takes place. An EOC may be a temporary facility or may be located in a more central or permanently established facility, perhaps at a higher level of organization within a jurisdiction. EOCs may be organized by major functional disciplines (e.g., fire, law enforcement, medical services), by jurisdiction (e.g., Federal, State, regional, tribal, city, county), or by some combination thereof.

Emergency Support Function (ESF) Annexes: The NRF includes 15 ESFs, which are managed by FEMA at the national level by the National Response Coordination Center (NRCC) or at the regional level by the Regional Response Coordination Center (RRCC) or Joint Field Office (JFO). The ESFs provide the structure for coordinating Federal interagency support for a Federal response to an incident. They are mechanisms for grouping functions most frequently used to provide Federal support to States and Federal-to-Federal support, both for declared disasters and emergencies under the Stafford Act and for non-Stafford Act incidents. During a response, FEMA employs ESFs as a critical mechanism to coordinate functional capabilities and resources provided by Federal departments and agencies, along with certain private-sector and nongovernmental organizations. They represent an effective way to bundle and funnel resources and capabilities to local, tribal, State, and other responders. These functions are generally coordinated by a single Department or Agency but may rely on several agencies that provide resources for each functional area. Organizing support by ESFs provides the greatest possible access to capabilities of the Federal Government regardless of which agency has those capabilities. ESFs comprise a wide range of operational-level mechanisms to provide assistance in functional areas such as transportation, communication, public works and engineering, firefighting, mass care, housing, human services, public health and medical services, search and rescue, response to oil and hazardous materials releases, law enforcement and public safety, agriculture and natural resources, and energy.

Emergency Support Function (ESF) Coordinator: The entity with management oversight for that particular ESF. The coordinator has ongoing responsibilities throughout the preparedness, response, and recovery phases of incident management. The role of the ESF coordinator is carried out through a "unified command" approach as agreed upon collectively by the designated primary agencies and, as appropriate, support agencies.

Emergency Support Function (ESF) Primary Agency: A Federal agency with significant authorities, roles, resources, or capabilities for a particular function within an ESF. A Federal agency designated as an ESF primary agency serves as a Federal executive agent under the Federal Coordinating Officer (or Federal Resource Coordinator for non-Stafford Act incidents) to accomplish the ESF mission. The ESF #9 Primary Agencies are DHS/FEMA, DHS/Coast Guard, DoD and DOI/NPS.

Emergency Support Function (ESF) Support Agency: An entity with specific capabilities or resources that support the primary agencies in executing the mission of the ESF.

Emergency Support Functions (ESFs): Used by the Federal Government and many State governments as the primary mechanism at the operational level to organize and provide assistance. ESFs align categories of resources and provide strategic objectives for their use. ESFs utilize standardized resource management concepts such as typing, inventorying, and tracking to facilitate the dispatch, deployment, and recovery of resources before, during, and after an incident.

Epicenter: The point on the earth's surface vertically above the hypocenter (or focus), point in the crust where a seismic rupture begins.

Evacuation: Organized, phased, and supervised withdrawal, dispersal, or removal of persons from dangerous or potentially dangerous areas, and their reception and care in safe areas.

Exposure (Radiation): The level of radiation flux to which a material or living tissue is exposed. The actual dose of radiation from the exposure depends on many factors including length of exposure time, the distance from the radiation source, and the amount of shielding between the radiation source and the exposed object.

Eye: The roughly circular area of comparatively light winds that encompasses the center of a severe tropical cyclone. The eye is either completely or partially surrounded by the eyewall cloud.

Eyewall/Wall Cloud: An organized band or ring of cumulonimbus clouds that surround the eye, or light-wind center of a tropical cyclone. Eyewall and wall cloud are used synonymously.

Fallout: The process or phenomenon of the descent to the earth's surface of particles contaminated with radioactive material from the radioactive cloud. The term is also applied in a collective sense to the contaminated particulate matter itself.

Fault: A fracture along which the blocks of crust on either side have moved relative to one another parallel to the fracture.

Federal Agency: Any department, independent establishment, Government corporation, or other agency of the executive branch of the Federal Government, including the U.S. Postal Service, but shall not include the American National Red Cross.

Federal Coordinating Officer (FCO): The official appointed by the President to execute Stafford Act authorities, including the commitment of Federal Emergency Management Agency (FEMA) resources and mission assignment of other Federal departments or agencies. In all cases, the FCO represents the FEMA Administrator in the field to discharge all FEMA responsibilities for the response and recovery efforts underway. For Stafford Act events, the FCO is the primary Federal representative with whom the State Coordinating Officer and other State, Tribal, Territorial, and local response officials interface to determine the most urgent needs and set objectives for an effective response in collaboration with the Unified Coordination Group. The FCO is responsible for coordinating the timely delivery of Federal disaster assistance resources and programs to the affected State and local governments, individual victims, and the private sector. The FCO issues MAs through the FEMA support team at the NRCC, RRCC, or JFO.

FEMA Task Force Search and Rescue Marking System: Distinct markings made with international orange spray paint near a collapsed structure's most accessible point of entry.

Finance/Administration Section: (1) Incident Command: Section responsible for all administrative and financial considerations surrounding an incident. (2) Joint Field Office (JFO): Section responsible for the financial management, monitoring, and tracking of all Federal costs relating to the incident and the functioning of the JFO while adhering to all Federal laws and regulations.

Flash Flood: A rapid and extreme flow of high water into a normally dry area, or a rapid water level rise in a stream or creek above a predetermined flood level, beginning within six hours of the causative event (e.g., intense rainfall, dam failure, ice jam). However, the actual time threshold may vary in different parts of the country. Ongoing flooding can intensify to flash flooding in cases where intense rainfall results in a rapid surge of rising flood waters.

Flash Flood Warning: Issued to inform the public, emergency management, and other cooperating agencies that flash flooding is in progress, imminent, or highly likely.

Flash Flood Watch: Issued to indicate current or developing hydrologic conditions that are favorable for flash flooding in and close to the watch area, but the occurrence is neither certain or imminent.

Flood: Any high flow, overflow, or inundation by water which causes or threatens damage.

Flood Warning: (FLW) In hydrologic terms, a release by the NWS to inform the public of flooding along larger streams in which there is a serious threat to life or property. A flood warning will usually contain river stage (level) forecasts.

Flood Watch: Issued to inform the public and cooperating agencies that current and developing hydrometeorological conditions are such that there is a threat of flooding, but the occurrence is neither certain nor imminent.

Frostbite: Human tissue damage caused by exposure to intense cold.

Function: The five major activities in the Incident Command System: Command, Operations, Planning, Logistics, and Finance/Administration. A sixth function, Intelligence/Investigations, may be established, if required, to meet incident management needs. The term function is also used when describing the activity involved (e.g., the planning function).

General Staff: A group of incident management personnel organized according to function and reporting to the Incident Commander. The General Staff normally consists of the Operations Section Chief, Planning Section Chief, Logistics Section Chief, and Finance/Administration Section Chief. An Intelligence/Investigations Chief may be established, if required, to meet incident management needs.

Governor: The chief executive of any State.

Group: An organizational subdivision established to divide the incident management structure into functional areas of operation. Groups are composed of resources assembled to perform a special function not necessarily within a single geographic division. See Division.

Half-Life: The time required for the activity of a given radioactive element to decrease to half of its initial value due to radioactive decay. The physical half-life is a characteristic property of each radioactive element and is independent of its amount or physical form. The effective or biological half-life of a given isotope in the body is the time in which the quantity in the body decreases to half because of both radioactive decay and biological elimination.

H-Hour: H-hour is a hurricane planning factor used to develop plans for pre-hurricane actions and differs from hurricane landfall in that it is based on the onset of tropical storm force winds, not the arrival of the hurricane eye wall ashore. H-Hour provides a determination for when tropical storm force winds arrive ashore, normally requiring the reduction or cessation of operations due to bridge closures, crosswind limits for flight operations, sheltering requirements, and other safety related considerations. Based on the size and intensity of a hurricane, there can be an 8 to 16 hour time difference between H-hour and hurricane landfall.

Hazardous Material: Any material that is flammable, corrosive, an oxidizing agent, explosive, toxic, poisonous, radioactive, nuclear, unduly magnetic, or chemical agent, biological research material, compressed gas, or any other material that, because of its quantity, properties, or packaging, may endanger life or property. For the purposes of ESF #1, hazardous material is a substance or material, including a hazardous substance that has been determined by the Secretary of Transportation to be capable of posing an unreasonable risk to health, safety, and property when transported in commerce, and which has been so designated (See reference (h)). For the purposes of ESF #10 and the Oil and Hazardous Materials Incident Annex, the term is intended to mean hazardous substances, pollutants, and contaminants as defined by the National Oil and Hazardous Substances Pollution Contingency Plan (NCP).

Household Pet: A domesticated animal, such as a dog, cat, bird, rabbit, rodent, or turtle that is traditionally kept in the home for pleasure rather than for commercial purposes, can travel in commercial carriers, and be housed in temporary facilities.

Hurricane/Typhoon: A tropical cyclone in which the maximum sustained surface wind (using the U.S. 1-minute average) is 64 kt (74 mph or 119 km/hr) or more. The term hurricane is used for Northern Hemisphere tropical cyclones east of the International Dateline to the Greenwich Meridian. The term typhoon is used for Pacific tropical cyclones north of the Equator west of the International Dateline.

Hurricane Season: The portion of the year having a relatively high incidence of hurricanes. The hurricane season in the Atlantic, Caribbean, and Gulf of Mexico runs from June 1 to November 30. The hurricane season in the Eastern Pacific basin runs from May 15 to November 30. The hurricane season in the Central Pacific basin runs from June 1 to November 30.

Hurricane Warning: An announcement that hurricane conditions (sustained winds of 74 mph or higher) are *expected* somewhere within the specified coastal area. Because hurricane preparedness activities become difficult once winds reach tropical storm force, the hurricane warning is issued 36 hours in advance of the anticipated onset of tropical-storm-force winds.

Hurricane Watch: An announcement that hurricane conditions (sustained winds of 74 mph or higher) are *possible* within the specified coastal area. Because hurricane preparedness activities become difficult once winds reach tropical storm force, the hurricane watch is issued 48 hours in advance of the anticipated onset of tropical-storm-force winds.

Hypothermia: A rapid, progressive mental

and physical collapse that accompanies the lowering of body temperature.

Improvised Nuclear Device (IND): An illicit nuclear weapon that is bought, stolen, or otherwise obtained from a nuclear State, or a weapon fabricated by a terrorist group from illegally obtained fissile nuclear weapons material and produces a nuclear explosion. The nuclear yield achieved by an IND produces extreme heat, powerful shockwaves, and prompt radiation that would be acutely lethal for a significant distance. An IND also produces radioactive fallout, which may spread and deposit over very large areas. If a nuclear yield is not achieved, the result would likely resemble an RDD in which fissile weapons material was utilized.

Incident: An occurrence, natural or manmade, that requires a response to protect life or property. Incidents can, for example, include major disasters, emergencies, terrorist attacks, terrorist threats, civil unrest, wildland and urban fires, floods, hazardous materials spills, nuclear accidents, aircraft accidents, earthquakes, hurricanes, tornadoes, tropical storms, tsunamis, war-related disasters, public health and medical emergencies, and other occurrences requiring an emergency response.

Incident Action Plan (IAP): An oral or written plan containing general objectives reflecting the overall strategy for managing an incident. It may include the identification of operational important information for management of the incident during one or more operational periods, resources and assignments. It may also include attachments that provide direction and important information for management of the incident during one or more operational periods.

Incident Command: The Incident Command System organizational element responsible for overall management of the incident and consisting of the Incident Commander (either single or unified command structure) and any assigned supporting staff.

Incident Command Post (ICP): The field location where the primary functions are performed. The ICP may be co-located with the Incident Base or other incident facilities.

Incident Command System (ICS): A standardized on-scene emergency management construct specifically designed to provide an integrated organizational structure that reflects the complexity and demands of single or multiple incidents, without being hindered by jurisdictional boundaries. ICS is the combination of facilities, equipment, personnel, procedures, and communications operating within a common organizational structure, designed to aid in the management of resources during incidents. It is used for all kinds of emergencies and is applicable to small as well as large and complex incidents. ICS is used by various jurisdictions and functional agencies, both public and private, to organize field-level incident management operations.

Incident Commander (IC): The individual responsible for all incident activities, including the development of strategies and tactics and the ordering and release of resources. The IC has overall authority and responsibility for conducting incident operations and is responsible for the management of all incident operations at the incident site.

Indian Tribes: The United States recognizes Indian tribes as domestic dependent nations under its protection and recognizes the right of Indian tribes to self-government. As such, tribes are responsible for coordinating tribal resources to address actual or potential incidents. When their resources are exhausted, tribal leaders seek assistance from States or even the Federal Government.

Incident Management: The broad spectrum of activities and organizations providing effective and efficient operations, coordination, and support applied at all levels of government, utilizing both governmental and nongovernmental resources to plan for, respond to, and recover from an incident, regardless of cause, size, or complexity.

Incident Management Assistance Team (IMAT): An interagency national- or regional-based team composed of subject-matter experts and incident management professionals from multiple Federal departments and agencies.

Incident Objectives: Statements of guidance and direction needed to select appropriate strategy(s) and the tactical direction of resources. Incident objectives are based on realistic expectations of what can be accomplished when all allocated resources have been effectively deployed. Incident objectives must be achievable and measurable, yet flexible enough to allow strategic and tactical alternatives.

Instrument Flight Rules (IFR): Rules governing the procedures for conducting instrument flight. Also a term used by pilots and controllers to indicate type of flight plan.

Inundation: The process of covering normally dry areas with flood waters.

Isolated Person: In an ESF #9 incident only, any non-distressed person or persons stranded within a specific area or residence by incident conditions where *immediate* assistance is determined not to be required.

Joint Field Office (JFO): The primary Federal incident management field structure. The JFO is a temporary Federal facility that provides a central location for the coordination of Federal, State, tribal, and local governments and private-sector and nongovernmental organizations with primary responsibility for response and recovery. The JFO structure is organized, staffed, and managed in a manner consistent with National Incident Management System principles. Although the JFO uses an Incident Command System structure, the JFO does not manage on-scene operations. Instead, the JFO focuses on providing support to on-scene efforts and conducting broader support operations that may extend beyond the incident site.

Joint Information Center (JIC): A facility established to coordinate all incident-related public information activities. It is the central point of contact for all news media. Public information officials from all participating agencies should co-locate at the JIC.

Jurisdiction: A range or sphere of authority. Public agencies have jurisdiction at an incident related to their legal responsibilities and authority. Jurisdictional authority at an incident can be political or geographical (e.g., Federal, State, tribal, and local boundary lines) or functional (e.g., law enforcement, public health).

Landfall: The intersection of the surface center of a tropical cyclone with a coastline. Because the strongest winds in a tropical cyclone are not located precisely at the center, it is possible for a cyclone's strongest winds to be experienced over land even if landfall does not occur. Similarly, it is possible for a tropical cyclone to make landfall and have its strongest winds remain over the water. Compare direct hit, indirect hit, and strike.

Lilly Pad: A lily pad is an interim stopping point during rescue operations where survivors can be accounted for, possibly have initial basic needs cared for, and from which they can be transported to a place of safety.

Liquefaction: In geology, liquefaction refers to the process by which saturated, unconsolidated sediments are transformed into a substance that acts like a liquid. Earthquakes can cause soil liquefaction where loosely packed, water-logged sediments come loose from the intense shaking of the earthquake. The term liquefaction is commonly misapplied to the displaced, saturated sediment as opposed to the process by which it was formed.

Livestock (Farm Animals): Livestock refers to horses, mares, mules, jacks, jennies, colts, cows, calves, yearlings, bulls, oxen, sheep, goats, lambs, kids, hogs, shoats and pigs. Mainly provide food for human or animal consumption.

Local Government: Public entities responsible for the security and welfare of a designated area as established by law: (1) a county, municipality, city, town, township, local public authority, school district, special district, intrastate district, council of governments (regardless of whether the council of governments is incorporated as a nonprofit corporation under State law), regional or interstate government entity,

or agency or instrumentality of a local government; (2) an Indian tribe or authorized tribal entity, or in Alaska a Native Village or Alaska Regional Native Corporation; and (3) a rural community, unincorporated town or village, or other public entity. See Section 2 (10), Homeland Security Act of 2002, Pub. L. 107-296, 116 Stat. 2135 (2002).

Logistics Section: (1) Incident Command: Section responsible for providing facilities, services, and material support for the incident. (2) Joint Field Office (JFO): Section that coordinates logistics support to include control of and accountability for Federal supplies and equipment; resource ordering; delivery of equipment, supplies, and services to the JFO and other field locations; facility location, setup, space management, building services, and general facility operations; transportation coordination and fleet management services; information and technology systems services; administrative services such as mail management and reproduction; and customer assistance.

Major Disaster: Under the Robert T. Stafford Disaster Relief and Emergency Assistance Act, any natural catastrophe (including any hurricane, tornado, storm, high water, wind-driven water, tidal wave, tsunami, earthquake, volcanic eruption, landslide, mudslide, snowstorm, or drought) or, regardless of cause, any fire, flood, or explosion in any part of the United States that, in the determination of the President, causes damage of sufficient severity and magnitude to warrant major disaster assistance under the Stafford Act to supplement the efforts and available resources of States, local governments, and disaster relief organizations in alleviating the damage, loss, hardship, or suffering caused thereby.

Major Hurricane: A hurricane that is classified as Category 3 or higher.

Manager: Individual within an Incident Command System organizational unit who is assigned specific managerial responsibilities (e.g., Staging Area Manager or Camp Manager).

Mass Rescue Operations (MRO): Civil search and rescue services characterized by the need for immediate response to large numbers of persons in distress, such that the capabilities normally available to search and rescue authorities are inadequate.

Mission Assignment: The mechanism used to support Federal operations in a Stafford Act major disaster or emergency declaration. It orders immediate, short-term emergency response assistance when an applicable State or local government is overwhelmed by the event and lacks the capability to perform, or contract for, the necessary work.

Mobilization: The process and procedures used by all organizations-Federal, State, tribal, and local-for activating, assembling, and transporting all resources that have been requested to respond to or support an incident.

Mobilization Guide: Reference document used by organizations outlining agreements, processes, and procedures used by all participating agencies/organizations for activating, assembling, and transporting resources.

Multiagency Coordination (MAC) Group: A group of administrators or executives, or their appointed representatives, who are typically authorized to commit agency resources and funds. A MAC Group can provide coordinated decision making and resource allocation among cooperating agencies, and may establish the priorities among incidents, harmonize agency policies, and provide strategic guidance and direction to support incident management activities. MAC Groups may also be known as multiagency committees, emergency management committees, or as otherwise defined by the Multiagency Coordination System.

Multiagency Coordination System (MACS): A system that provides the architecture to support coordination for incident prioritization, critical resource allocation, communications systems integration, and information coordination. MACS assist agencies and organizations responding to an incident. The elements of a MACS include facilities, equipment, personnel, procedures, and communications. Two of the most commonly used elements are Emergency Operations Centers and MAC Groups.

Multijurisdictional Incident: An incident requiring action from multiple agencies that each have jurisdiction to manage certain aspects of an incident. In the Incident Command System, these incidents will be managed under Unified Command.

Mutual Aid Agreement or Assistance Agreement: Written or oral agreement between and among agencies/ organizations and/or jurisdictions that provides a mechanism to quickly obtain emergency assistance in the form of personnel, equipment, materials, and other associated services. The primary objective is to facilitate rapid, short-term deployment of emergency support prior to, during, and/or after an incident.

National Essential Functions: A subset of government functions that are necessary to lead and sustain the Nation during a catastrophic emergency and that, therefore, must be supported through continuity of operations and continuity of government capabilities.

National Incident Management System (NIMS): A set of principles that provides a systematic, proactive approach guiding government agencies at all levels, nongovernmental organizations, and the private sector to work seamlessly to prevent, protect against, respond to, recover from, and mitigate the effects of incidents, regardless of cause, size, location, or complexity, in order to reduce the loss of life or property and harm to the environment.

National Response Framework (NRF): A guide to how the U.S. conducts all-hazards response. It is built upon scalable, flexible, and adaptable coordinating structures to align key roles and responsibilities across the Nation, linking all levels of government, nongovernmental organizations, and the private sector. The NRF is intended to capture specific authorities and best practices for managing incidents that range from the serious but purely local, to large-scale terrorist attacks or catastrophic natural disasters.

National Search and Rescue Committee (NSARC): Federal committee comprised of the Departments of Homeland Security, Defense, Transportation, Interior, and Commerce, the Federal Communications Commission, National Aeronautics and Space Administration. Established to oversee the National Search and Rescue Plan and act as a coordinating forum for national search and rescue matters.

National Search and Rescue Plan (NSP): An interagency agreement providing national arrangements for coordination of search and rescue services to meet domestic needs and international commitments.

National Special Security Event (NSSE): A designated event that, by virtue of its political, economic, social, or religious significance, may be the target of terrorism or other criminal activity.

National Urban Search and Rescue Response System: Specialized teams that locate, rescue (extricate), and provide initial medical stabilization of victims trapped in confined spaces.

National Voluntary Organizations Active in Disaster (National VOAD): A consortium of more than 30 recognized national organizations active in disaster relief. Their organizations provide capabilities to incident management and response efforts at all levels. During major incidents, National VOAD typically sends representatives to the National Response Coordination Center to represent the voluntary organizations and assist in response coordination.

Natural Disaster: Any hurricane, tornado, storm, flood, high water, wind-driven water, tidal wave, tsunami, earthquake, volcanic eruption, landslide, mudslide, snowstorm, drought, fire, or other catastrophe in any part of the U.S. which causes, or which may cause, substantial damage or injury to civilian property or persons.

Nerve Agent: The most toxic chemical warfare agents. Nerve agents are absorbed into the body through breathing, by injection, or absorption through the skin. They affect the nervous and the respiratory systems and various body functions.

Nongovernmental Organization (NGO): An entity with an association that is based on interests of its members, individuals, or institutions. It is not created by a government, but it may work cooperatively with government. Such organizations serve a public purpose, not a private benefit. Examples of NGOs include faith-based charity organizations and the American Red Cross. NGOs, including voluntary and faith-based groups, provide relief services to sustain life, reduce physical and emotional distress, and promote the recovery of disaster victims. Often these groups provide specialized services that help individuals with disabilities. NGOs and voluntary organizations play a major role in assisting emergency managers before, during, and after an emergency.

Nuclear Blast: An explosion with intense light and heat, a damaging pressure wave and widespread radioactive material that can contaminate the air, water and ground surfaces for miles around.

Nuclear Detonation: A nuclear explosion resulting from fission or fusion reactions in nuclear materials, such as from a nuclear weapon.

Nuclear Radiation: Particulate and electromagnetic radiation emitted from atomic nuclei in various nuclear processes. The important nuclear radiations, from the weapons effects standpoint, are alpha and beta particles, gamma rays, and neutrons.

Officer: The Incident Command System title for a person responsible for one of the Command Staff positions of Safety, Liaison, and Public Information.

On Scene Coordinator (OSC): A person designated to coordinate search and rescue operations within a specified search area. See also **Incident Commander (IC)**.

Operational Period: The time scheduled for executing a given set of operation actions, as specified in the Incident Action Plan. Operational periods can be of various lengths, although usually they last 12 to 24 hours.

Operations Section: (1) Incident Command: Responsible for all tactical incident operations and implementation of the Incident Action Plan. In the Incident Command System, it normally includes subordinate Branches, Divisions, and/or Groups. (2) Joint Field Office: Coordinates operational support with on-scene incident management efforts. Branches, divisions, and groups may be added or deleted as required, depending on the nature of the incident. The Operations Section is also responsible for coordinating with other Federal facilities that may be established to support incident management activities.

Person in Distress: There is reasonable certainty that a person is threatened by grave and imminent danger and requires immediate assistance.

Personal Protective Equipment: Clothing and/or equipment worn by workers (including first responders and first receivers) to prevent or mitigate job-related illness or injury. Individual PPE elements can include respiratory and percutanous protective equipment.

Place of Safety: Location where rescue operations are considered to terminate and where: 1) The survivor's safety or life is no longer threatened; 2) basic human needs (such as food, shelter and medical needs) can be met; and (3) transportation arrangements can be made for the survivor's next or final destination.

Planning Section: (1) Incident Command: Section responsible for the collection, evaluation, and dissemination of operational information related to the incident, and for the preparation and documentation of the Incident Action Plan. This Section also maintains information on the current and forecasted situation and on the status of resources assigned to the incident. (2) Joint Field Office: Section that collects, evaluates, disseminates, and uses information regarding the threat or incident and the status of Federal resources. The Planning Section prepares and documents Federal support actions and develops unified action, contingency, long-term, and other plans.

Pre-Scripted Mission Assignment (PSMA): A mechanism used by the Federal Government to facilitate rapid Federal resource response. Pre-scripted mission assignments identify resources or capabilities that Federal departments and agencies, through various Emergency Support Functions (ESFs), are commonly called upon to provide during incident response. Pre-scripted mission assignments allow primary and supporting ESF agencies to organize resources that will be deployed during incident response.

Primary Agency: See Emergency Support Function (ESF) Primary Agency.

Primary Mission Essential Functions: Government functions that must be performed in order to support or implement the performance of National Essential Functions before, during, and in the aftermath of an emergency.

Principal Federal Official (PFO): May be appointed to serve as the Secretary of Homeland Security's primary representative to ensure consistency of Federal support as well as the overall effectiveness of the Federal incident management for catastrophic or unusually complex incidents that require extraordinary coordination.

Private Sector: Organizations and individuals that are not part of any governmental structure. The private sector includes for-profit and not-for-profit organizations, formal and informal structures, commerce, and industry.

Protective Action Guide (PAG): A radiation exposure level or range (or level of other hazard) established by appropriate Federal or State agencies beyond which protective action should be considered.

Public Information: Processes, procedures, and systems for communicating timely, accurate, and accessible information on an incident's cause, size, and current situation; resources committed; and other matters of general interest to the public, responders, and additional stakeholders (both directly affected and indirectly affected).

Public Information Officer: A member of the Command Staff responsible for interfacing with the public and media and/or with other agencies with incident-related information requirements.

Radiation: Energy moving in the form of particles or waves. Familiar radiations are heat, light, radio waves, and microwaves. Ionizing radiation is a very high-energy form of electromagnetic radiation.

Radiation Absorbed Dose (rad): A unit expressing the absorbed dose of ionizing radiation. Absorbed dose is the energy deposited per unit mass of matter. The units of rad and gray are the units in two different systems for expressing absorbed dose. 1 rad = 0.01 gray (Gy); 1 Gy = 100 rad.

Radioactive Contamination: The deposition of unwanted radioactive material on the surfaces of structures, areas, objects, or people. It can be airborne, external, or internal.

Radiological Incident: An event or series of events, deliberate or accidental, leading to the release, or potential release, in to the environment of radioactive material in sufficient quantity to warrant consideration of protective actions. Use of an RDD or IND is an act of terror that results in a radiological incident.

Radiological Dispersal Device (RDD): An event or series of events, deliberate or accidental, leading to the release, or potential release, into the environment of radioactive material in sufficient quantity to warrant consideration of protective actions. Use of an RDD is an act of terror that results in a radiological incident.

Recovery: The development, coordination, and execution of service- and site-restoration plans; the reconstitution of government operations and services; individual, private-sector, nongovernmental, and public assistance programs to provide housing and to promote restoration; long-term care and treatment of affected persons; additional measures for social, political, environmental, and economic restoration; evaluation of the incident to identify lessons learned; post-incident reporting; and development of initiatives to mitigate the effects of future incidents.

Regional Response Coordination Center (RRCC): Located in each Federal Emergency Management Agency (FEMA) region, these multiagency agency coordination centers are staffed by Emergency Support Functions in anticipation of a serious incident in the region or immediately following an incident. Operating under the direction of the FEMA Regional Administrator, the RRCCs coordinate Federal regional response efforts and maintain connectivity with State emergency operations centers, State fusion centers, Federal Executive Boards, and other Federal and State operations and coordination centers that have potential to contribute to development of situational awareness.

Rem: A unit of absorbed dose that accounts for the relative biological effectiveness of ionizing radiations in tissue (also called equivalent dose). Not all radiation produces the same biological effect, even from the same amount of absorbed dose; rem relates the absorbed dose in human tissue to the effective biological damage of the radiation. The units of rem and Sievert are the units in two different systems of expressing equivalent dose. 1 rem = 0.01 Sieverts (Sv); 1 Sv = 100 rem.

Rescue: An operation to retrieve persons in distress, provide for their medical or other needs, and deliver them to a place of safety.

Responder: A person designated by a responsible authority, that is trained, equipped, and qualified in a position to perform a specific task.

Response: Activities that address the short-term, direct effects of an incident. Response includes immediate actions to save lives, protect property, and meet basic human needs. Response also includes the execution of emergency operations plans and of mitigation activities designed to limit the loss of life, personal injury, property damage, and other unfavorable outcomes. As indicated by the situation, response activities include applying intelligence and other information to lessen the effects or consequences of an incident; increased security operations; continuing investigations into nature and source of the threat; ongoing public health and agricultural surveillance and testing processes; immunizations, isolation, or quarantine; and specific law enforcement operations aimed at preempting, interdicting, or disrupting illegal activity, and apprehending actual perpetrators and bringing them to justice.

Richter scale: The Richter magnitude scale was developed in 1935 by Charles F. Richter of the California Institute of Technology as a mathematical device to compare the size of earthquakes. The magnitude of an earthquake is determined from the logarithm of the amplitude of waves recorded by seismographs. Adjustments are included for the variation in the distance between the various seismographs and the epicenter of the earthquakes. On the Richter scale, magnitude is expressed in whole numbers and decimal fractions. For example, a magnitude 5.3 might be computed for a moderate earthquake, and a strong earthquake might be rated as magnitude 6.3. Because of the logarithmic basis of the scale, each whole number increase in magnitude represents a tenfold increase in measured amplitude; as an estimate of energy, each whole number step in the magnitude scale corresponds to the release of about 31 times more energy than the amount associated with the preceding whole number value.

Robert T. Stafford Disaster Relief and Emergency Assistance Act ("Stafford Act"; PL 100-707): Stafford Act was signed into law November 23, 1988; amended the Disaster Relief Act of 1974, PL 93-288. This Act describes the programs and processes by which the Federal Government provides disaster and emergency assistance to State and local governments, tribal nations, eligible private nonprofit organizations, and individuals affected by a declared major disaster or emergency. The Stafford Act covers all hazards, including natural disasters and terrorist events.

Roentgen (R): A unit of gamma or x-ray exposure in air. It is the primary standard of measurement used in the emergency responder community in the U.S. For the purpose of this guidance, one R of exposure is approximately equal to one rem of whole-body external dose.

Safety Officer: A member of the Command Staff responsible for monitoring incident operations and advising the Incident Commander on all matters relating to operational safety, including the health and safety of emergency responder personnel.

Saffir-Simpson Hurricane Wind Scale: The Saffir-Simpson Hurricane Wind Scale is a 1 to 5 categorization based on the hurricane's intensity at the indicated time. The scale provides examples of the type of damage and impacts in the United States associated with winds of the indicated intensity.

Search: An operation, normally coordinated by an Incident Commander, RCC or RSC, using available personnel and facilities to locate persons in distress.

Search Action Plan: Message, normally developed by the SMC, for passing instructions to SAR facilities and agencies participating in a SAR mission. See also **incident action plan**.

Search and Rescue Coordinator (SC): One or more persons or agencies within an Administration with overall responsibility for establishing and providing SAR services, and ensuring that planning for those services is properly coordinated.

Search and Rescue Facility: (1) Any mobile resource, including designated SRUs, used to conduct SAR operations. (2) Fixed facilities for the incident; may include the Incident Base, feeding areas, sleeping areas, sanitary facilities, etc.

Search and Rescue Mission Coordinator (SMC): The official temporarily assigned to coordinate response to an actual or apparent distress situation.

Search and Rescue Region (SRR): An area of defined dimensions, associated with an RCC, within which SAR services are provided.

Secretary of Defense: Responsible for homeland defense and Defense Support of Civil Authorities for domestic incidents when requested by civil authorities or when directed by the President. Federal military forces always remain under the command of the Secretary of Defense.

Secretary of Homeland Security: Serves as the principal Federal official for domestic incident management, which includes coordinating both Federal operations within the United States and Federal resources used in response to or recovery from terrorist attacks, major disasters, or other emergencies. The Secretary of Homeland Security is by Presidential directive and statutory authority also responsible for coordination of Federal resources utilized in the prevention of, preparation for, response to, or recovery from terrorist attacks, major disasters, or other emergencies, excluding law enforcement responsibilities otherwise reserved to the Attorney General.

Secretary of State: Responsible for managing international preparedness, response, and recovery activities relating to domestic incidents and the protection of U.S. citizens and U.S. interests overseas.

Section: The Incident Command System organizational level having responsibility for a major functional area of incident management (e.g., Operations, Planning, Logistics, Finance/Administration, and Intelligence/Investigations (if established). The Section is organizationally situated between the Branch and the Incident Command.

Service Animal: Any guide dog, signal dog, assistive dog, seizure dog, or other animal individually trained to do work or perform tasks for the benefit of an individual with a disability, including but not limited to guiding individuals with impaired vision, alerting individuals with impaired hearing to intruders or sounds, providing minimal protection or rescue work, pulling a wheelchair, or fetching dropped items. Service animals shall be treated as required by laws such as the Americans with Disabilities Act.

Situation Report (SITREP): Document that contains confirmed or verified information and explicit details (who, what, where, and how) relating to an incident.

Special Needs Population: A population whose members may have additional needs before, during, and after an incident in functional areas, including but not limited to: maintaining independence, communication, transportation, supervision, and medical care. Individuals in need of additional response assistance may include those who have disabilities; who live in institutionalized settings; who are elderly; who are children; who are from diverse cultures, who have limited English proficiency, or who are non-English-speaking; or who are transportation disadvantaged.

Staging Area: Temporary location for available resources. A Staging Area can be any location in which personnel, supplies, and equipment can be temporarily housed or parked while awaiting operational assignment.

State: When capitalized, refers to any State of the U.S., the District of Columbia, the Commonwealth of Puerto Rico, the Virgin Islands, Guam, American Samoa, the Commonwealth of the Northern Mariana Islands, and any possession of the United States. See Section 2 (14), Homeland Security Act of 2002, Pub. L. 107-296, 116 Stat. 2135 (2002).

Storm Surge: An abnormal rise in sea level accompanying a hurricane or other intense storm, and whose height is the difference between the observed level of the sea surface and the level that would have occurred in the absence of the cyclone. Storm surge is usually estimated by subtracting the normal or astronomic high tide from the observed storm tide.

Strike-Slip Fault: Vertical (or nearly vertical) fractures where the blocks have mostly moved horizontally. If the block opposite an observer looking across the fault moves to the right, the slip style is termed right lateral; if the block moves to the left, the motion is termed left lateral.

Supervisor: The Incident Command System title for an individual responsible for a Division or Group.

Support Agency: See Emergency Support Function (ESF) Support Agency.

Support Annex: Describes how Federal Departments and Agencies, the private sector, volunteer organizations, and nongovernmental organizations coordinate and execute the common support processes and administrative tasks required during an incident. The actions described in the Support Annexes are not limited to particular types of events, but are overarching in nature and applicable to nearly every type of incident.

Swift Water: Water moving at a rate greater than one knot (1.15 mph).

Swift Water Rescue: Swiftwater rescue (also called "whitewater rescue") is a subset of technical rescue dealing in whitewater river conditions. Due to the added pressure of moving water, swiftwater rescue involves the use of specially trained personnel, ropes and mechanical advantage systems that are often much more robust than those used in standard rope rescue. The main goal is to use or deflect the water's power to assist in the rescue of the endangered person(s), as in most situations there is no easy way to overcome the power of the water.

Tactics: The deployment and directing of resources on an incident to accomplish the objectives designated by strategy.

Task Force: Any combination of resources assembled to support a specific mission or operational need. All resource elements within a Task Force must have common communications and a designated leader.

Temporary Flight Restriction (TFR): A TFR is a regulatory action issued via the U.S. Notice to Airmen (NOTAM) system to restrict certain aircraft from operating within a defined area, on a temporary basis, to protect persons or property in the air or on the ground.

Territory: Territories are one type of political division of the U.S., overseen directly by the Federal Government and not any part of a U.S. State. Under the Stafford Act, Puerto Rico, the U.S. Virgin Islands, Guam, American Samoa, and the Commonwealth of the Northern Mariana Islands are included in the definition of "State." Stafford Act assistance is also available to two sovereign nations under the compact of free association: (1) Federated States of Micronesia (FSM); and the (2) Republic of the Marshall Islands (RMI).

Terrorism: As defined in the Homeland Security Act of 2002, activity that involves an act that is dangerous to human life or potentially destructive of critical infrastructure or key resources; is a violation of the criminal laws of the U.S. or of any State or other subdivision of the U.S.; and appears to be intended to intimidate or coerce a civilian population, to influence the policy of a government by intimidation or coercion, or to affect the conduct of a government by mass destruction, assassination, or kidnapping.

Threat: Natural or manmade occurrence, individual, entity, or action that has or indicates the potential to harm life, information, operations, the environment, and/or property.

Tornado A violently rotating column of air, usually pendant to a cumulonimbus, with circulation reaching the ground. It nearly always starts as a funnel cloud and may be accompanied by a loud roaring noise. On a local scale, it is the most destructive of all atmospheric phenomena.

Tornado Warning: Issued when a tornado is indicated by radar or sighted by spotters; therefore, people in the affected area should seek safe shelter immediately. Tornado Warnings can be: issued without a Tornado Watch being already in effect; are usually issued for a duration of approximately 30 minutes; and issued by the local National Weather Service office (NWFO). Tornado Warnings will include where the tornado was located and what towns will be in its path. If the tornado will affect the near shore or coastal waters, it will be issued as the combined product: Tornado Warning and Special Marine Warning. If the thunderstorm which is causing the tornado is also producing torrential rains, this warning may also be combined with a Flash Flood Warning. After the Tornado Warning is issued, the affected NWFO will issue Severe Weather Statements that contain updated information on the tornado and when warning is no longer in effect.

Tornado Watch: Issued by the National Weather Service when conditions are favorable for the development of tornadoes in and close to the watch area. The Tornado Watch size can vary depending on the weather situation, but are usually issued for a duration of 4 to 8 hours. Tornado Watches are normally issued well in advance of the actual occurrence of severe weather. During the watch, people should review tornado safety rules and be prepared to move a place of safety if threatening weather approaches.

Tribal Assistance Coordination Group (TAC-G): The TAC-G is: Federal Government representatives that support Emergency Support Function #15 (External Affairs) and the Tribal Relations Support Annex of the NRF during disaster response operations under the Stafford Act; provides liaison officers to coordinate between Tribal

Governments and the Incident Command; and comprised of Indian Affairs, Indian Health Services, and Federal Emergency Management Agency representatives.

Tribal Government (Indian): Any Federally-recognized governing body of an Indian or Alaska Native tribe, band, nation, pueblo, village, or community that the Secretary of Interior acknowledges to exist as an Indian Tribe under the Federally Recognized Tribe List Act of 1994, 25 U.S.C 479a (does not include Alaska Native corporations, the ownership of which is vested in private individuals).

Tribal Leader (Indian): Individual responsible for the public safety and welfare of the people of that tribe.

Tribe (Indian): An Indian Tribe, band, nation, or other organized group or community, including a native village, regional corporation, or village corporation, as those terms are defined in Section 3 of the Alaska Native Claims Settlement Act (43 USC 1602), which is recognized as eligible for the special programs and services provided by the United States to Indians because of their status as Indians (36 CFR Part 800).

Tropical Cyclone: A warm-core non-frontal synoptic-scale cyclone, originating over tropical or subtropical waters, with organized deep convection and a closed surface wind circulation about a well-defined center. Once formed, a tropical cyclone is maintained by the extraction of heat energy from the ocean at high temperature and heat export at the low temperatures of the upper troposphere. In this they differ from extratropical cyclones, which derive their energy from horizontal temperature contrasts in the atmosphere (baroclinic effects).

Tropical Depression: A tropical cyclone in which the maximum sustained surface wind speed (using the U.S. 1-minute average) is 33 kt (38 mph or 62 km/hr) or less.

Tropical Disturbance: A discrete tropical weather system of apparently organized convection -- generally 100 to 300 mi. in diameter -- originating in the tropics or subtropics, having a nonfrontal migratory character, and maintaining its identity for 24 hours or more. It may or may not be associated with a detectable perturbation of the wind field.

Tropical Storm: A tropical cyclone in which the maximum sustained surface wind speed (using the U.S. 1-minute average) ranges from 34 kt (39 mph or 63 km/hr) to 63 kt (73 mph or 118 km/hr).

Tropical Storm Warning: An announcement that tropical storm conditions (sustained winds of 39 to 73 mph) are *expected* somewhere within the specified coastal area within 36 hours.

Tropical Storm Watch: An announcement that tropical storm conditions (sustained winds of 39 to 73 mph) is *possible* within the specified coastal area within 48 hours.

Tropical Wave: A trough or cyclonic curvature maximum in the trade-wind easterlies. The wave may reach maximum amplitude in the lower middle troposphere.

Tsunami: A series of long-period waves (on the order of tens of minutes) that are usually generated by an impulsive disturbance that displaces massive amounts of water, such as an earthquake occurring on or near the sea floor. Underwater volcanic eruptions and landslides can also cause tsunamis. While traveling in the deep oceans, tsunami have extremely long wavelengths, often exceeding 50 nm, with small amplitudes (a few tens of centimeters) and negligible wave steepness, which in the open ocean would cause nothing more than a gentle rise and fall for most vessels, and possibly go unnoticed. Tsunamis travel at very high speeds, sometimes in excess of 400 knots. As tsunamis reach the shallow waters near the coast, they begin to slow down while gradually growing steeper, due to the decreasing water depth. The building walls of destruction can become extremely large in height, reaching tens of meters 30 feet or more as they reach the shoreline. The effects can be further amplified where a bay, harbor, or lagoon funnels the waves as they move inland. Large tsunamis have been known to rise to over 100 feet. The amount of water and energy contained in tsunami can have devastating effects on coastal areas.

Tsunami Advisory: 1) Products of the Pacific Tsunami Warning Center (PTWC - Pacific (except Alaska, British Columbia and Western States) Hawaii, Caribbean (except Puerto Rico, Virgin Is.), and Indian Ocean): The third highest level of tsunami alert. Advisories are issued to coastal populations within areas not currently in either warning or watch status when a tsunami warning has been issued for another region of the same ocean. An Advisory indicates that an area is either outside the current warning and watch regions or that the tsunami poses no danger to that area. The Center will continue to monitor the event, issuing updates at least hourly. As conditions warrant, the Advisory will either be continued, upgraded to a watch or warning, or ended. 2) Products of the West Coast/Alaska Tsunami Warning Center (WC/ATWC - Alaska, British Columbia and Western States, Canada, Eastern and Gulf States, Puerto Rico, U.S Virgin Islands): A tsunami advisory is issued due to the threat of a potential tsunami which may produce strong currents or waves dangerous to those in or near the water. Coastal regions historically prone to damage due to strong currents induced by tsunamis are at the greatest risk. The threat may continue for several hours after the arrival of the initial wave, but significant widespread inundation is not expected for areas under an advisory. Appropriate actions to be taken by local officials may include closing beaches, evacuating harbors and marinas, and the repositioning of ships to deep waters when there is time to safely do so. Advisories are normally updated to continue the advisory, expand/contract affected areas, upgrade to a warning, or cancel the advisory.

Tsunami Information Statement: Issued to inform emergency management officials and the public that an earthquake has occurred, or that a tsunami warning, watch, or advisory has been issued for another section of the ocean. In most cases, information statements are issued to indicate there is no threat of a destructive tsunami and to prevent unnecessary evacuations as the earthquake may have been felt in coastal areas. An information statement may, in appropriate situations, caution about the possibility of destructive local tsunamis. Information statements may be re-issued with additional information, though normally these messages are not updated. However, a watch, advisory or warning may be issued for the area, if necessary, after analysis and/or updated information becomes available.

Tsunami Warning: 1) Products of the Pacific Tsunami Warning Center

(PTWC - Pacific (except Alaska, British Columbia and Western States) Hawaii, Caribbean (except Puerto Rico, Virgin Is.), and Indian Ocean): The highest level of tsunami alert. Warnings are issued due to the imminent threat of a tsunami from a large undersea earthquake or following confirmation that a potentially destructive tsunami is underway. They may initially be based only on seismic information as a means of providing the earliest possible alert. Warnings advise that appropriate actions be taken in response to the tsunami threat. Such actions could include the evacuation of low-lying coastal areas and the movement of boats and ships out of harbors to deep water. Warnings are updated at least hourly or as conditions warrant to continue, expand, restrict, or end the warning. 2) Products of the West Coast/Alaska Tsunami Warning Center (WC/ATWC - Alaska, British Columbia and Western States, Canada, Eastern and Gulf States, Puerto Rico, U.S Virgin Islands): A tsunami warning is issued when a potential tsunami with significant widespread inundation is imminent or expected. Warnings alert the public that widespread, dangerous coastal flooding accompanied by powerful currents is possible and may continue for several hours after arrival of the initial wave. Warnings also alert emergency management officials to take action for the entire tsunami hazard zone. Appropriate actions to be taken by local officials may include the evacuation of low-lying coastal areas, and the repositioning of ships to deep waters when there is time to safely do so. Warnings may be updated, adjusted geographically, downgraded, or canceled. To provide the earliest possible alert, initial warnings are normally based only on seismic information.

Tsunami Watch: 1) Products of the Pacific Tsunami Warning Center (PTWC - Pacific (except Alaska, British Columbia and Western States) Hawaii, Caribbean (except Puerto Rico, Virgin Is.), and Indian Ocean): The second highest level of tsunami alert. Watches are issued based on seismic information without confirmation that a destructive tsunami is underway. Issued as a means of providing an advance alert to areas that could be impacted by destructive tsunami waves. Watches are updated at least hourly to continue, expand their coverage, upgrade to a Warning, or end the alert. 2) Products of the West Coast/Alaska Tsunami Warning Center (WC/ATWC - Alaska, British Columbia and Western States, Canada, Eastern and Gulf States, Puerto Rico, U.S Virgin Islands): A tsunami watch is issued to alert emergency management officials and the public of an event which may later impact the watch area. The watch area may be upgraded to a warning or advisory or canceled based on updated information and analysis. Therefore, emergency management officials and the public should prepare to take action. Watches are normally issued based on seismic information without confirmation that a destructive tsunami is underway.

Type: An Incident Command System resource classification that refers to capability. Type 1 is generally considered to be more capable than Types 2, 3, or 4, respectively, because of size, power, capacity, or (in the case of Incident Management Teams) experience and qualifications.

Typhoon: A tropical cyclone in the Western Pacific Ocean in which the maximum 1-

minute sustained surface wind is 64 knots (74 mph) or greater.

Typhoon Season: The part of the year having a relatively high incidence of tropical cyclones. In the western North Pacific, the typhoon season is from July 1 to December 15. Tropical cyclones can occur year-round in any basin.

Unified Area Command: A Unified Area Command is established when incidents or accidents under an Area Command are multijurisdictional (see Area Command).

Unified Command (UC): An Incident Command System application used when more than one agency has incident jurisdiction or when incidents cross political jurisdictions. Agencies work together through the designated members of the UC, often the senior persons from agencies and/or disciplines participating in the UC, to establish a common set of objectives and strategies and a single Incident Action Plan.

Unified Coordination Group (UCG): Provides leadership within the Joint Field Office. The Unified Coordination Group is comprised of specified senior leaders representing State and Federal interests, and in certain circumstances Tribal Governments, local jurisdictions, the private sector, or nongovernmental organizations. The Unified Coordination Group typically consists of the Principal Federal Official (if designated), Federal Coordinating Officer, State Coordinating Officer, and senior officials from other entities with primary statutory or jurisdictional responsibility and significant operational responsibility for an aspect of an incident (e.g., the Senior Health Official, Department of Defense representative, or Senior Federal Law Enforcement Official if assigned). Within the Unified Coordination Group,

the Federal Coordinating Officer is the primary Federal official responsible for coordinating, integrating, and synchronizing Federal response activities.

Unit: The organizational element with functional responsibility for a specific incident planning, logistics, or finance/administration activity.

Unit Leader: The individual in charge of managing Units within an Incident Command System (ICS) functional Section. The Unit can be staffed by a number of support personnel providing a wide range of services. Some of the support positions are preestablished within ICS (e.g., Base/Camp Manager), but many others will be assigned as technical specialists.

Unity of Command: An Incident Command System principle stating that each individual involved in incident operations will be assigned to only one supervisor.

Urban Search and Rescue (US&R): Operational activities that include locating, extricating, and providing on-site medical treatment to victims trapped in collapsed structures.

Urban Search and Rescue (US&R) Incident Support Team (IST): US&R IST is a FEMA support team that provides Federal, State, and local officials with technical assistance in the acquisition and utilization of Emergency Support Function (ESF) #9 resources through advice, incident command assistance, management and coordination of US&R task forces, and obtaining ESF-9 logistic support.

Urban Search and Rescue (US&R) Task Forces: A framework for structuring local emergency services personnel into integrated disaster response task forces.

The 28 National US&R Task Forces, complete with the necessary tools, equipment, skills, and techniques, can be deployed by the Federal Emergency Management Agency to assist State and local governments in rescuing victims of structural collapse incidents or to assist in other search and rescue missions.

Vector: An agent, such as an insect or rat, capable of transferring a pathogen from one organism to another.

Visual Flight Rules (VFR): Rules governing procedures for conducting flight under visual meteorological conditions. In addition, used by pilots and controllers to indicate type of flight plan. (The term "VFR" is also used in the U. S. to indicate weather conditions equal to or greater than minimum VFR requirements.)

Volunteer: For purposes of the National Incident Management System, any individual accepted to perform services by the lead agency (which has authority to accept volunteer services) when the individual performs services without promise, expectation, or receipt of compensation for services performed. See 16 U.S.C. 742f(c) and 29 CFR 553.10.

Warning: A warning is issued when a hazardous weather or hydrologic event is occurring, is imminent, or has a very high probability of occurring. A warning is used for conditions posing a threat to life or property.

Watch: A watch is used when the risk of a hazardous weather or hydrologic event has increased significantly, but its occurrence, location, and/or timing is still uncertain. It is intended to provide enough lead time so that those who need to set their plans in motion can do so.

Waterspout: In general, a tornado occurring over water. Specifically, it normally refers to a small, relatively weak rotating column of air over water beneath a cumulonimbus or towering cumulus cloud. Waterspouts are most common over tropical or subtropical waters. The exact definition of waterspout is debatable. In most cases the term is reserved for small vortices over water that are not associated with storm-scale rotation. But there is sufficient justification for calling virtually any rotating column of air a waterspout if it is in contact with a water surface.

Weapons of Mass Destruction (WMD): 1. WMD are defined in US Code 2332a(c)2 as: (A) any destructive device as defined in section 921 of this title: a. The term "destructive device" means any explosive, incendiary, or poison gas bomb, grenade, rocket having a propellant charge of more than four ounces, missile having an explosive or incendiary charge of more than one-quarter ounce, mine, or similar device; (B) any weapon that is designed or intended to cause death or serious bodily injury through the release, dissemination, or impact of toxic or poisonous chemicals or their precursors; (C) any weapon involving a biological agent, toxin, or vector (as those terms are defined in section 179 of this title); or (D) any weapon that is designed to release radiation or radioactivity at a level dangerous to human life. 2. DoD uses the following approved definition of WMD as "chemical, biological, radiological, or nuclear weapons capable of a high order of a high order of destruction or causing mass casualties and exclude the means of transporting or propelling the weapon where such means is a separable and divisible part from the weapon."

Weapons of Mass Destruction - Civil Support Team (WMD-CST): Joint Army and Air National Guard teams that provide chemical, biological, and radiological initial survey and assessment operations for domestic WMD incidents. The 57 WMD-CSTs are designed to provide a specialized capability to respond to a CBRNE incident primarily in a Title 32 status within the United States and its territories. The mission of the WMD-CST is to support civil authorities at domestic CBRNE incident sites by identifying CBRNE agents and substances, assessing current and projected consequences, advising on response measures, and assisting with appropriate requests for additional support. This includes incidents involving intentional or unintentional release of CBRNE and natural or manmade disasters that result or could result in the catastrophic loss of life or property in the United States.

List of Acronyms

AARC	Airspace Access Response Cell	DEN	Domestic Events Network
ACO	Aircraft Coordinator	DHS	Department of Homeland Security
ADA	Americans with Disabilities Act	DNDO	Domestic Nuclear Detection Office
AFRCC	Air Force Rescue Coordination Center	DoD	Department of Defense
		DOI	Department of Interior
AHJ	Authority Having Jurisdiction	DOS	Department of State
AKRCC	Alaska Rescue Coordination Center	DSC	Dual Status Commander
ALARA	As Low As Reasonably Achievable	DSCA	Defense Support of Civil Authorities
ANS	Air Navigation Services		
AOB	Air Operations Branch	EICC	Emergency Incident Coordination Center
AOR	Area of Responsibility		
ATC	Air Traffic Control	ELT	Emergency Locator Transmitter
ATM	Air Traffic Management	EMAC	Emergency Management Assistance Compact
ATO	Air Traffic Organization		
BOB	Boat Operations Branch	EMP	Electromagnetic Pulse
C2CRE	Command and Control CBRN Response Enterprise	EMS	Emergency Medical System
		EMT	Emergency Medical Technician
CBRN	Chemical, Biological, Radiological and Nuclear	EOC	Emergency Operations Center
		EPIRB	Emergency Position Indicating Radio Beacon
CBRNE	Chemical, Biological, Radiological, Nuclear, and High-Yield Explosives	EPLO	Emergency Preparedness Liaison Officer
CDC	Centers for Disease Control		
CERFP	CBRNE Emergency Response Force Package	ESF	Emergency Support Function
		FAA	Federal Aviation Administration
CFS	Cubic Feet per Second	FCO	Federal Coordinating Officer
CIKR	Critical Infrastructure and Key Resources	FEMA	Federal Emergency Management Agency
CIS	Catastrophic Incident Supplement	FSARCG	Federal SAR Coordination Group
CISAR	Catastrophic Incident Search and Rescue	GA	General Aviation
		GARS	Global Area Reference System
COP	Common Operating Picture	GIS	Geospatial Information System
CPG	Comprehensive Preparedness Guide	GPS	Global Positioning System
CSP	Commence Search Point	HAV	Hepatitis A Virus
CST	Civil Support Team	HEPA	High-Efficiency Particulate Air
DCO	Defense Coordination Officer	HRF	Homeland Response Force
DCRF	Defense CBRN Response Force	HSPD	Homeland Security Presidential Directive
DECO	Disaster Emergency Communications Officer	IA	Indian Affairs

IAA	Incident Assessment and Awareness	NRF	National Response Framework
IAMSAR	International Aeronautical and Maritime Search and Rescue	NSP	National Search and Rescue Plan
IAP	Incident Action Plan	NSS	National Search and Rescue Supplement
IC	Incident Commander	NSSE	National Security Special Events
ICISF	International Critical Incident Stress Foundation	OSC	On Scene Coordinator
ICS	Incident Command System	PDD	Presidential Disaster Declaration
IFR	Instrument Flight Rules	PFD	Personal Flotation Device
IHS	Indian Health Service	PIC	Pilot in Command
IMC	Instrument Meteorological Conditions	PIO	Public Information Officer
IND	Improvised Nuclear Device	PLB	Personal Locator Beacon
INSARAG	International Search and Rescue Advisory Group	PPE	Personal Protective Equipment
		PR	Personnel Recovery
IST	Incident Support Team	RCC	Rescue Coordination Center
JAC	Joint Analysis Center	RDD	Radiological Dispersal Device
JFO	Joint Field Office	REAC	Radiation Emergency Assistance Center
JIC	Joint Information Center	RFA	Request for Assistance
JPRC	Joint Personnel Recovery Center	RFF	Request for Forces
JTF	Joint Task Force	RRCC	Regional Response Coordination Center
kHz	Kilohertz		
MA	Mission Assignment	SAP	Search Action Plan
MHz	Megahertz	SAR	Search and Rescue
MOA	Memorandum of Agreement	SBTF	Small Boat Task Force
MOG	Maximum on Ground	SC	Search and Rescue Coordinator
MOU	Memorandum of Understanding	SECDEF	Secretary of Defense
MRO	Mass Rescue Operation	SGS	Strategic Guidance Statement
NAD	North American Datum	SITREP	Situation Report
NAS	National Airspace System	SMC	Search and Rescue Mission Coordinator
NDART	National Disaster Animal Response Team	SOSC	System Operations Support Center
NFPA	National Fire Protection Association	SRR	Search and Rescue Region
NG	National Guard	SRT	Special Response Team
NGO	Nongovernmental Organization	TAC-G	Tribal Assistance Coordination Group
NIFOG	National Interoperability Field Operations Guide	TFR	Temporary Flight Restriction
		UCG	Unified Coordinating Group
NIMS	National Incident Management System	UAS	Unmanned Aerial System
		US&R	Urban Search and Rescue
NOK	Next-of-kin	USC	United States Code
NPS	National Park Service	USCG	United States Coast Guard
NRCC	National Response Coordinator Center	USNG	United States National Grid

USNORTHCOM United States Northern Command

USPACOM United States Pacific Command

VFR Visual Flight Rule

VHF Very High Frequency

VIP Very Important Person

VMC Visual Meteorological Conditions

VOAD Voluntary Organizations Active in Disasters

UHF Ultra High Frequency

US&R Urban Search and Rescue

WGS World Geodetic System

WMD Weapons of Mass Destruction

WMD-CST WMD Civil Support Team

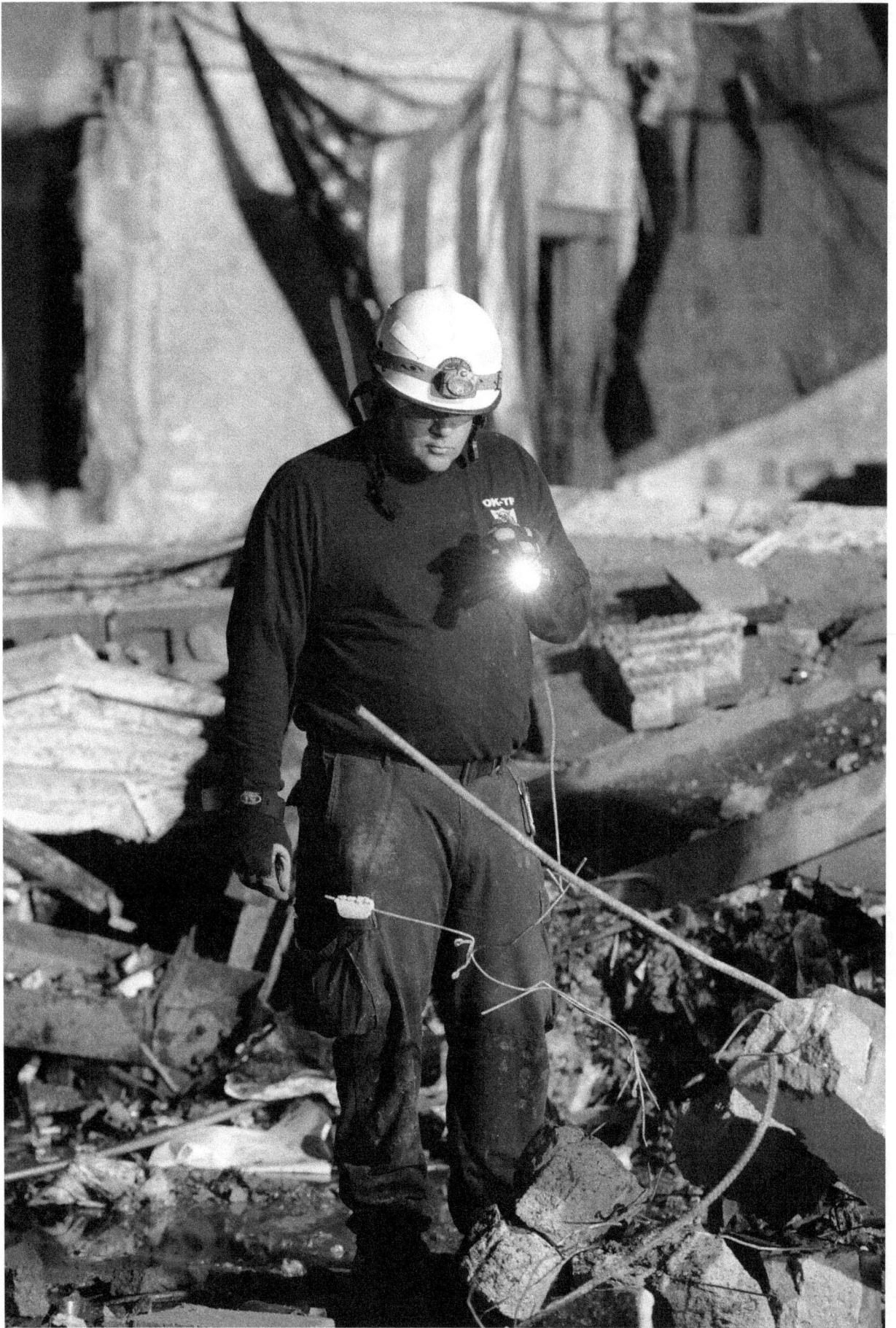

Introduction:
CISAR Addendum (Version 3.0)

Introduction

Purpose

Limited Scope

Potential Large Scale

Effective CISAR Planning

Introduction

In recent years, our nation has faced a remarkable series of disasters and emergencies. As a result, our national response structures are evolving to meet these threats. The *National Response Framework (NRF)* is the next step in this evolution by defining how we respond as a nation to these challenges.

Based on best practices and lessons learned, the NRF and its associated annexes have been developed to provide guiding principles that enable all Federal response partners to prepare for and provide a unified national response to disasters and emergencies – from the smallest incident to the largest catastrophe.

The NRF's key principles include:

- Engaged partnership;
- Tiered response;
- Scalable, flexible, and adaptable operational capabilities; and
- Unity of effort through unified command,

In support of these principles, the National Search and Rescue Committee (NSARC)

has completely revised the *Catastrophic Incident Search and Rescue (CISAR) Addendum (Version 3.0)* to the *National Search and Rescue Supplement (NSS)*. This *CISAR Addendum (Version 3.0)* includes over one hundred additional pages of information and has revised the existing text with the latest policies and procedures in the conduct of disaster search and rescue (SAR) operations. Its purpose is to address the continuing process of national preparedness and response to catastrophic incidents and Emergency Support Function (ESF) #9 (SAR) operations in support of Federal, State, Tribal, Territorial, and local authorities.

Purpose

This CISAR Addendum (Version 3.0) is intended to provide a description of the:

- Federal Government's civil SAR response to catastrophic incidents and other ESF #9 operations;
- Guide Federal authorities involved in the SAR response; and
- Inform States, Tribes, and Territories on what to expect when Federal SAR responders are requested to assist.

Limited Scope

This CISAR Addendum (Version 3.0) provides Federal, interagency guidance in the conduct of SAR operations during catastrophic incidents and other ESF #9 operations. CISAR consists of SAR operations carried out as all or part of the response to an emergency or disaster declared by the President, under provisions of the NRF and Emergency Support Function (ESF) #9.

This Addendum does not supersede other Federal, State, Tribal, or Territorial SAR plans.

Potential Large Scale

Depending on the nature of a catastrophic incident (e.g., earthquake, hurricane, terrorist attack, etc.), CISAR operations may be required. If CISAR operations are conducted and ESF #9 is implemented, operations may be either a minor or major aspect of the overall incident response. Even if CISAR operations are conducted, it may

very well be that State, Tribal, Territorial, and local authorities can conduct CISAR operations using their own SAR resources. This Addendum recognizes the need for a Federal, scalable, coordinated SAR effort that can be tailored to the incident.

An effective response to a major catastrophic incident typically requires immediate, well-planned, and coordinated large-scale actions and use of resources from multiple organizations.

Effective CISAR Planning

Successful large-scale CISAR operations depend on the development of contingency plans, as well as effective cooperation and coordination between Federal, State, Tribal, Territorial, and local SAR authorities *before* an event occurs. Knowing your SAR partners, developing interagency plans, and exercising those plans is critical for the conduct of mass rescue operations in a disaster.

Part 1: Organization

The Stafford Act

Robert T. Stafford
Disaster Relief and Emergency
Assistance Act, *as Amended*

April 2013

FEMA

Section 1-1: CISAR Primary Reference and Guidance Documents

Table 1-1-1 below provides a summary of primary references and guidance documents for CISAR operations.

Table 1-1-1: Key CISAR References	
Robert T Stafford Disaster Relief and Emergency Assistance Act ("Stafford Act")	Robert T. Stafford Disaster Relief and Emergency Assistance Act (P.L. 100-707): Provides the statutory framework for a Presidential declaration of an emergency or a declaration of a major disaster, and describes the programs and processes by which the Federal Government provides disaster and emergency assistance to State and local governments, tribal nations, eligible private nonprofit organizations, and individuals affected by a declared major disaster or emergency. The Stafford Act structure for the declaration process reflects the fact that federal resources under this act supplement state and local resources for disaster relief and recovery. The Stafford Act covers all hazards, including natural disasters and terrorist events.
Flood Control and Coastal Emergencies Act, 33 USC 701n (2007) (commonly referred to as Pub. L. 84-99)	Authorizes an emergency fund for preparation for emergency response to, among other things, natural disasters, flood fighting and rescue operations, repair or restoration of flood control and hurricane protection structures, temporary restoration of essential public facilities and services, and provision of emergency supplies of water.
Economy Act (31 U.S.C. 1535)	Authorizes Federal agencies to provide goods or services, on a reimbursable basis, to other Federal agencies when more specific statutory authority does not exist.
Executive Order 12148, 44 Fed. Reg. 43239 (1979), as amended, by Executive Order 13286, 68 Fed. Reg. 10619 (2003)	Designates DHS as the primary agency for coordination of Federal disaster relief, emergency assistance, and emergency preparedness. The order also delegates the President's relief and assistance functions under the Stafford Act to the Secretary of Homeland Security, with the exception of the declaration of a major disaster or emergency.
Homeland Security Presidential Directive (HSPD) 5: Management of Domestic Incidents	HSPD-5 serves to enhance the ability of the United States to manage domestic incidents by establishing a single, comprehensive national incident management system. This management system is designed to cover the prevention, preparation, response, and recovery from terrorist attacks, major disasters, and other emergencies. The directive gives further detail on which government officials oversee and have authority for various parts of the national incident management system.
Presidential Policy Directive (PPD) 8: National Preparedness	PPD 8 is aimed at strengthening the security and resilience of the United States through systematic preparation for the threats that pose the greatest risk to the security of the Nation, including acts of terrorism, cyber attacks, pandemics, and catastrophic natural disasters.

Table 1-1-1 is continued on the next page.

Table 1-1-1: Key CISAR References (continued)	
National Response Framework (NRF)	The NRF presents the guiding principles that enable all response partners to prepare for and provide a unified national response to disasters and emergencies. It establishes a comprehensive, national, all-hazards approach to domestic incident response.
Emergency Support Function (ESF) #9 – Search and Rescue (SAR) Annex	ESF #9 details the Federal Government's SAR responsibilities, identifies FEMA as the ESF #9 Coordinator, and explains the duties of the Federal Agency(s) assigned as Lead Primary Agency for a specific incident requiring Federal SAR assistance (See Section 1-3: Emergency Support Function #9).
Catastrophic Incident Annex	The Catastrophic Incident Annex to the National Response Framework (NRF-CIA) establishes the context and overarching strategy for implementing and coordinating an accelerated, proactive national response to a catastrophic incident.
National Search and Rescue Plan (NSP)	The NSP is an interagency agreement that constitutes the primary authority and policy guidance for involvement of Federal Agencies (including the military), in coordinating, providing, or supporting civil SAR services so that the United States can meet both domestic needs and international commitments.
National Search and Rescue Supplement (NSS)	The NSS is a Federal manual on civil SAR that, together with its various addenda, provides extensive guidance for implementation of the NSP.
International Aeronautical and Maritime Search and Rescue Manual (IAMSAR Manual)	The IAMSAR Manual is a three-volume Manual used worldwide for aeronautical and maritime civil SAR. In the United States, the IAMSAR Manual is supplemented by the NSS and this Addendum.

Section 1-2: Catastrophic Incident SAR (CISAR)

Catastrophic Incident Search and Rescue (CISAR)

Homeland Security Presidential Directive 5 (HSPD-5)

CISAR Legal Considerations

Catastrophic Incident Search and Rescue (CISAR)

Civil SAR is composed of search operations, rescue operations, and associated civilian services provided to assist persons and property in potential or actual distress in a non-hostile environment.

Catastrophic Incident Search and Rescue (CISAR) consists of civil SAR operations carried out as all or part of the response to an emergency or disaster declared by the President under provisions of the NRF and ESF #9 – SAR (Appendix A).

Catastrophic Incident

"A catastrophic incident is any natural or manmade incident, including terrorism, which results in extraordinary levels of mass causalities, damage, or disruption severely affecting the population, infrastructure, environment, economy, national morale, and/or government functions."

National Response Framework, page 42

The nature of CISAR could range from normal SAR operations (limited number of persons in distress) to the conduct of mass rescue operations. Two criteria must be met for an incident to be identified as CISAR. First, the response must be associated with a Presidential Declaration. Second, ESF #9 must be implemented. Clear delineation between normal SAR, mass rescue

operations, and CISAR may not be apparent. However, it is important to understand that flexible response options are available for these progressive or potentially overwhelming events. This Addendum attempts to provide standardized and flexible options for any type of CISAR operation.

Provisions of the NSP, NSS, and relevant addenda always apply to civil SAR regardless of whether the operations are CISAR. If the operations are CISAR under ESF #9, then provisions of the NRF and its relevant supporting documents also apply and are intended to *support* the provisions of the NSP.

Homeland Security Presidential Directive 5 (HSPD-5)

HSPD-5 explains the Federal Government's policy on responding to disasters. HSPD-5 states the following:

To prevent, prepare for, respond to, and recover from terrorist attacks, major disasters, and other emergencies, the United States Government shall establish a single, comprehensive approach to domestic incident management. The objective of the United States Government is to ensure that all levels of government across the Nation have the capability to work efficiently and effectively together, using a national approach to domestic incident management.

In these efforts, with regard to domestic incidents, the United States Government treats crisis management and consequence management as a single, integrated function, rather than as two separate functions.

The Secretary of Homeland Security is the principal Federal official for domestic incident management. Pursuant to the Homeland Security Act of 2002, the Secretary is responsible for coordinating Federal operations within the United States to prepare for, respond to, and recover from terrorist attacks, major disasters, and other emergencies. The Secretary shall coordinate the Federal Government's resources utilized in response to or recovery from terrorist attacks, major disasters, or other emergencies if and when any one of the following four conditions applies:

(1) A Federal Department or Agency acting under its own authority has requested the assistance of the Secretary;

(2) The resources of State and local authorities are overwhelmed and Federal assistance has been requested by the appropriate State and local authorities;

(3) More than one Federal department or agency has become substantially involved in responding to the incident; or

(4) The Secretary has been directed to assume responsibility for managing the domestic incident by the President.

Figure 1-2-1 below details a typical State request for Federal assistance.

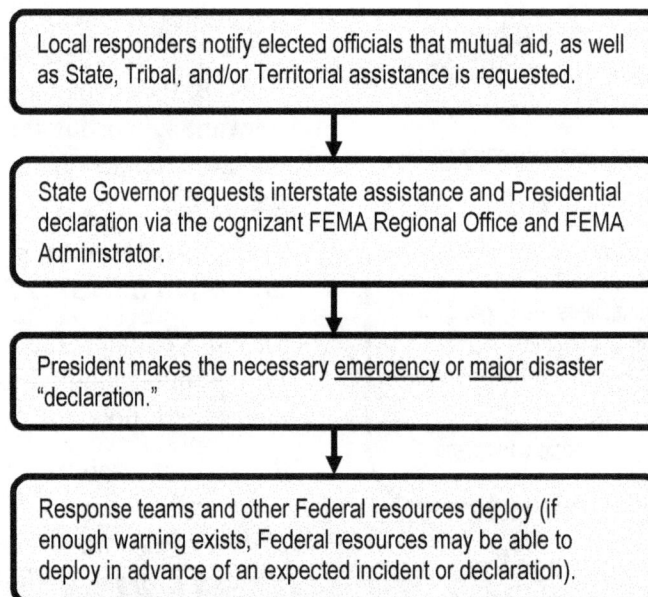

```
┌─────────────────────────────────────────────────────────────┐
│ Local responders notify elected officials that mutual aid, as │
│ well as State, Tribal, and/or Territorial assistance is        │
│ requested.                                                     │
└─────────────────────────────────────────────────────────────┘
                            ↓
┌─────────────────────────────────────────────────────────────┐
│ State Governor requests interstate assistance and Presidential │
│ declaration via the cognizant FEMA Regional Office and FEMA    │
│ Administrator.                                                 │
└─────────────────────────────────────────────────────────────┘
                            ↓
┌─────────────────────────────────────────────────────────────┐
│ President makes the necessary emergency or major disaster      │
│ "declaration."                                                │
└─────────────────────────────────────────────────────────────┘
                            ↓
┌─────────────────────────────────────────────────────────────┐
│ Response teams and other Federal resources deploy (if          │
│ enough warning exists, Federal resources may be able to        │
│ deploy in advance of an expected incident or declaration).     │
└─────────────────────────────────────────────────────────────┘
```

Figure 1-2-1: Emergency or Major Disaster Declaration

CISAR Legal Considerations

The Robert T. Stafford Disaster Relief and Emergency Assistance Act (the "Stafford Act") authorizes and describes various programs and processes by which the Federal Government provides disaster and emergency assistance to State and local Governments, tribal nations, eligible private nonprofit organizations, and individuals affected by a declared major disaster or emergency. The Stafford Act covers all hazards, including natural disasters and terrorist events. Generally, Stafford Act assistance is rendered upon request from a State Governor, provided certain conditions are met, primarily that the Governor certifies that the State lacks the resources and capabilities to manage the consequences of the event without Federal assistance.

Federal SAR resources deploying in support of a Stafford Act Declaration must balance their legal authority under the Stafford Act and other applicable authorities with the concurrent authority of State and local jurisdictions. The Stafford Act provides authority to Federal SAR resources, when it is essential to meet immediate threats to life and property resulting from a major disaster, to perform on public or private lands or waters any work or services essential to saving lives and protecting or preserving property or public health and safety including debris removal, search and rescue and demolition of unsafe structures which endanger the public (42 USC 5170b(a)(3); Similar authority exists for Emergency Declarations under 42 USC 5192). In addition, States must agree to assist Federal agencies in all support and local jurisdictional matters (44 CFR 206.208). However, State and local authorities continue to maintain their inherent authority over their jurisdictions. This means Federal SAR resources normally must adhere to the instructions of State and local authorities even when deployed in support of a declaration under the Stafford Act.

Section 1-3: Emergency Support Function #9 – Search and Rescue (SAR)

Overview: Emergency Support Functions (ESFs)

ESF #9 – Search and Rescue

FEMA: ESF #9 Coordinator

ESF Primary Agency (General Requirements)

ESF #9 Overall Primary Agency

ESF #9 Support Agencies

Mission Assignments

ESF #9 Planning and Coordination

Overview: Emergency Support Functions (ESFs)

As detailed in the NRF, Federal, State, Tribal, Territorial, and local governments, as well as other organizations, organize resources and capabilities under 15 ESFs (Table 1-3-1 on the next page). ESFs align categories of resources and provide strategic objectives for their use.

Additionally, ESFs provide the structure for coordinating Federal interagency support for a Federal response to an incident. ESFs are mechanisms for grouping functions most frequently used to provide Federal support to States and Federal-to-Federal support, both for declared disasters and emergencies under the Stafford Act and for non-Stafford Act incidents.

Each ESF Annex identifies the ESF Coordinator, as well as the Primary and Support Agencies with the resources and capabilities to support the specific functional activities of the particular ESF. Several ESFs incorporate multiple components, with Primary Agencies designated for each component to ensure seamless integration of and transition between preparedness, response, and recovery activities.

ESFs with multiple Primary Agencies (e.g., ESF #9 – SAR) designate an ESF Coordinator for the purposes of pre-incident planning and coordination of Primary and Support Agency efforts throughout the incident.

ESFs may be selectively activated for both Stafford Act and non-Stafford Act incidents where State, Tribal, or Territorial authorities request Department of Homeland Security (DHS) assistance, or under other circumstances as defined in HSPD-5.

Not every incident results in the activation of all ESFs. Activation depends on the nature and magnitude of the event, the suddenness of onset, and for ESF #9, the knowledge or expectation that the incident may result in a request for an integrated SAR response and the capabilities available to local SAR authorities may be inadequate or exceeded.

Table 1-3-1: ESF Roles and Responsibilities (Reference: National Response Framework)	
ESF	**Scope**
ESF #1 – Transportation	• Aviation/airspace management and control • Transportation safety • Restoration/recovery of transportation infrastructure • Movement restrictions • Damage and impact assessment
ESF #2 – Communications	• Coordination with telecommunications and information technology industries • Restoration and repair of telecommunications infrastructure • Protection, restoration, and sustainment of national cyber and information technology resources • Oversight of communications within the Federal incident management and response structures
ESF #3 – Public Works and Engineering	• Infrastructure protection and emergency repair • Infrastructure restoration • Engineering services and construction management • Emergency contracting support for life-saving and life-sustaining services
ESF #4 – Firefighting	• Coordination of Federal firefighting activities • Support to wildland, rural, and urban firefighting operations
ESF #5 – Emergency Management	• Coordination of incident management and response efforts • Issuance of mission assignments • Resource and human capital • Incident action planning • Financial management
ESF #6 – Mass Care, Emergency Assistance, Housing, and Human Services	• Mass care • Emergency assistance • Disaster housing • Human services
ESF #7 – Logistics Management and Resource Support	• Comprehensive, national incident logistics planning, management, and sustainment capability • Resource support (facility space, office equipment and supplies, contracting services, etc.)

Table 1-3-1 continued on the next page.

Table 1-3-1: ESF Roles and Responsibilities (continued) (Reference: National Response Framework)	
ESF #8 – Public Health and Medical Services	• Public health • Medical • Mental health services • Mass fatality management
ESF #9 – Search and Rescue	• Life-saving assistance • Search and rescue operations
ESF #10 – Oil and Hazardous Materials Response	• Oil and hazardous materials (chemical, biological, radiological, etc.) response • Environmental short- and long-term cleanup
ESF #11 – Agriculture and Natural Resources	• Nutrition assistance • Animal and plant disease and pest response • Food safety and security • Natural and cultural resources and historic properties protection and restoration • Safety and well-being of household pets
ESF #12 – Energy	• Energy infrastructure assessment, repair, and restoration • Energy industry utilities coordination • Energy forecast
ESF #13 – Public Safety and Security	• Facility and resource security • Security planning and technical resource assistance • Public safety and security support • Support to access, traffic, and crowd control
ESF #14 – Long-Term Community Recovery	• Social and economic community impact assessment • Long-term community recovery assistance to States, local governments, and the private sector • Analysis and review of mitigation program implementation
ESF #15 – External Affairs	• Emergency public information and protective action guidance • Media and community relations • Congressional and international affairs • Tribal and insular affairs

ESF #9 – Search and Rescue

Federal assistance under ESF #9 (Appendix A), coordinates the provisioning of Federal SAR resources in the conduct of lifesaving operations in support of the State or other Federal Agency requesting the resources. Assistance under ESF #9 is scalable to meet the specific needs of each incident, based upon the nature and magnitude of the event and the capabilities of State, Tribal, Territorial, and local SAR resources.

Figure 1-3-1 on the next page conceptually represents the planning, interagency coordination, and conduct of ESF #9 SAR operations.

1-11

Figure 1-3-1: Stafford Act Declaration/Emergency Support Function #9 Concept

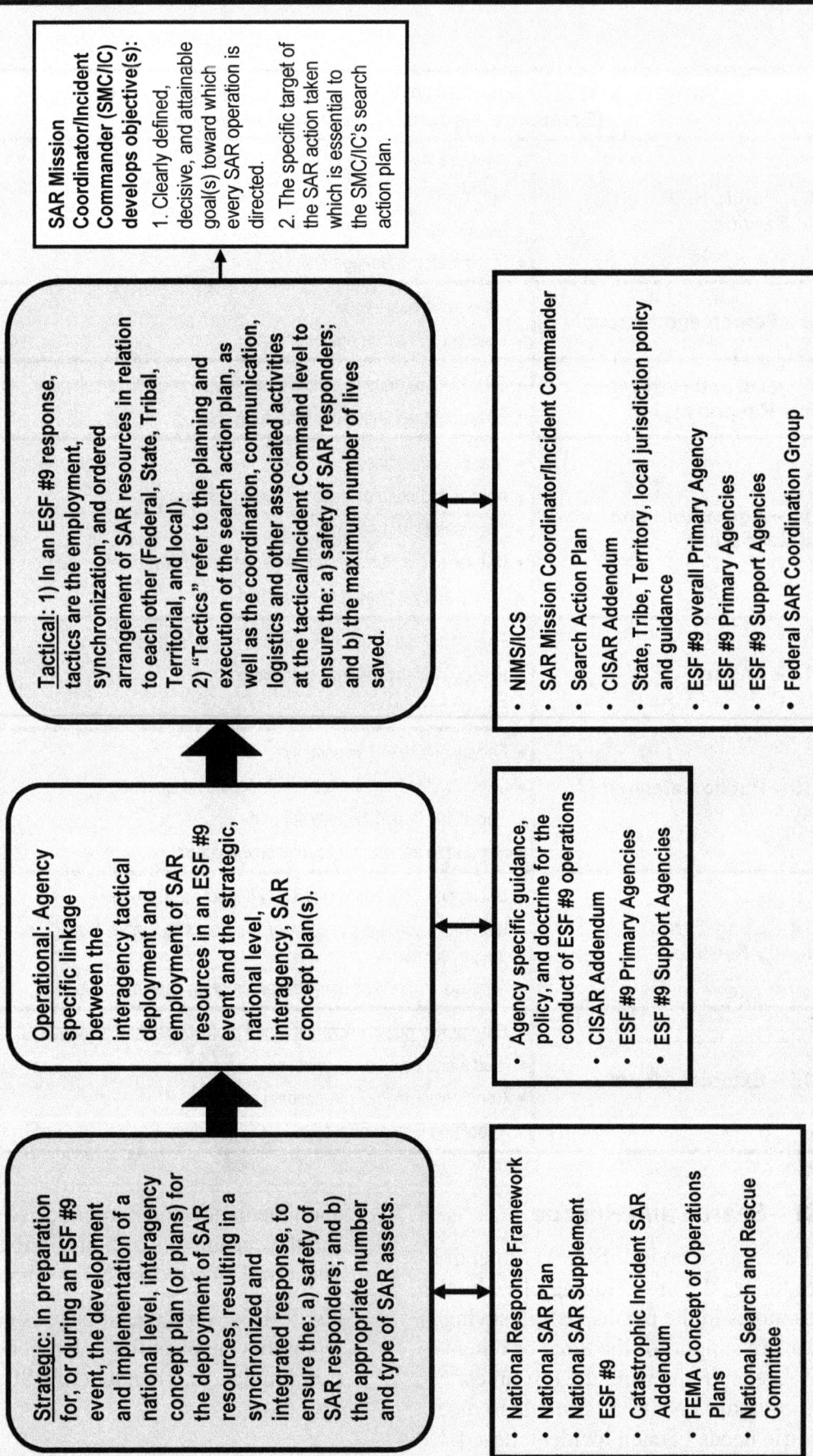

Strategic: In preparation for, or during an ESF #9 event, the development and implementation of a national level, interagency concept plan (or plans) for the deployment of SAR resources, resulting in a synchronized and integrated response, to ensure the: a) safety of SAR responders; and b) the appropriate number and type of SAR assets.

- National Response Framework
- National SAR Plan
- National SAR Supplement
- ESF #9
- Catastrophic Incident SAR Addendum
- FEMA Concept of Operations Plans
- National Search and Rescue Committee

Operational: Agency specific linkage between the interagency tactical deployment and employment of SAR resources in an ESF #9 event and the strategic, national level, interagency, SAR concept plan(s).

- Agency specific guidance, policy, and doctrine for the conduct of ESF #9 operations
- CISAR Addendum
- ESF #9 Primary Agencies
- ESF #9 Support Agencies

Tactical: 1) In an ESF #9 response, tactics are the employment, synchronization, and ordered arrangement of SAR resources in relation to each other (Federal, State, Tribal, Territorial, and local). 2) "Tactics" refer to the planning and execution of the search action plan, as well as the coordination, communication, logistics and other associated activities at the tactical/Incident Command level to ensure the: a) safety of SAR responders; and b) the maximum number of lives saved.

- NIMS/ICS
- SAR Mission Coordinator/Incident Commander
- Search Action Plan
- CISAR Addendum
- State, Tribe, Territory, local jurisdiction policy and guidance
- ESF #9 overall Primary Agency
- ESF #9 Primary Agencies
- ESF #9 Support Agencies
- Federal SAR Coordination Group

SAR Mission Coordinator/Incident Commander (SMC/IC) develops objective(s):

1. Clearly defined, decisive, and attainable goal(s) toward which every SAR operation is directed.
2. The specific target of the SAR action taken which is essential to the SMC/IC's search action plan.

While the Federal Emergency Management Agency (FEMA) coordinates the overall planning and conduct of disaster response operations under the Stafford Act, the *strategic*, national level, interagency ESF #9 planning and guidance development is coordinated through the National Search and Rescue Committee (NSARC).

In the NSARC, the Federal Agencies responsible for the planning and conduct of ESF #9 operations are:

- FEMA Urban Search and Rescue (US&R);

- U.S. Coast Guard (USCG);

- National Park Service (NPS); and

- Department of Defense (DoD)/ Commander, U.S. Northern Command (CDRUSNORTHCOM)/Commander, U.S. Pacific Command (CDRUSPACOM)

At the *operational* level, FEMA US&R, USCG, NPS, and CDRUSNORTHCOM/ CDRUSPACOM (the ESF #9 Primary Agencies) are:

- Responsible for the conduct and implementation of ESF #9 operations; and

- Development of agency specific plans, doctrine, training, tactics, techniques, and procedures for the conduct of ESF #9 operations.

At the *tactical* level, the ESF #9 Primary Agencies conduct CISAR operations at the request of the State/SAR Mission Coordinator (SMC)/Incident Commander (IC). Federal SAR assistance is normally conducted under an ESF #9 Mission Assignment (MA).

ESF #9: Purpose

Emergency Support Function (ESF) #9 – Search and Rescue rapidly deploys Federal SAR resources to provide lifesaving assistance to State, Tribal, Territorial, and local SAR Coordinator(s), SAR Mission Coordinator(s), and other authorities when a request for assistance is made or anticipated from an authority recognized by the NRF for incidents or potential incidents requiring an integrated SAR response.

ESF #9 – Search and Rescue Annex

FEMA: ESF #9 Coordinator

The ESF Coordinator is the entity with management oversight for a particular ESF. DHS/FEMA is the ESF #9 Coordinator.

Responsibilities of the ESF Coordinator include:

- Coordination before, during, and after an incident, including pre-incident planning and coordination;

- Maintaining ongoing contact with ESF primary and support agencies;

- Conduct periodic ESF meetings and conference calls;

- Coordinate efforts with corresponding private-sector organizations; and

- Coordinate ESF activities relating to catastrophic incident planning and critical infrastructure preparedness, as appropriate.

For ESF #9 in particular, DHS/FEMA will:

- Activate ESF #9 when an incident is anticipated or occurs that may result in a request for an integrated SAR response to an impacted area;

- Designate the overall Primary Agency for an ESF #9 SAR response (based on incident circumstances, SAR environment, type of response required, etc.); and

- Coordinate with other ESFs to ensure the most expedient and efficient resources are mobilized.

ESF Primary Agency (General Requirements)

An ESF Primary Agency is a Federal Agency with significant authorities, roles, resources, or capabilities for a particular function within an ESF.

A Federal Agency designated as an ESF Primary Agency serves as a Federal executive agent under the Federal Coordinating Officer (or Federal Resource Coordinator for non-Stafford Act incidents) to accomplish the ESF mission.

As described in the *ESF Annex Introduction (http://www.fema.gov/pdf/emergency/nrf/nrf -esf-intro.pdf),* when an ESF is activated in response to an incident, the Primary Agency (including all ESF #9 Primary Agencies), are responsible for:

- Supporting the ESF Coordinator and coordinating closely with the other Primary and Support Agencies;

- Orchestrating Federal support within their functional area for an affected State;

- Providing staff for the operations functions at fixed and field facilities;

- Notifying and requesting assistance from Support Agencies;

- Managing MAs;

- Coordinating with Support Agencies, as well as appropriate State officials, operations centers, and agencies;

- Working with appropriate private-sector organizations to maximize use of all available resources;

- Supporting and keeping other ESFs and organizational elements informed of ESF operational priorities and activities;

- Conducting situational and periodic readiness assessments;

- Executing contracts and procuring goods and services as needed;

- Ensuring financial and property accountability for ESF activities;

- Planning for short- and long-term incident management and recovery operations;

- Maintaining trained personnel to support interagency emergency response and support teams; and

- Identifying new equipment or capabilities required to prevent or respond to new or emerging threats and hazards, or to improve the ability to address existing threats.

ESF #9 has four Primary Agencies. Table 1-3-2 on the next page lists the ESF #9 Primary Agencies, their SAR environment responsibility, and an operational overview of each respective Primary Agency.

Table 1-3-2: ESF #9 Primary Agencies		
Type of SAR	**Primary Agency**	**Operational Overview**
Structure Collapse (Urban) SAR (US&R)	**DHS/Federal Emergency Management Agency (FEMA)**	Includes operations for natural and man-made disasters and Catastrophic Incidents as well as other structural collapse operations that primarily require DHS/FEMA US&R task force operations. The National US&R Response System integrates DHS/FEMA US&R task forces, Incident Support Teams (ISTs), and technical specialists. The Federal structural collapse SAR response integrates DHS/FEMA task forces in support of integrated SAR operations conducted in accordance with the NSP.
Maritime/Coastal/ Waterborne SAR	**DHS/U.S. Coast Guard (USCG)**	Includes operations for natural and man-made disasters that primarily require DHS/USCG air, ship, boat, and response team operations. The Federal maritime/coastal/waterborne SAR response integrates DHS/USCG resources in support of integrated SAR operations conducted in accordance with the NSP.
Land SAR	**DOI/National Park Service (NPS) and Department of Defense (DoD)**	Includes operations that require aviation and ground forces to meet mission objectives other than maritime/coastal/waterborne and structural collapse SAR operations as described above. Land SAR Primary Agencies will integrate their efforts to provide an array of diverse capabilities under ESF #9.

ESF #9 (Appendix A) lists the duties and responsibilities of each Primary Agency.

ESF #9 Overall Primary Agency

The purpose of the ESF #9 overall Primary Agency is to coordinate the provisioning and response of Federal SAR resources (both Primary and Support Agencies) to the State, Tribe, Territory, or other Federal Agency requesting those resources as rapidly as possible.

In order to expedite resource requests, the ESF #9 overall Primary Agency should have representatives at key response coordination locations (e.g., NRCC, RRCC, JFO, IMAT and State, Tribal, Territorial and local Incident Commands (SAR Subject Matter Expert), EOCs, or SAR Branches).

The ESF #9 overall Primary Agency representatives:

- Are critical for assisting in coordinating which ESF #9 Primary Agency will provide the requested SAR resources;

- Should understand how ESF #9 resources are requested, deployed, and employed under the NSP as well as through the MA process; and

- Have available ESF #9 Primary and Support Agency contact information for coordinating the availability and support of Federal SAR resources.

The Federal Agency designated ESF #9 overall Primary Agency by DHS/FEMA (ESF #9 Coordinator) will:

- Determine what SAR resources are available and required from the other ESF #9 Primary Agencies. Recommend if additional resources are needed to FEMA;

- Provide a consolidated (Federal, State and local, as available) ESF #9 SAR resource laydown to FEMA and the other ESF #9 Primary Agencies on a daily basis, or as required;

- Provide incident reports, assessments, and situation reports on a daily basis, or as required;

- Coordinate integration of Federal SAR resources in support of the State, Tribe, or Territory requesting the support;

- Coordinate the provisioning of additional support assets;

- As required, provide representation at FEMA's National Response Coordination Center (NRCC), Joint Field Office (JFO), and State, Tribal, Territorial, and local EOCs; and

The other ESF #9 Primary Agencies will provide support to the designated overall Primary Agency, as required.

Other information concerning ESF #9 Primary Agency and overall Primary Agency is provided in Table 1-3-3 below.

Table 1-3-3: Other ESF #9 Overall Primary Agency Considerations
1. ESF#9 Federal SAR resources *always* operate in support of the authority with local jurisdiction. State, local, Tribal or Territorial authorities. Take care to remember you are there *on request*, and to support and assist.
2. Anticipate the affected local/State's will assume responsibility as the SAR Mission Coordinator (SMC) for the disaster response. If the ESF #9 overall Primary Agency or the overall Primary Agency's representative does not know who the SMC (or whomever is coordinating SAR resources) is, find out quick, and establish a professional working rapport.
3. Each local/State has different SAR capabilities and response plans. Some States are better prepared/equipped than others. Each ESF #9 Primary Agency should begin coordination and planning with their respective State partners before an event occurs, to understand their potential disaster scenarios and what SAR resources they would likely request from the ESF #9 Primary Agencies
4. Anticipate States not having the experience or the plans for coordinating a unified Federal SAR response during a major disaster/CISAR. Appropriately champion the proven efficiencies of an integrated planning cell (State, Federal, and local ESF#9 agencies) with the authority having jurisdiction (normally the State SC).
5. When Federal SAR resources are requested by a State(s), FEMA Region, or other Federal Agency, anticipate using the Catastrophic Incident SAR Addendum as operational guidance during a CISAR operation.
6. Generally, ESF #9 Primary Agencies have a 72-96 hour lifesaving window in a catastrophic incident response. Plan accordingly.
7. Preplanning and coordination between the ESF #9 Primary Agencies (at all organizational levels) BEFORE an event occurs must be conducted to ensure lifesaving resources are rapidly provided when requested during a disaster.
8. Do not let paperwork slow the provisioning of SAR resources (Remember: 72-96 hour lifesaving window). The processing of MAs is important. However, disagreements about funding can occur after lives are saved.
9. Anticipate State(s) not understanding how to request Federal SAR resources (especially in the high stress environment of a disaster/mass rescue response). FEMA Mission Assignment (MA) Managers, ESF #9 Primary Agency and Defense Coordinating Officers' (DCO) representatives located at the Joint Field Office/IMAT, Incident Command, State EOC, and/or in the FEMA RRCC can assist in expediting this process.

Table 1-3-3 is continued on the next page.

Table 1-3-3: Other ESF #9 Overall Primary Agency Considerations (continued)

10. The Coast Guard will normally be designated ESF #9 overall Primary Agency for all flooding (e.g., "waterborne") events where Federal SAR support is requested.

11. The Coast Guard will normally be designated the ESF #9 overall Primary Agency for all hurricane landfall response operations. However, if there is minimal flooding and significant structural/wind damage, DHS/FEMA, as the ESF #9 Coordinator, can designate another overall Primary Agency. This is the exception rather than the rule. The Coast Guard should always assume they will be the ESF #9 overall Primary Agency for any hurricane landfall that requires ESF #9 Federal SAR support.

12. FEMA US&R will normally be designated ESF #9 overall Primary Agency for structural collapse disasters, such as earthquakes.

13. DoD/NPS ESF #9 overall Primary Agency land SAR responsibilities are generally understood to include SAR not captured under FEMA US&R and Coast Guard overall Primary Agency responsibilities (e.g., wide-area land SAR, wilderness SAR, volcanic eruption, CBRNE event, etc.).

14. FEMA US&R Task Forces have no airlift capability. While ESF #1 – Transportation, may seem to be the preferred means to airlift SAR personnel to the disaster site, anticipate using and/or dedicating other ESF #9 Primary Agency airlift resources to rapidly deploy FEMA US&R Task Force and Incident Support Team (IST) personnel to the scene (Remember: 72-96 hour lifesaving window).

15. Each ESF #9 Primary Agency should plan for: 1) "notice" (e.g., hurricanes, tsunamis (sometimes), etc.); and 2) "no-notice" events (e.g., earthquakes, tsunamis, crisis management events that require consequence management action (terrorist attack, etc.)).

16. Under ESF #9, the NPS has the only authorized, Federal swift water SAR capability.

17. The designated ESF #9 overall Primary Agency, as well as the other ESF #9 Primary Agencies, should provide representatives to key coordination locations to assist in the SAR resource coordination process. Representatives should have a good understanding of NIMS/ICS, ESF #9, the MA process, etc., and have the communications capability to contact the NRCC and other ESF #9 Primary Agency points of contact (Coast Guard Districts/Areas, CDRUSNORTHCOM, NPS Emergency Services, FEMA US&R Task Forces and Incident Support Teams, etc.).

    a. The ESF #9 overall Primary Agency will assist FEMA US&R in standing the watch at the FEMA NRCC ESF #9 Desk. Each ESF #9 Primary Agency must plan for this 24 hour watchstander requirement.

    b. The ESF #9 Primary Agencies should anticipate sending representatives to the affected State(s) EOC(s) and the FEMA RRCC(s), and/or JFO.

18. FEMA will normally provide MAs for ESF #9 Primary Agency representatives.

19. Coordinating and responding to a disaster, where large numbers of persons may be in need of lifesaving resources, is a difficult, high stress effort. In many instances, the person(s) planning the SAR response may not be experienced in planning large scale, mass rescue operations (very few are). Any assistance the ESF #9 Primary Agencies can provide, whether experienced SAR planners, or even liaison officers with limited knowledge of disaster SAR operations, can be of help. In many instances, just having extra people available to assist may be of considerable value. Also, do not become part of the problem. For example, if a State/FEMA ESF #9 SAR MA is not perfectly completed, but "good enough," then press on (It's about saving lives).

20. Each FEMA Region processes MAs differently. Each ESF #9 Primary Agency should plan with their respective State(s)/FEMA Region(s) how an ESF #9 MA will be processed before a disaster occurs (Remember: 72-96 hour lifesaving window).

21. Under the Stafford Act, the ESF #9 Primary Agencies may be required to provide SAR resources to Puerto Rico, U.S. Virgin Islands, Guam, American Samoa, the Commonwealth of the Northern Mariana Islands, Federated States of Micronesia, and the Republic of the Marshall Islands (see Section 1-11: U.S. Territories). Plan accordingly.

Table 1-3-3 is continued on the next page.

Table 1-3-3: Other ESF #9 Overall Primary Agency Considerations (continued)

22. Federal SAR resources contracted to the State under an MA will continue to follow Agency specific policies, procedures, and doctrine in the conduct of SAR operations. The only difference is that the State(s) has assumed SMC and will coordinate the SAR response within their respective State(s).

23. The FEMA NRCC ESF #9 Desk must be notified of any ESF #9 MAs requested and accepted by any ESF #9 Primary Agency.

24. ESF #9 Primary Agencies must keep the FEMA NRCC ESF #9 Desk appraised of any ESF #9 operations being conducted.

25. The ESF #9 overall Primary Agency must keep the FEMA NRCC ESF #9 Desk appraised of ESF #9 Primary Agency force lay down (e.g., where and what type of Federal SAR resources are conducting SAR operations).

26. The ESF #9 overall Primary Agency concludes Federal SAR support when ALL three of the following criteria are met:

    a. No Federal ESF #9 SAR resources are being utilized by the affected State(s)/FEMA Region(s);

    b. Authorities having jurisdiction no longer require or anticipate the use of Federal SAR resources; and

    c. The ESF #9 overall Primary Agency, in consultation with FEMA and the other ESF #9 Primary Agencies, concurs with concluding Federal ESF #9 SAR support for the affected State(s), Tribe, Territory, or other Federal Agency, and the affected FEMA Region(s).

27. The planning and application of effective contingency air traffic and airspace management measures, including disaster TFRs, is often critical to safe and efficient aerial SAR missions. The Incident Command (and ESF#9 Primary Agencies) must coordinate closely with FAA operations liaisons deployed to the NRCC and activated JFOs as CISAR aerial operations are planned and carried out.

28. If in doubt, "launch"! When lives are at stake, a bias towards action will always be better than grappling with indecision and/or bureaucracy (Remember: 72-96 hour lifesaving window).

Figure 1-3-2 on the next page provides a general description of an ESF #9 response, detailing the responsibilities of the overall Primary Agency, other Primary Agencies and the general flow of the response.

Each ESF #9 event is different. Figure 1-3-2 only provides a general flow of the information and resourcing of SAR assets to the requesting government authority.

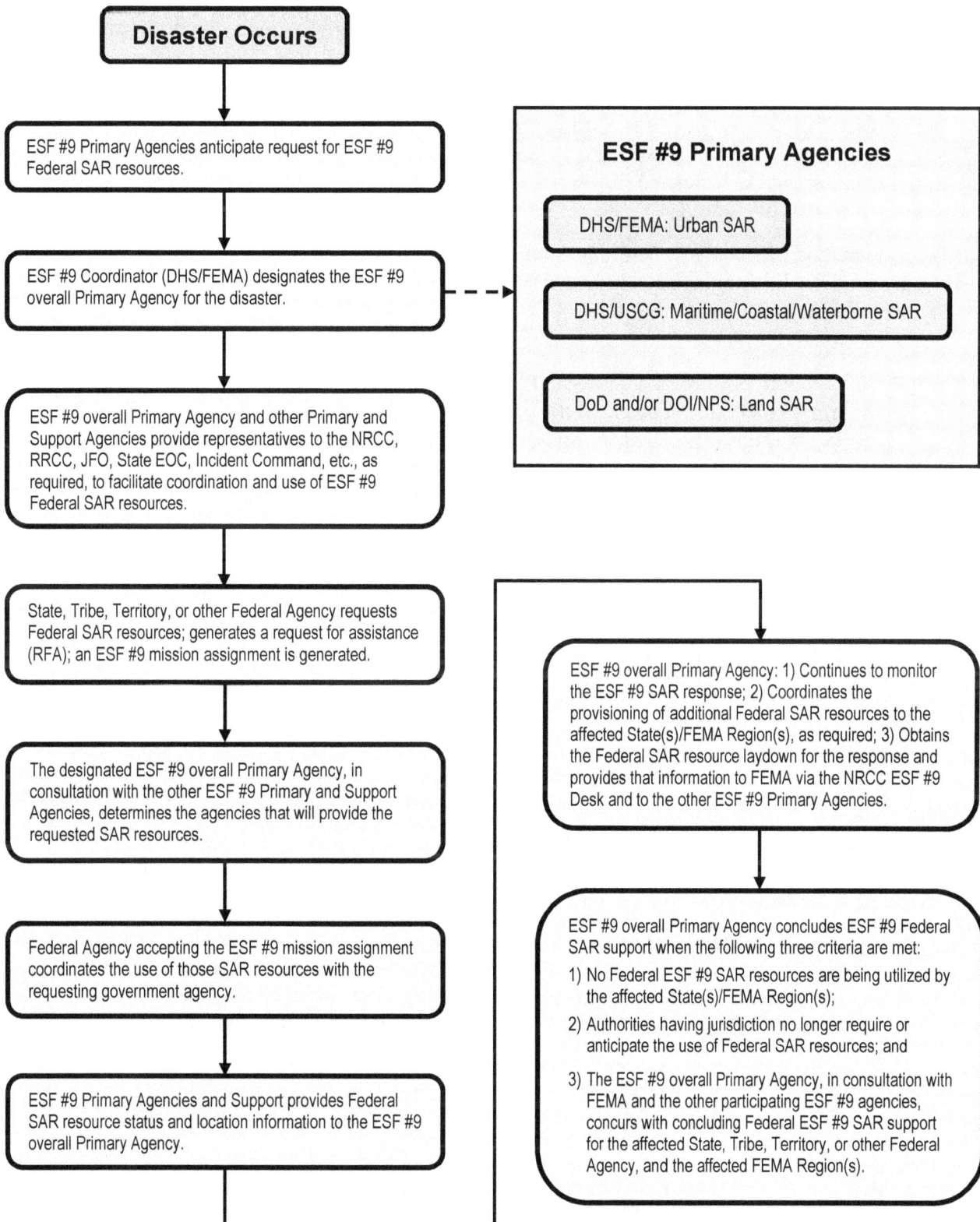

Disaster Occurs

ESF #9 Primary Agencies anticipate request for ESF #9 Federal SAR resources.

ESF #9 Coordinator (DHS/FEMA) designates the ESF #9 overall Primary Agency for the disaster.

ESF #9 overall Primary Agency and other Primary and Support Agencies provide representatives to the NRCC, RRCC, JFO, State EOC, Incident Command, etc., as required, to facilitate coordination and use of ESF #9 Federal SAR resources.

State, Tribe, Territory, or other Federal Agency requests Federal SAR resources; generates a request for assistance (RFA); an ESF #9 mission assignment is generated.

The designated ESF #9 overall Primary Agency, in consultation with the other ESF #9 Primary and Support Agencies, determines the agencies that will provide the requested SAR resources.

Federal Agency accepting the ESF #9 mission assignment coordinates the use of those SAR resources with the requesting government agency.

ESF #9 Primary Agencies and Support provides Federal SAR resource status and location information to the ESF #9 overall Primary Agency.

ESF #9 Primary Agencies

DHS/FEMA: Urban SAR

DHS/USCG: Maritime/Coastal/Waterborne SAR

DoD and/or DOI/NPS: Land SAR

ESF #9 overall Primary Agency: 1) Continues to monitor the ESF #9 SAR response; 2) Coordinates the provisioning of additional Federal SAR resources to the affected State(s)/FEMA Region(s), as required; 3) Obtains the Federal SAR resource laydown for the response and provides that information to FEMA via the NRCC ESF #9 Desk and to the other ESF #9 Primary Agencies.

ESF #9 overall Primary Agency concludes ESF #9 Federal SAR support when the following three criteria are met:

1) No Federal ESF #9 SAR resources are being utilized by the affected State(s)/FEMA Region(s);

2) Authorities having jurisdiction no longer require or anticipate the use of Federal SAR resources; and

3) The ESF #9 overall Primary Agency, in consultation with FEMA and the other participating ESF #9 agencies, concurs with concluding Federal ESF #9 SAR support for the affected State, Tribe, Territory, or other Federal Agency, and the affected FEMA Region(s).

Figure 1-3-2: General ESF #9 SAR response.

(Note: Figure 1-3-2 above provides a general response for a "notice" event. For a "no notice" event, Federal SAR resources will respond as required, followed by the generation of an MA.)

Support Agencies

Support Agencies are those entities with specific capabilities or resources that support the Primary Agency in executing the mission of the ESF. When an ESF is activated, Support Agencies are responsible for:

- Conducting operations, when requested by DHS or the designated ESF Primary Agency, consistent with their own authority and resources;

- Participating in planning for short- and long-term incident management and recovery operations and the development of supporting operational plans, SOPs, checklists, or other job aids, in concert with existing first-responder standards;

- Assisting in the conduct of situational assessments;

- Furnishing available personnel, equipment, or other resource support as requested by DHS or the ESF primary agency;

- Providing input to periodic readiness assessments;

- Maintaining trained personnel to support interagency emergency response and support teams; and

- Identifying new equipment or capabilities required to prevent or respond to new or emerging threats and hazards, or to improve the ability to address existing threats.

ESF #9 (Appendix A) lists the ESF #9 Support Agencies and their responsibilities.

Mission Assignments (MAs)

The Stafford Act provides that FEMA may issue MAs to any Federal Agency, with or without reimbursement, to utilize its authorities and the resources granted to it under Federal law in support of disaster relief efforts.

The FEMA Disaster Relief Fund (DRF) is available for purposes of the Stafford Act. Reimbursement may be provided from the DRF for activities conducted pursuant to these actions. However, the DRF is *not* available for activities:

- Not authorized by the Stafford Act;

- Undertaken under other authorities or agency missions; or

- That are non-Stafford Act incidents requiring a coordinated Federal response.

MAs are a work order issued by FEMA directing an agency to provide a specified service and/or goods for a specified amount of time and reimbursable funding. MAs can be issued by FEMA's NRCC, an RRCC, or a Joint Field Office (JFO) established after a Presidential Disaster Declaration (PDD) has been issued (RFAs may be submitted and MAs can also be issued through the FEMA surge account in anticipation of a PDD).

Specifically, FEMA issues MAs to Federal Agencies to:

- Address a State's request for Federal Assistance to meet unmet emergency needs; or

- Support overall Federal operations pursuant to, or in anticipation of, a Stafford Act declaration.

ESF #9 Primary Agencies providing SAR resources during a disaster SAR response can be issued an MA for SAR resources provided to the requesting State(s).

Processes and procedures (e.g., Pre-Scripted Mission Assignments) for expedited issuance of ESF #9 MAs for Federal SAR resources should be coordinated by each Primary Agency with DHS/FEMA (ESF #9 Coordinator).

ESF #9 Planning and Coordination

Successful, Federal ESF #9 SAR support to the State, Tribe, Territory, or Federal Agency requesting the resources, begins with effective interagency planning and coordination *before* a disaster occurs. Knowing and understanding each ESF #9 Primary Agency's points of contact, procedures for coordinating SAR resources, and capabilities is critical for providing the correct, requested SAR resources in a timely manner.

ESF #9 planning and coordination needs to occur at all ESF #9 Primary Agency organization levels. The challenge is that each ESF #9 Primary Agency organizes and coordinates SAR operations and resources differently.

Table 1-3-4 below is a planning and coordination matrix that describes at which organization level each ESF #9 Primary Agency should engage their respective peers for ongoing disaster planning and SAR resource coordination.

Table 1-3-4: ESF #9 Primary Agency Planning and Coordination Matrix				
	DoD	**USCG**	**NPS**	**FEMA**
National, Interagency ESF #9 Policy Development/ Coordination	• Office of Secretary of Defense • Joint Staff	• Office of Search and Rescue, USCG HQ	• Deputy Chief, Emergency Services NPS HQ	• US&R Branch Chief, FEMA HQ
National NRCC ESF #9 level planning and coordination	• Joint Staff • USNORTHCOM/ USPACOM • National Guard Bureau	• Office of Search and Rescue, USCG HQ	• Deputy Chief, Emergency Services NPS HQ	• US&R Branch Chief, FEMA HQ
Regional level planning and coordination	• USNORTHCOM/ USPACOM • Air National Guard (ANG)/ Army National Guard (ARNG)	• LANTAREA/ PACAREA	• Deputy Chief, Emergency Services NPS HQ	• FEMA Region ESF #9 Reps
JFO level planning and coordination	• USNORTHCOM/ USPACOM • National Guard	• District • LANTAREA/ PACAREA	• Regional Emergency Coordinator NPS	• FEMA Region ESF #9 Reps
State level planning and coordination	• USNORTHCOM (NC RCC)/USPACOM (as required) • National Guard	• Sector and District	• Regional Emergency Coordinator NPS	• FEMA Region ESF #9 Reps
Local/IC level planning and coordination	• USNORTHCOM (NC RCC)/USPACOM (as required) • National Guard	• Sector	• Park Emergency Services Coordinator	• FEMA Region ESF #9 Reps

This page intentionally left blank.

Section 1-4: Federal SAR Responsibilities

General SAR Provisions

ESF #9 SAR Operations vs. Routine SAR

NSP: SAR Coordinator (SC)

Summary of Federal Responsibilities

General SAR Provisions

The NRF provides an overview of key Federal roles and responsibilities in disaster response operations. In particular, the ESF #9 Annex to the NRF designates the four Federal Agencies as Primary Agencies, based on their respective SAR environment responsibility for each specific event.

ESF #9 SAR operations are normally coordinated by the SMC/IC as part of the overall response to the disaster.

By comparison, the NSP provides an overview of key Federal roles and responsibilities for routine (non-ESF #9) SAR operations. Federal SAR Coordinators (SC), responsible for the provisioning and equipping of SAR services within their respective SAR regions, are identified.

Routine (non-ESF #9) SAR operations under the NSP are coordinated by an SMC within that specific SAR region.

ESF #9 SAR Operations vs. Routine SAR

ESF #9 SAR operations are conducted when a State, Tribe, Territory, or other Federal Agency requests Federal SAR assistance. ESF #9 Federal SAR resources can be requested for both Stafford Act disasters and non-Stafford SAR operations. Per the NRF, the NSP and its associated documents are identified as key policy and guidance documents for implementing ESF #9 operations.

Table 1-4-1 on the next page highlights the differences between CISAR operations and routine SAR.

Table 1-4-1: ESF #9 Operations vs. Routine SAR		
	ESF #9 SAR Operations	**Routine (non-ESF #9) SAR**
What type of SAR is conducted?	Federal ESF #9 assistance consists of SAR operations carried out as all or part of the response to an emergency or disaster declared by the President under provisions of the NRF and ESF #9.	Any SAR operations not conducted under ESF #9.
What guidance applies?	The NRF, ESF #9, IAMSAR Manual, NSP, NSM, this Addendum, and Agency specific policy and doctrine.	IAMSAR Manual, NSP, NSM, and Agency specific policy and doctrine.
Who is responsible for coordinating the SAR mission?	As per ESF #9, the Federal overall Primary Agency coordinates the provisioning of Federal SAR resources requested by a government authority (State, Tribe, Territory, or other Federal Agency). Designation as ESF #9 overall Primary Agency is based on the type of SAR environment (e.g., DHS/FEMA: Urban; DHS/U.S. Coast Guard: Maritime/Coastal/ Waterborne; DOI/NPS or DoD: Land SAR). The ESF #9 overall Primary Agency will be designated by the ESF #9 Coordinator (FEMA US&R). When Federal SAR resources are provided to the requesting government authority, command relationships will normally be in support of FEMA and SMC/IC.	As per the NSP, the Federal SC is the person or Agency with overall responsibility for establishing and providing SAR services for a particular SAR region (SRR) in which the U.S. has primary responsibility (e.g., CDRUSNORTHCOM: Continental U.S.; CDRUSPACOM: Alaska; Coast Guard oceanic SAR Regions, Hawaii and navigable waters over which the U.S. has jurisdiction). The SMC is the official temporarily assigned to coordinate a SAR response to an actual or apparent distress situation within the SAR region.
What Command and Control system will be used?	NIMS/ICS	As per the NSP, various command and control systems are used based on agency specific policies and procedures. In CONUS and Alaska, SAR operations are normally conducted using NIMS/ICS. In the oceanic SAR regions, the Coast Guard will normally use the international SAR system as detailed in the IAMSAR Manual.

NSP: Federal SAR Coordinator (SC)

According to the NSP, a Federal SC is one or more persons or agencies with overall responsibility for establishing and providing SAR services, and for ensuring that planning for those services is properly coordinated. Similar to the ESF #9 Primary Agencies, the Federal SCs identified in the NSP are associated with certain types of SAR, but also with responsibilities for certain geographic areas known as SAR regions (SRRs).

(Note: Details about SRRs and associated responsibilities are provided in the NSS.)

Federal SC responsibilities apply to all relevant types of SAR covered by the NSP within each SRR. Additionally, certain SAR responsibilities for SAR services have been assumed by Federal SCs according to agreements signed with States.

Summary of Federal Responsibilities

Refer to ESF #9 for information about Primary and Support Agencies, the types of CISAR operations, and the responsible overall Primary Agency for each.

Refer to the NSP concerning Federal SC responsibilities.

It is important to understand that the NSP continues to apply even when ESF #9 is implemented. As such, both the Federal SC and ESF #9 Primary Agencies will coordinate providing SAR resources for a CISAR incident. However, the designated ESF #9 Primary Agency has lead responsibility in coordinating Federal SAR resources in support of the requesting State, Tribe, or Territory for the particular incident.

Table 1-4-2 below explains how ESF #9 Primary Agency and Federal SC responsibilities compare.

Table 1-4-2: Federal Responsibilities: Routine SAR and ESF #9 SAR Operations	
ESF #9 SAR Operations	**Routine SAR (non-ESF #9) Operations**
ESF #9: Primary Agencies	**NSP: Federal SAR Coordinators (SCs)**
Generally, a Primary Agency is the Federal Agency with significant authorities, roles, resources, or capabilities for a particular function within an ESF. Under ESF #9 in particular, the overall Primary Agency is designated by the ESF #9 Coordinator (FEMA US&R) based on the nature of the CISAR environment (Table 1-3-2). The other Primary and Support Agencies assist the overall Primary Agency by providing SAR resources as required.	SAR Coordinator: One or more persons or agencies with overall responsibility for establishing and providing SAR services, and for ensuring that planning for those services is properly coordinated.
Per ESF #9, the Primary Agencies are responsible for the following SAR environments:	Per the NSP, the U.S. Federal SCs are responsible for the following SRRs:
FEMA US&R: Urban SAR	**U.S. Northern Command:** Continental U.S.
USCG: Maritime/Coastal/Waterborne SAR	**U.S. Pacific Command:** Alaska
DoD/CDRUSNORTHCOM and NPS: Land SAR	**USCG:** Oceanic SAR Regions, Hawaii and navigable waters over which the U.S. has jurisdiction.
(Note: The NSP remains applicable, even during CISAR operations. Federal SAR Coordinators will continue to oversee the conduct SAR operations in their respective area of responsibility.)	**National Park Service:** National Parks

Section 1-5: Federal Emergency Management Agency (FEMA)

FEMA, ESF #9, and Urban SAR (US&R) Responsibilities

National US&R Response System

Type I US&R Task Force

US&R Incident Support Team (IST)

FEMA Regional Offices

Military Support to FEMA US&R

FEMA, ESF #9, and Urban SAR (US&R) Responsibilities

FEMA is the ESF #9 Coordinator. In addition, FEMA is the ESF #9 Primary Agency for Structural Collapse SAR, which includes operations for natural and man-made disasters and catastrophic incidents as well as other structural collapse operations that primarily require FEMA US&R task force operations under ESF #9.

The FEMA US&R Branch develops national US&R policy, provides planning guidance and coordination assistance, standardizes task force procedures, evaluates task force operational readiness, funds special equipment and training within available appropriations, and reimburses, as appropriate, task force costs incurred as a result of ESF #9 deployment.

National US&R Response System

FEMA manages the National US&R Response System, which is a framework for organizing State and local partner emergency response teams as integrated Federal US&R Task Forces, located

throughout the continental U.S. (Figure 1-5-1 on the next page), are equipped with the necessary tools, equipment, skills, and techniques, and can be deployed by FEMA to assist State and local governments in rescuing victims of structural collapse incidents or to assist in other SAR operations in support of ESF #9.

Upon activation under the NRF, FEMA US&R Task Forces:

- Are considered Federal assets under the Robert T. Stafford Disaster Relief and Emergency Assistance Act and other applicable authorities;

- Are staffed primarily by emergency services personnel who are trained and have experience in collapsed structure SAR operations and possess specialized expertise and equipment;

- Can be activated and deployed by FEMA to a disaster area to provide assistance in structural collapse rescue, or may be pre-positioned when a major disaster threatens a community; and

- Will have all personnel and equipment at the embarkation point within six hours of activation so that it can be dispatched and en route to the disaster location in a matter of hours.

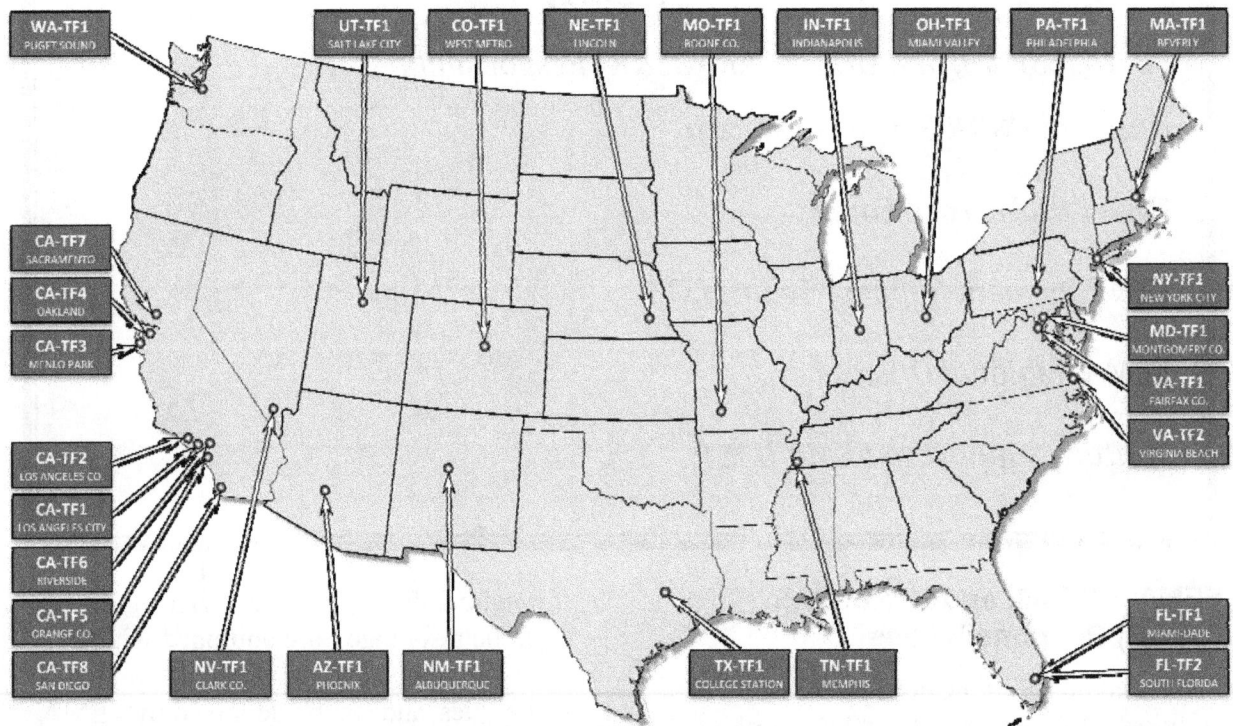

Figure 1-5-1: FEMA US&R Task Force Locations

Type I US&R Task Force

A Type I Task Force is made up of 70 multi-faceted, cross-trained personnel who serve in six major functional areas, including search, rescue, medical, hazardous materials, logistics and planning and are supported by canines. A Task Force is able to conduct physical search and heavy rescue operations in damaged or collapsed reinforced concrete buildings. Each task force can be divided into two 35-member teams to provide 24-hour search and rescue operations. Self-sufficient for the initial 72 hours, the task forces are equipped with convoy vehicles to support over the road deployments and can be configured into Light Task Forces to support weather events such as hurricanes and tornadoes and other similar incidents (Figure 1-5-2 on the next page).

FEMA US&R Task Force (Type I)

- 70 personnel (+10 drivers/mechanics if transported by ground;
- Physical SAR operations;
- Emergency medical care for entrapped victims;
- Reconnaissance to assess damage to infrastructure;
- Hazardous materials surveys/shut off utilities to buildings;
- Structure/hazard evaluations of buildings; and
- Hazmat detection, monitoring decontamination capability.

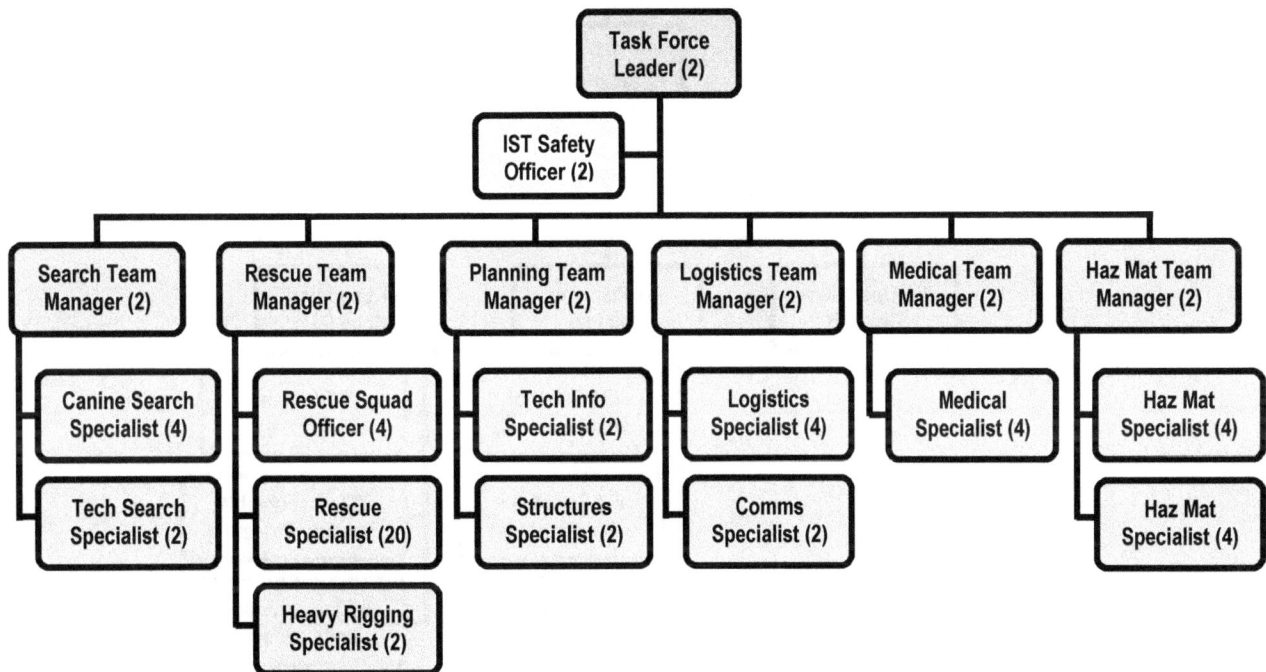

Task Force Leader (2)

IST Safety Officer (2)

Search Team Manager (2)	Rescue Team Manager (2)	Planning Team Manager (2)	Logistics Team Manager (2)	Medical Team Manager (2)	Haz Mat Team Manager (2)
Canine Search Specialist (4)	Rescue Squad Officer (4)	Tech Info Specialist (2)	Logistics Specialist (4)	Medical Specialist (4)	Haz Mat Specialist (4)
Tech Search Specialist (2)	Rescue Specialist (20)	Structures Specialist (2)	Comms Specialist (2)		Haz Mat Specialist (4)
	Heavy Rigging Specialist (2)				

Figure 1-5-2: FEMA US&R Task Force – Type I

US&R Task Force capabilities include the following:

- Conduct physical SAR operations in damaged/collapsed structures;

- Provide reconnaissance to assess damage and needs, and to report results to appropriate officials;

- Render emergency medical care for trapped victims, US&R personnel and search canines;

- Survey and evaluate hazardous materials threats;

- Assess and shut off utilities to homes and other buildings;

- Operate in a known or suspected weapons-of-mass-destruction environment;

- Provide structural and hazard evaluations of buildings; and

- Stabilize damaged structures, including shoring and cribbing.

US&R Incident Support Team (IST)

FEMA US&R Incident Support Teams (ISTs) provide coordination and logistical support to US&R Task Forces during emergency operations. The initial deployment normally consists of an IST-Advance (IST-A) team, which can then be expanded as the incident response requires (Figure 1-5-3 below). The IST also conducts needs assessments and provides technical advice and assistance to State, Tribal, Territorial, and local government emergency managers.

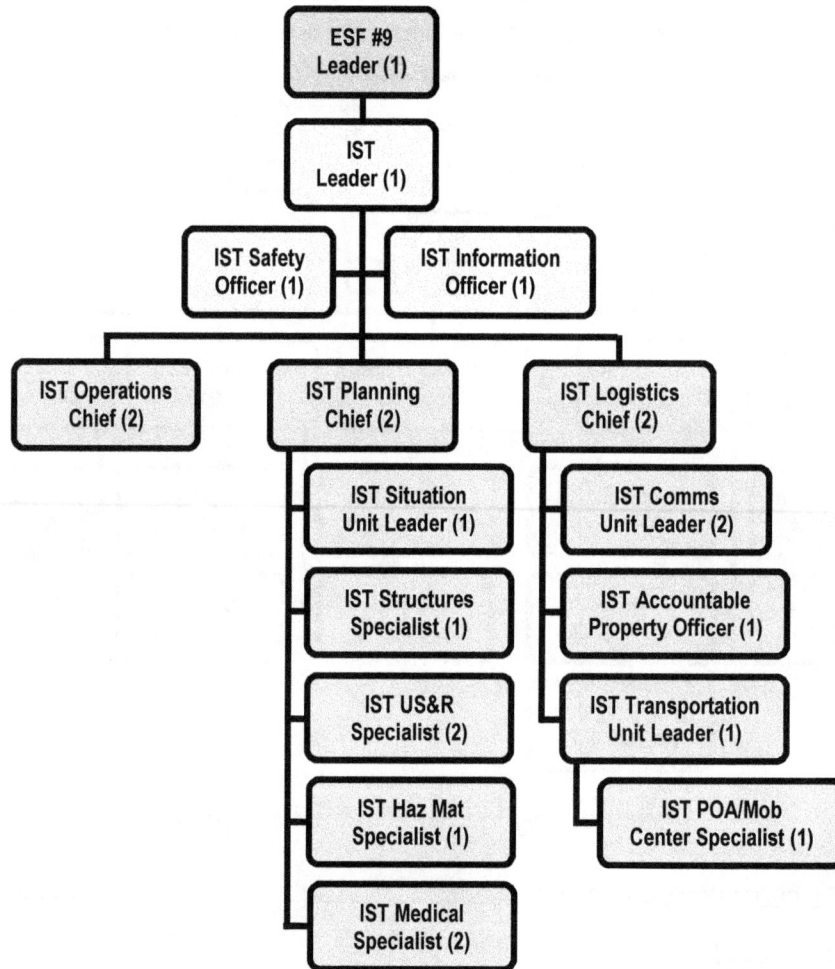

Figure 1-5-3: FEMA US&R Incident Support Team – Advance (IST-A)

FEMA Regional Offices

FEMA has ten Regional Offices (Figure 1-5-4 below and Table 1-5-1 on the next page) that:

- Support development of NRF-related response plans;

- Assist States, Tribes, Territories, and local communities improve readiness; and

- Mobilize FEMA assets and evaluation teams.

Military Support to FEMA US&R

In a catastrophic incident, FEMA's 28 US&R Task Forces can be augmented by DoD military personnel in order to meet the high demand for timely, efficient, and effective structural collapse SAR capability.

(Note: See Section 1-6: Department of Defense, for further information on military support in the conduct of US&R operations.)

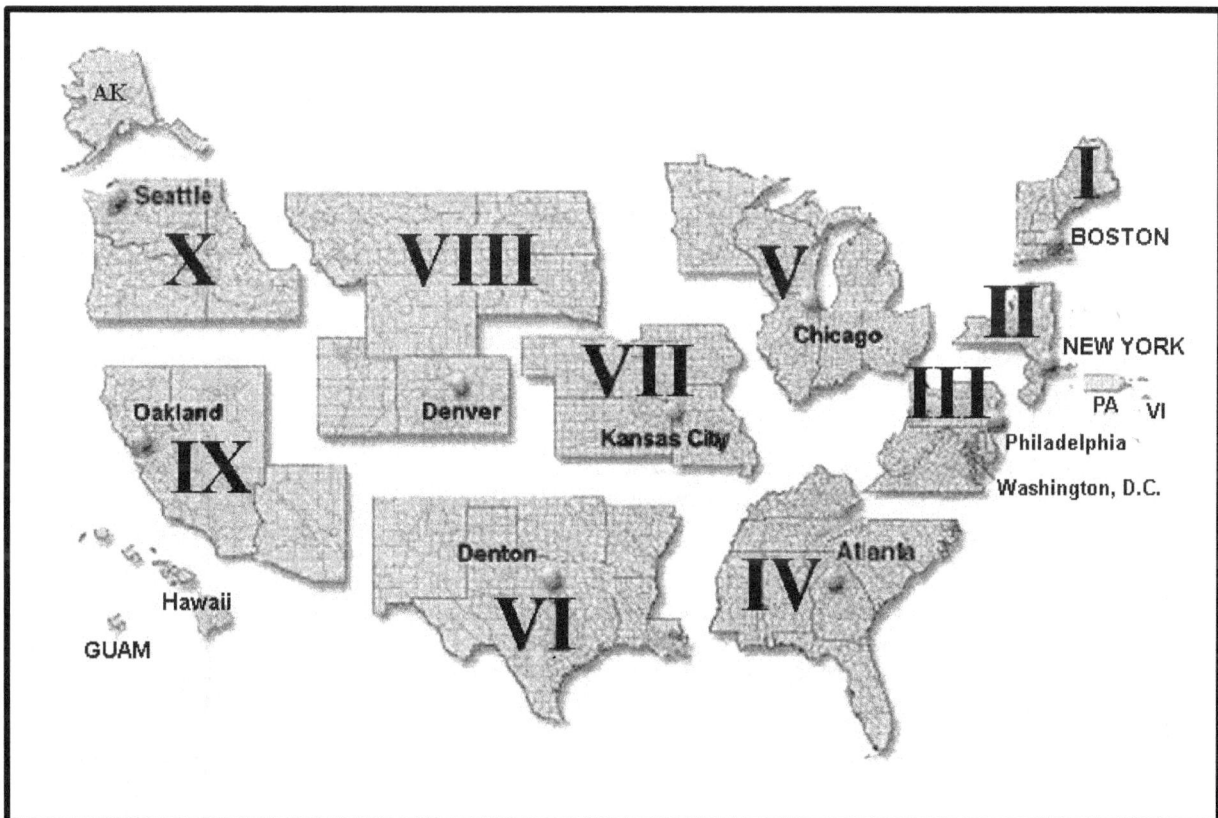

Figure 1-5-4: FEMA Regions

Table 1-5-1 FEMA Regions

Region	States	Address	Telephone
Region I	Connecticut, Maine, Massachusetts, New Hampshire, Rhode Island, Vermont	FEMA 99 High Street, 6th Floor Boston, MA 02110-2320	(617) 956-7501
Region II	New Jersey, New York, Puerto Rico, Virgin Islands	FEMA 26 Federal Plaza, Room 1337, New York, NY 10278-0002	(212) 225-7209
Region III	Delaware, District of Columbia, Maryland, Pennsylvania, Virginia, West Virginia	FEMA One Independence Mall, 6th Floor 615 Chestnut Street Philadelphia, PA 19106-4404	(215) 931-5608
Region IV	Alabama, Florida, Georgia, Kentucky, Mississippi, North Carolina, South Carolina, Tennessee	FEMA 3003 Chamblee-Tucker Road Atlanta, GA 30341	(770) 220-5200
Region V	Illinois, Indiana, Michigan, Minnesota, Ohio, Wisconsin	FEMA 536 South Clark Street, 6th Floor Chicago, IL 60605	(312) 408-5500
Region VI	Arkansas, Louisiana, New Mexico, Oklahoma, Texas	FEMA Federal Regional Center 800 N. Loop 288 Denton, TX 76201-3698	(940) 898-5104
Region VII	Iowa, Kansas, Missouri, Nebraska	FEMA 9221, Ward Parkway, Suite 300 Kansas City, MO 64114	(816) 283-7061
Region VIII	Colorado, Montana, North Dakota, South Dakota, Utah, Wyoming	FEMA Denver Federal Building Building 710, Box 25267 Denver, CO 80225-0267	(303) 235-4812
Region IX	American Samoa, Arizona, California, Guam, Hawaii, Nevada, Commonwealth of the Northern Mariana Islands, Federated States of Micronesia, Republic of the Marshall Islands	FEMA 1111 Broadway Suite 120 Oakland, CA 94607-4052	(510) 627-7100
Region X	Alaska, Idaho, Oregon, Washington	FEMA Federal Regional Center 130 228th Street, S.W. Bothell, WA 98021-9796	(425) 487-4600

Section 1-6: Department of Defense (DoD)

DoD Support to Civil SAR

DoD Policy

Immediate Response Authority

Military Support to US&R

DoD Responsibilities

Resources

Defense Support of Civil Authorities (DSCA)

Legal Restrictions

DoD Support Structure for CISAR

National Guard

Title 10 SAR Support

DoD Support for Civil SAR

The Department of Defense (DoD) has significant capabilities that can support civil SAR operations. Whether responding to planned events or crisis incidents, SAR providers at all levels should consider using DoD capabilities as early as possible.

Under provisions of the NSP and existing Memoranda of Understanding (MOUs)/ Memoranda of Agreement (MOAs), DoD Components maintain active, reserve, and other DoD SAR capabilities that can support civil authorities.

Federal military commanders and responsible DoD officials are also authorized by DoD policy to provide assistance to save lives, prevent human suffering, or mitigate great property damage under imminently serious conditions within the United States. This support is known as "immediate response" (See DoD Directive 3025.18).

SAR operations required for an actual or potential Mass Rescue Operations (MROs) may require a larger Federal response under the NRF with the activation of ESF #9. In these situations, FEMA will coordinate all SAR operations. Requests for assistance from FEMA for DoD capabilities supporting ESF #9 will be validated by the appropriate Defense Coordinating Officer (DCO) in the affected FEMA region or by the DoD Liaison at the FEMA National Response

Coordination Center, depending on where the request originates. Any request generated by the FEMA NRCC will be coordinated with the affected FEMA region. DoD will be reimbursed for all support in accordance with the authorities it was provided.

DoD resources for support to civil SAR are not limited to traditional SAR aviation platforms. CISAR operations may require requests for small boats, swift water rescue operations, planners, and specially trained rescue personnel, as well as general purpose troops. For example, in the event of an overwhelming collapsed structure event, when requested, USNORTHCOM is responsible for executing the military support to FEMA US&R mission, and would provide military augmentation to FEMA US&R Task Forces.

CISAR operations require a high-degree of planning and cooperation in order to be successful. In accordance with the NSP and national response doctrine, DoD Components must provide support with a spirit of unity of effort and cooperation with their civil counterparts at all levels.

DoD Policy

Per DoD Instruction 3003.01 (26 September 2011), DoD will support domestic civil authorities by providing civil SAR services to the fullest extent practicable on a non-interference basis with primary military duties; such services are provided according to applicable national directives, plans, guidelines, and agreements and under the authority of and consistent with the NSP. For day-to-day SAR, a DoD response to a request under the NSP is normally on a non-reimbursable basis. However, DoD capabilities provided under the NSP during the onset of a catastrophic incident may be reimbursed later in accordance with the Economy Act (Title 31 USC 1535–1536) or the Robert T. Stafford Act Disaster Relief

and Emergency Assistance Act (Title 42 USC 5121–5206). In accordance with the NRF, and to ensure a fully coordinated and integrated DoD response, all requests for DoD support of CISAR operations should be submitted to the Secretary of Defense from either the FCO through the DCO at the Joint Field Office or from the FEMA NRCC.

As discussed in the NSP, DoD may also, if requested, support civil SAR operations anywhere in the world, consistent with its expertise, capabilities, and legal authority.

Immediate Response Authority

Federal military commanders, Heads of DoD Components, and/or responsible DoD civilian officials (hereafter referred to collectively as "DoD officials") have independent Immediate Response Authority that allows them to provide needed assistance after an emergency, under certain narrow circumstances.

In response to a request for assistance from a civil authority, under imminently serious conditions and if time does not permit approval from higher authority, DoD officials may provide an immediate response by temporarily employing the resources under their control (subject to any supplemental direction provided by higher headquarters) to save lives, prevent human suffering, or mitigate great property damage within the United States. DoD Directive 3025.18, "Defense Support of Civil Authorities (DSCA)," December 29, 2010, governs the use of this authority by DoD officials.

Numerous legal considerations apply to the use of Immediate Response Authority and in situations where DoD forces have been asked to provide support to civilian authorities, so DoD officials should consult with their legal advisor when providing such support or assistance.

Military Support to US&R

For a large-scale US&R event, FEMA may request military augmentation for its twenty-eight (28) US&R Task Forces.

Developed in close coordination with FEMA, the DoD US&R concept of operations facilitates the augmentation of military forces with FEMA US&R Task Forces with specially trained search and extraction elements from the National Guard. National Guard CBRN Emergency Response Force Packages (CERFP) are the primary military force option for DoD's responsibility in providing military support to FEMA US&R Task Forces when requested by FEMA and approved by Secretary of Defense.

Additionally, general purpose military personnel may also augment FEMA US&R Task Forces in a non-technical support capacity after receiving FEMA training prior to employment at a disaster site.

DoD Responsibilities

SC duties as detailed in the NSP are separate and distinct from NRF ESF #9 overall Primary Agency responsibilities. SC duties are assigned to the U.S. Northern Command for the continental United States other than Alaska, and to the U.S. Pacific Command for Alaska.

SC duties are assigned to DoD and approved by the Deputy Secretary of Defense in the NSP.

DoD and DOI/National Park Service:

ESF #9 Primary Agencies - Land SAR

Operational Overview: Land SAR includes operations that require aviation and ground forces to meet mission objectives other than maritime/coastal/waterborne (Primary Agency: Coast Guard) and structural collapse SAR operations (Primary Agency: FEMA).

DoD, through USNORTHCOM and USPACOM, will coordinate facilities, resources, and special capabilities that conduct and support air, land, and maritime SAR operations according to applicable directives, plans, guidelines and agreements. Per the NSP, the U.S. Air Force and USPACOM provide resources for the organization and coordination of civil SAR services and operations with their assigned SAR regions and when requested, assist Federal, State, Tribe, Territory, and local authorities.

Resources

Civil authorities may use existing MOUs/MOAs under authorities provided in the NSP to facilitate the immediate use of DoD resources for civil SAR.

AFRCC

The US Air Force Rescue Coordination Center (AFRCC) maintains MOUs/MOAs with each State and has extensive resource files of available DoD and civilian SAR assets. At the State's request, the AFRCC coordinates arrangements for their use.

Civil authorities requiring an immediate response from DoD for civil SAR within the 48 contiguous states should contact the AFRCC at 1-800-851-3051 as soon as a need is anticipated or identified.

This authority is applicable only to DoD support provided within the 50 United States, the District of Columbia, the Commonwealth of Puerto Rico, the U.S. Virgin Islands, Guam, American Samoa, the Commonwealth of the Northern Mariana Islands, and any possession of the United States or any political subdivision thereof.

A CISAR event may involve large numbers of persons needing assistance, and response priority must be given to saving human lives. Lifesaving efforts must be immediate (within 48 hours or less, depending on the circumstances), which often requires deploying resources before they are requested under the authority of the NSP. The AFRCC/AKRCC and/or Joint Personnel Recovery Center (JPRC) can provide critical planning and deployment of aeronautical SAR assets to support State, Tribe, or Territory CISAR plans.

Defense Support of Civil Authorities (DSCA)

DoD's primary mission is national defense; because of this critical role, resources are normally committed to NRF-related operations only after approval by the Secretary of Defense or at the direction of the President. DoD Support of Civil Authorities is referred to as DSCA. The relevant DoD Directive regarding DSCA policy is DoDD 3025.18, *Defense Support to Civil Authorities*.

DSCA may involve Federal military forces, DoD civilians, contractor personnel, and DoD agencies and components.

Department of Defense

Nothing in this directive impairs or otherwise affects the authority of the Secretary of Defense over the Department of Defense, including the chain of command for military forces from the President as Commander in Chief, to the Secretary of Defense, to the commander of military forces, or military command and control procedures. The Secretary of Defense shall provide military support to civil authorities for domestic incidents as directed by the President or when consistent with military readiness and appropriate under the circumstances and the law. The Secretary of Defense shall retain command of military forces providing civil support. The Secretary of Defense and the Secretary [of Homeland Security] shall establish appropriate relationships and mechanisms for cooperation and coordination between their two departments.

Homeland Security Presidential Directive 5 (HSPD-5), Paragraph 9

Legal Restrictions

Some CISAR situations may involve the presence of military personnel supporting SAR on scene when law enforcement operations may need to be carried out concurrently in the same location. The Posse Comitatus Act is a Federal criminal statute (18 USC 1385) that prohibits most members of the Federal armed forces (Army, Air Force, Navy, Marine Corps, and National Guard forces when called into Federal service) from exercising law enforcement, police or peace officer powers that maintain "law and order" within the U.S., except where expressly authorized by the Constitution or Congress. The Posse Comitatus Act does not apply to the Coast Guard (14 USC 2).

This is relevant because some CISAR situations may involve the presence of military personnel supporting SAR on scene when law enforcement operations may need to be carried out concurrently in the same location.

DoD Support Structure for CISAR

U.S. Northern Command (USNORTHCOM).
USNORTHCOM has command and control of assigned personnel and resources providing DSCA in their area of responsibility (AOR). Additional forces can be made available, under the command and control of USNORTHCOM as authorized by Secretary of Defense order (includes resources authorized under the DSCA Execution Order).

U.S. Pacific Command (USPACOM).
USPACOM has command and control of assigned personnel and resources providing DSCA in their AOR. Additional forces can be made available, under the command and control of USPACOM as authorized by Secretary of Defense order (includes resources authorized under the DSCA Execution Order).

USNORTHCOM and USPACOM have established Joint Personnel Recover Centers (JPRC) to coordinate DoD SAR operations in their respective AORs

Air Force Rescue Coordination Center (AFRCC). For incidents in which DoD is the designated ESF #9 overall Primary Agency, the AFRCC will continue to handle day-to-day SAR and transfer CISAR responsibilities over to the USNORTHCOM JPRC.

Joint Personnel Recovery Center (JPRC).
For incidents in which DoD is the ESF #9 overall Primary Agency, the JPRC is the command and control node for DoD Title 10 SAR assets.

For CISAR operations, the JPRC:

- Commands and controls DoD SAR forces for CISAR operations based on the State, Tribe, or Territory SAR Plans and requests;

- Handles CISAR reports, assessments, and situation reports;

- Helps de-conflict demands for DoD SAR air assets; and

- Facilitates coordination among DoD and other Federal, State, and local response activities.

Defense Coordinating Officer (DCO).
USNORTHCOM and USPACOM use DCOs to coordinate with FEMA in their respective AORs. USNORTHCOM has permanently stationed DCOs in each of the 10 FEMA Regions. The DCO serves as DoD's single point of contact in a Joint Field Office for requesting DoD assistance.

Joint Task Force (JTF). Based on the complexity and type of incident, and the anticipated level of DoD resource involvement, USNORTHCOM or USPACOM may designate a JTF to command military activities in support of DSCA incident objectives. DoD civil SAR forces will normally be coordinated by a SAR or Joint Personnel Recovery Liaison within the JTF; these officers will liaise DoD support with the respective SAR Branches of the Federal Joint Field Office (JFO) and/or the State Emergency Operations Centers (if established).

(Note: Within the USPACOM AOR, JTF-Homeland Defense (JTF-HD, is permanently activated and serves as the designated supported commander for DSCA operations.)

Dual Status Commander – Joint Support Staff Element (DSC JSF-SE). As an option to improve unity of effort and ensure a rapid response to save lives, or in response to an emergency or major disaster within the U.S., a State may request Dual Status Command when Federal military forces and State military forces are employed, or anticipated to be employed simultaneously in support of civil authorities.

Emergency Preparedness Liaison Officer (EPLO) Program. EPLOs are senior

(Army/Air Force/Marine Colonel, Navy Captain) Title 10 U.S.C. reserve component officers from each military service and the U.S. Coast Guard, who provide information on appropriate military assistance to other Federal Agencies and State governments, and help inform and coordinate these DoD capabilities with the DCO.

National Guard (NG)

National Guard (NG) forces employed under State active duty or Title 32 status are under the command and control of the Governor of their respective State when providing support. Federal military forces, when requested and approved by the Secretary of Defense, will coordinate closely with National Guard forces to promote unity of effort.

In addition to their partnership with DoD in operating under DSC status, the National Guard, specifically the (S/E) of the NG CBRN Emergency Response Force Packages (CERFP), are the primary military force option for DoD's responsibility (when requested by FEMA and approved by Secretary of Defense) in providing military support to FEMA US&R Task Forces in response to a catastrophic, domestic incident.

National Guard Bureau – National Guard Coordination Center (NGB-NGCC)

The NGB-NGCC operates 24/7. Contact information:

Telephone: 703-607-8712/8713/0040 (fax)
DSN: 327-8712/8713/0040 (fax)
Secure Fax: 703-607-8740 (DSN prefix 327)
NIPRNET: ngbjoccmsgctr@ng.army.mil
SIPRNET: ngbjoccmsgctr@ng.army.smil.mil

Title 10 SAR Support

When requested by the State and approved by the SECDEF, additional Title 10 forces are available for SAR operations. The SAR elements in the Defense CBRN Response Force (DCRF) and the Command and Control (C2) CBRN Response Enterprise (C2CRE) are trained to a high standard in order to be certified to operate independently at an incident site or in support of the FEMA US&R Task force, under direction of the incident commander while operating under C2 of the DSC or Title 10 Commander. The DCRF and C2CRE SAR capability in the Initial Response Force are certified in Structural Collapse SAR at the Technician Level as defined by National Fire Protection Association (NFPA) 1670 standards. This definition incorporates Technician Level capabilities for Rope Rescue, Confined Space SAR, Vehicle SAR, Trench and Excavation SAR, and Machinery SAR as well as Awareness Level for Water SAR.

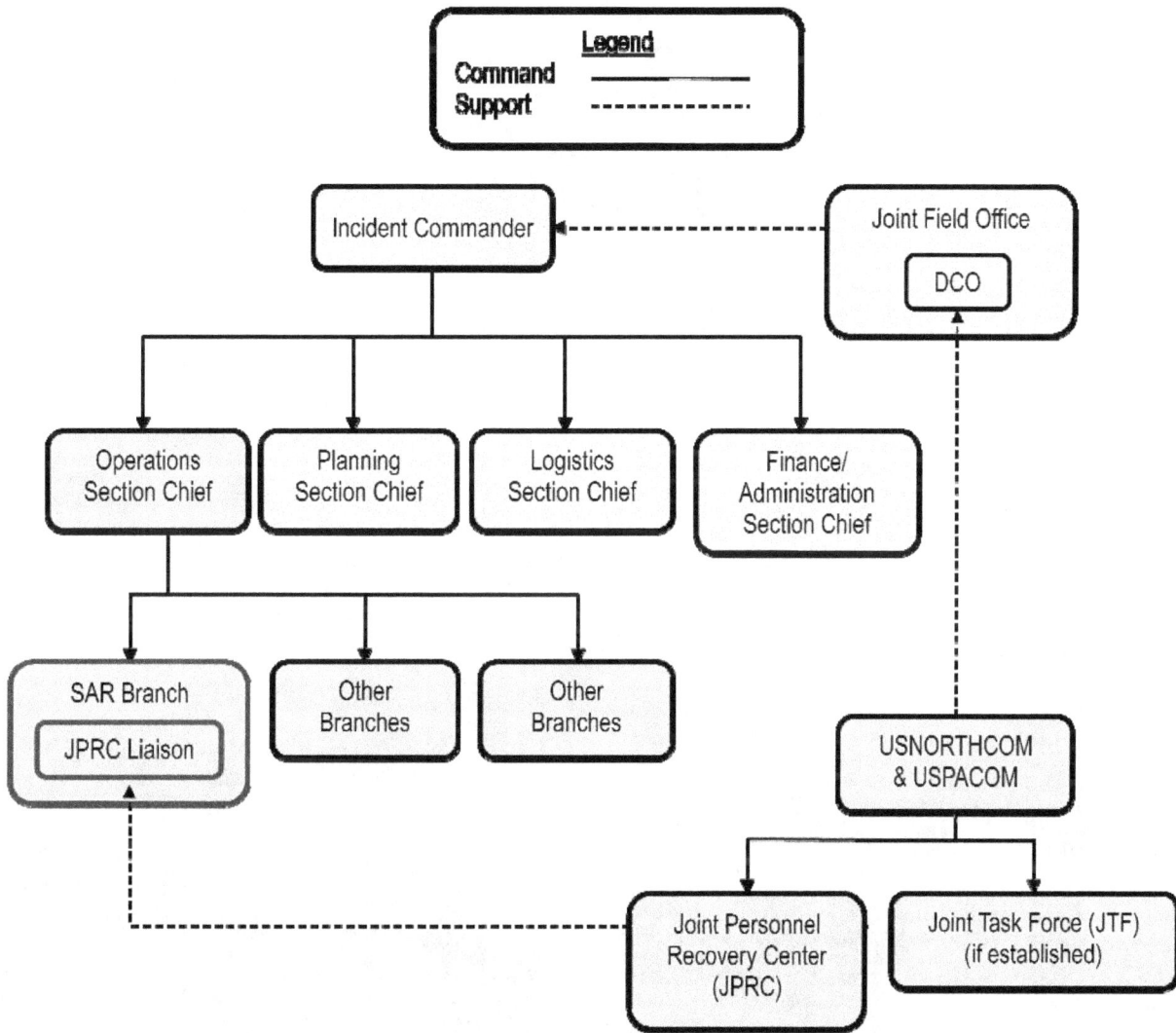

Figure 1-6-1: Simplified DoD Support Structure for Catastrophic Incident SAR

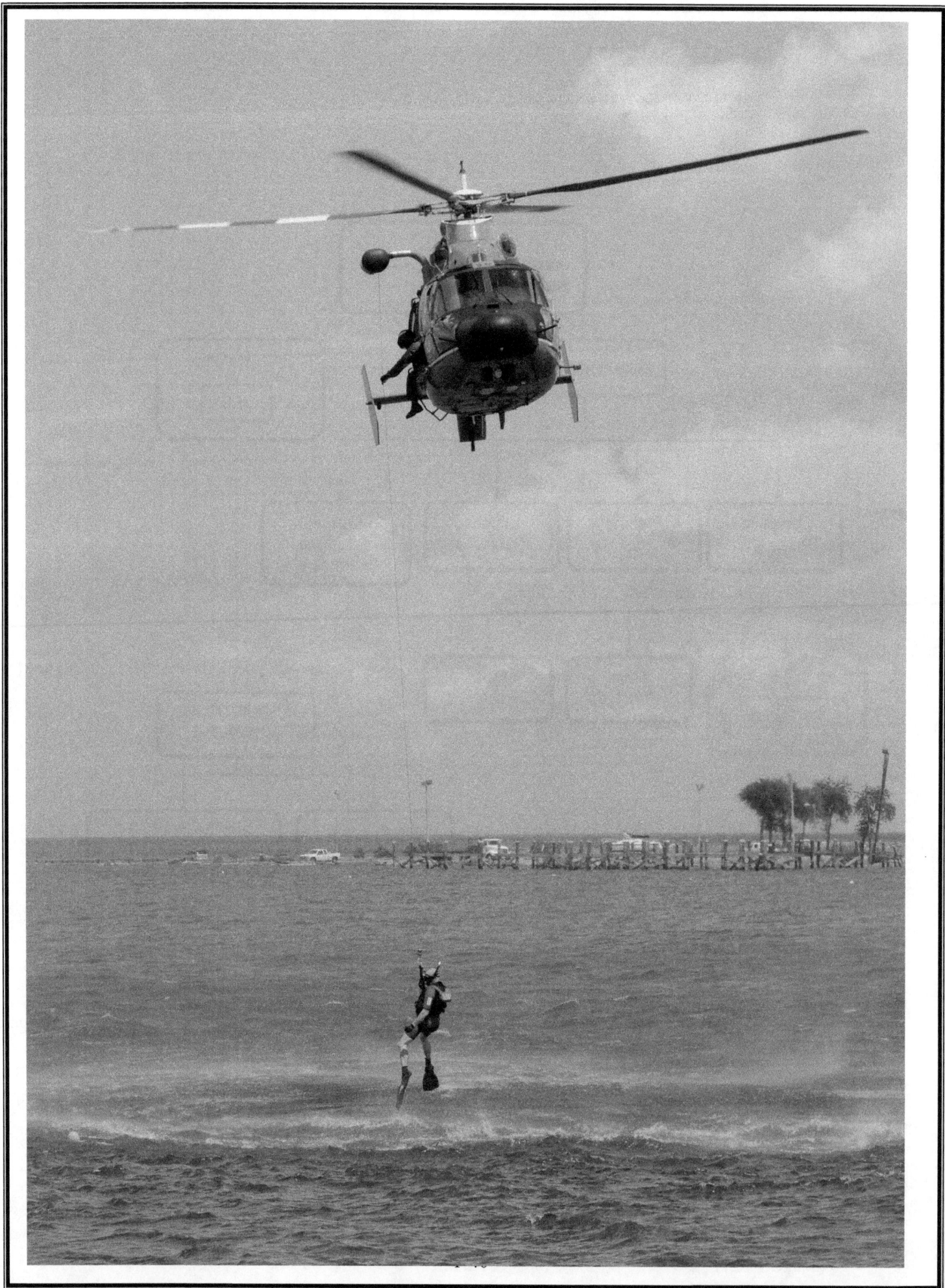

Section 1-7: United States Coast Guard (USCG)

USCG Authority

Routine USCG SAR Support

ESF #9 & CISAR

CIS Coordination

USCG: Overall Primary Agency (PA)

USCG Districts and FEMA Regions

References

USCG Authority

The USCG's statutory authority to conduct SAR missions is found in 14 USC 2, 88, and 141.

The USCG has many resources that may be immediately available to support civil SAR operations. Whether responding to notice or no-notice events, emergency managers should consider using these resources as early in the planning process as possible.

Routine USCG SAR Support

For routine/normal SAR support, USCG assistance does not require coordination by DHS through ESF #9 Mission Assignments (MA) and can be provided without a Presidential emergency or major disaster declaration.

In these instances the USCG provides SAR assistance to States, Tribes, Territories, and local jurisdictions, consistent with their authorities.

In these routine types of SAR operations, the Coast Guard Addendum, NSS and NSP apply.

ESF #9 SAR & CISAR

In CISAR operations, the response system shifts to providing SAR resources through the ICS, to ensure their most effective and safe use.

USCG and ESF #9

Under ESF #9, the USCG is the Primary Agency for Maritime/Coastal/Waterborne SAR operations. This includes operations for natural and man-made disasters that primarily require DHS/USCG air, ship, boat, and response team operations. The Federal maritime/coastal/waterborne SAR response integrates DHS/USCG resources in support of integrated SAR operations conducted in accordance with the NSP.

In ESF #9 SAR operations, USCG SAR responders assist and augment State, Tribal, Territorial, and local SAR capabilities in incidents requiring a coordinated Federal response. USCG SAR responders will coordinate SAR activities with other SAR response agencies within the Incident Command.

In contrast, routine, immediate SAR operations (non-ESF #9) are conducted in accordance with the IAMSAR Manual, NSP, NSS, and Coast Guard Addendum which define SAR responsibilities and provide guidance to Federal Agencies with civil SAR mandates.

For CISAR, the SMC will normally be placed within the Incident Command, typically as the SAR Branch Director or SAR Group Supervisor in the Operations Section.

USCG CISAR response assets will most likely be coordinated through the IC in support of the State, Tribe, Territory, or other Federal lead as dictated by ESF #9.

(Note: In order to support a multi-agency CISAR effort, Coast Guard responders must be NIMS/ICS qualified to ensure effective integration with other response organizations.)

USCG: CISAR vs. Routine SAR

The USCG may support CISAR operations with USCG assets that are organic to the region in which the catastrophic incident occurred. Other assets may also be deployed from other geographic locations.

In addition, the USCG will still continue to respond to other SAR operations within the U.S. maritime SAR regions and waters under the jurisdiction of the U.S. As such, this may require the USCG to retain SMC of USCG assets during these other SAR operations and may only be able to accept CISAR mission assignments on a not to interfere basis.

As CISAR operations are conducted, it is critical that the IC fully understands the USCG's continued responsibility to conduct other SAR operations in addition to supporting ongoing CISAR operations.

CISAR Coordination

Successful response requires unity of effort, which respects the chain of command of each participating organization while harnessing seamless coordination across jurisdictions in support of common lifesaving objectives.

This kind of support may well require USCG personnel to liaise/interact at the Federal, State, Tribal, Territorial, or local Emergency Operations Centers (EOC).

The EOC will most likely be staffed by personnel representing multiple Federal, State, Tribal, Territorial, or local emergency response agencies, the private sector, and nongovernmental organizations.

USCG: ESF #9 Overall Primary Agency (PA)

When a disaster occurs and ESF #9 Federal SAR resources are requested, the USCG will be the overall Primary Agency (PA) if the disaster is maritime/coastal/waterborne (e.g., flooding) event. For these types of disasters, either Commander, Atlantic Area or Commander, Pacific Area will be the overall PA for the duration of the ESF #9 assistance.

Overall PA responsibilities include the planning, coordination, and conduct of ESF #9 Federal Agency assistance in support of the State in catastrophic incidents such as hurricane response, disaster flooding, etc.

(Note: See, Section 1-3: Emergency Support Function #9, for further guidance on ESF #9 overall Primary Agency responsibilities.)

During these events, USCG personnel will be responsible for assisting in SAR planning and coordination efforts with the affected State(s) and other Federal Agencies. These coordination activities may include providing USCG liaisons at the effect State's EOC, FEMA's RRCC/JFO and the FEMA Headquarters NRCC.

Designation as overall PA does not necessarily mean the USCG will be required to provide Federal ESF #9 SAR assets in support of the State. DoD, National Guard, National Park Service (NPS), FEMA Urban SAR, etc., also have SAR assets available to meet a State's ESF #9 request. For example, if a State requests the USCG provide two helicopters for multi-day SAR standby for *potential* flooding, then the USCG, with a daily SAR response mission requirement, may *not* be the best resource to meet the State's ESF #9 SAR requirement.

This is why it is important for the USCG, as ESF #9 overall PA, request the State provide their *SAR requirement* and not request a specific type of asset (e.g., helicopters, personnel, etc.), that may be based on what the State considers their preferred Federal Agency. With the State's request, the ESF #9 overall PA, in consultation with the other ESF #9 PAs, can determine the best Federal Agency(s) to meet the State's requirement.

Over the past several years, experience has shown that early and continuous interagency planning and coordination between local, Tribal, Territorial, State and Federal SAR planners will ensure successful coordination and use of ESF #9 SAR assets during complex, multi-agency SAR operations. If the disaster is a "notice" event and provides the opportunity for early interagency planning and coordination (e.g., hurricane, flooding, etc.), then the USCG, if assigned as overall PA, is responsible for coordinating Federal support and assisting the State in planning and coordinating the disaster response.

USCG Districts and FEMA Regions

Figure 1-7-1 on the next page provides a pictorial overview of USCG Districts and FEMA Regions. In many instances, CISAR operations may very well overlap other USCG Districts as well as multiple FEMA Regions. When the USCG assumes ESF #9 overall Primary Agency for maritime/coastal/waterborne SAR operations, the challenge will be coordinating Federal SAR resources in support of several States and coordinating with other FEMA Regions, USCG Districts and even CDRUSNORTHCOM or CDRUSPACOM.

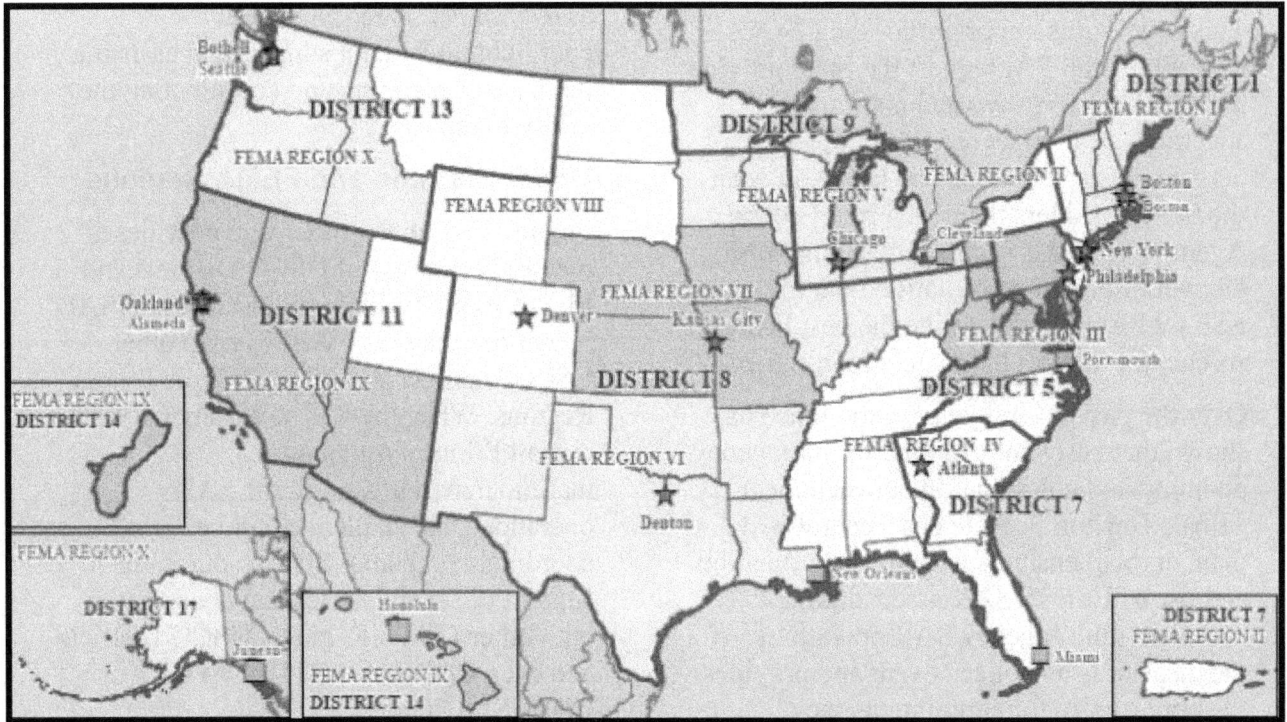

Figure 1-7-1: USCG Districts and FEMA Regions

References

The information in this Section was obtained from the following sources:

- *Coast Guard Addendum to the U.S. Search and Rescue Manual to the*

- *IAMSAR Manual (COMDTINST M16130.2E) September, 2009;*

- *Coast Guard Incident Management Handbook (August 2006)*; and

- *National Response Framework, January, 2008.*

National Command Center

The National Command Center is located at Coast Guard Headquarters in Washington, DC and can be reached at 202-372-2100.

Section 1-8: National Park Service (NPS)

NPS Authority

NPS Regional Offices

ESF #9 Land SAR Overall Primary Agency

NPS Authority

The National Park Service (NPS) is an agency within the Department of the Interior (DOI).

The NPS traditionally provides visitor protection services - including SAR - within areas of the National Park System. The provision of these services is in accordance with the Organic Act of August 25, 1916 (16 USC). This Act allows for the rendering of emergency rescue and law enforcement for related purposes outside of the National Park System.

NPS Regional Offices

Each of the seven regional NPS offices coordinates SAR resources and training through the NPS Regional SAR Coordinator. This Coordinator will facilitate the management of SAR resources and maintains equipment and supplies that can be available during CISAR operations.

ESF #9 Land SAR Overall Primary Agency

Under ESF #9, NPS is the overall Primary Agency for Land SAR.

NPS Land SAR Teams provide personnel, equipment, and supplies for conducting operations during a CISAR event. These teams are specially trained to operate in various roles including ground search, small boat operations, swiftwater rescue, helo-aquatic rescue, and other technical rescue disciplines.

In particular, NPS Small Boat Task Forces (SBTF) provide personnel and shallow draft vessels for use in situations where rescue and evacuation are necessary.

NPS maintains preconfigured teams that include personnel and equipment from the NPS, the U.S. Fish and Wildlife Service, U.S. Geological Survey, Bureau of Indian Affairs, and other DOI components in planning for ESF #9.

The National Park Service's Emergency Incident Coordination Center (EICC) operates 24/7 and provides resource management for routine law enforcement and SAR events. During a CISAR event, the EICC provides coordination of the agency's assets.

The EICC may be contacted at 540-999-3422.

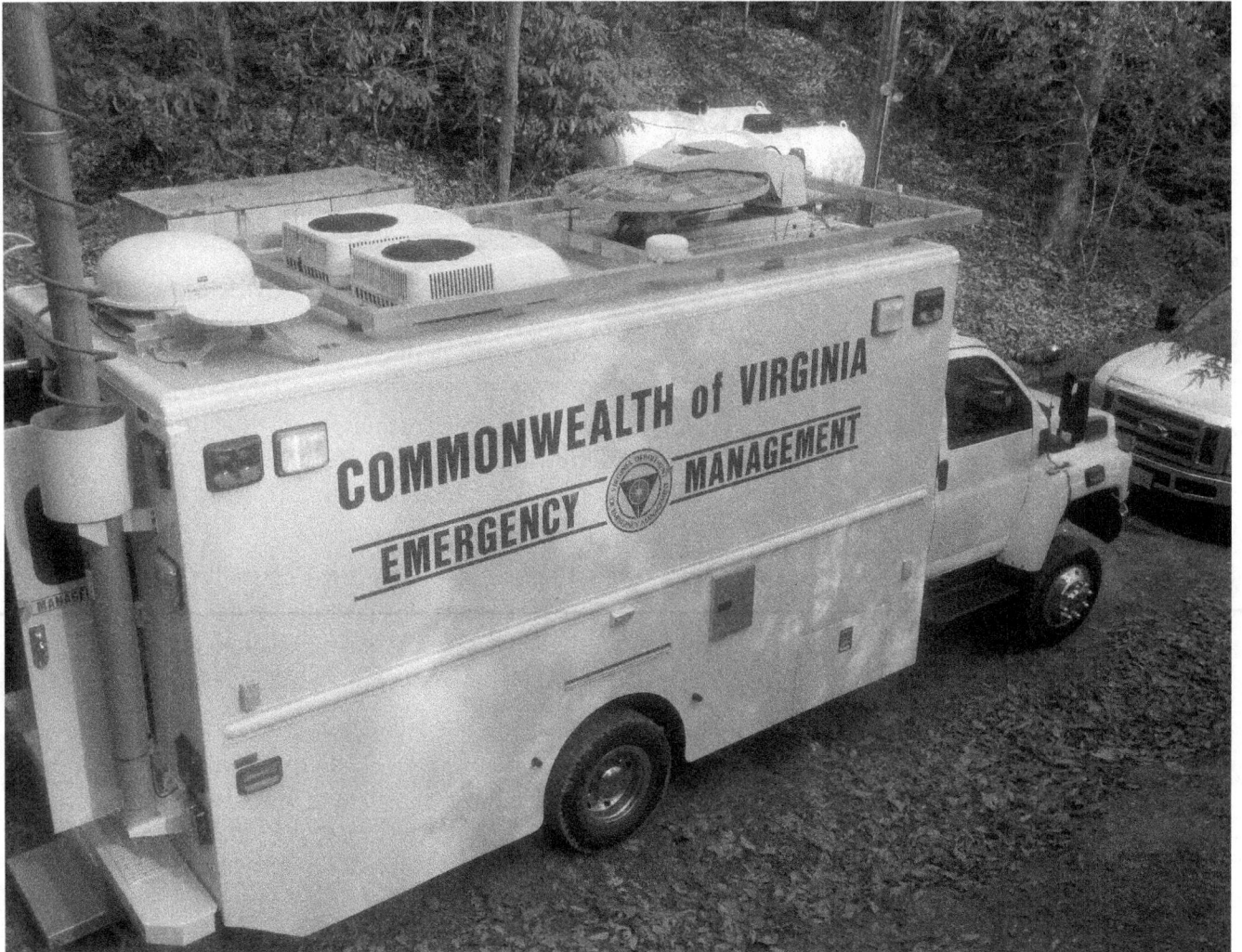

Section 1-9: States

Overview

Providing and Requesting Assistance

SAR Agreements and Plans

Emergency Management Assistance Compact (EMAC)

Disaster Declarations

    Preliminary Disaster Assessment (PDA)

    State Resources Overwhelmed

Disaster Declaration Types

President Directed Emergency Assistance

States

"The Federal Government recognizes the roles and responsibilities of State and local authorities in domestic incident management. Initial responsibility for managing domestic incidents generally falls on State and local authorities. The Federal Government will assist State and local authorities when their resources are overwhelmed, or when Federal interests are involved. The Secretary [of Homeland Security] will coordinate with State and local governments to ensure adequate planning, equipment, training, and exercise activities. The Secretary [of Homeland Security] will also provide assistance to State and local governments to develop all-hazards plans and capabilities, including those of greatest importance to the security of the United States, and will ensure that State, local, and Federal plans are compatible."

Homeland Security Presidential Directive 5 (HSPD-5), Paragraph 6

Overview

Key State-level SAR issues include:

- Designation of responsible State SAR Coordinator;

- An efficient and effective State SAR plan (including CISAR contingency and integrated unified SAR response plans);

- Effective relationships with intrastate and other State and Federal SAR resource providers and agencies;

- Training and maintaining qualified and proficient SAR resources;

- SAR support to local governments and Incident Commanders;

- Brokering interstate and intrastate SAR resources and enabling capabilities;

- Identifying capability gaps and requesting timely, additional assistance; and

- Maintaining necessary balance of multi-jurisdictional issues.

Providing and Requesting Assistance

The State provides assistance to local governments if requested. States have significant resources of their own that include emergency management and homeland security agencies, State Police, health agencies, transportation agencies, incident management teams, specialized teams, and the National Guard.

If additional resources are required, States may request assistance from other States through interstate mutual aid and assistance agreements such as the Emergency Management Assistance Compact (EMAC).

If an incident is beyond State and local capabilities, the Governor can seek Federal assistance. The State will collaborate with the impacted communities and the Federal Government to provide the assistance requested.

SAR Agreements and Plans

Most States have signed SAR agreements and/or plans with the Federal SC designated in the NSP responsible for their respective areas. For example, the Air Force, via the AFRCC, has an MOA with each Governor and MOUs with many State authorities responsible for SAR. In addition to these MOAs, States are encouraged to develop integrated plans that cover CISAR operations and the integration of Federal assets in support of these operations. A State request for assistance may also include Federal SAR planners to assist in the development or enhancement of their CISAR plan.

Emergency Management Assistance Compact (EMAC)

EMAC is administered by the National Emergency Management Association, which provides form and structure to the interstate

mutual aid and assistance process. All States belong to the EMAC system.

Most CISAR operations will involve National Guard forces. National Guard forces deployed through EMAC typically will remain under the control of their respective State.

Disaster Declarations

The Robert T. Stafford Disaster Relief and Emergency Assistance Act, 42 USC 5121-5207 ("Stafford Act") states in part that, "All requests for a declaration by the President that a major disaster exists shall be made by the governor of the affected state."

Preliminary Disaster Assessment (PDA)

The governor's request is made through the applicable FEMA Regional Office. State and Federal officials conduct a joint Federal, State, and local Preliminary Damage Assessment (PDA) to determine the extent of the disaster (or impending disaster) and its impact on individuals and public facilities. This information is included in the governor's request to show that the disaster is, or is forecasted to be, of such severity and magnitude that the effective response is beyond the capabilities of the State and local governments and supplemental Federal assistance is necessary. Normally, the PDA is completed prior to the submission of the governor's request. However, when an obviously severe or catastrophic event is anticipated or occurs, the State Governor's request may be submitted prior to the PDA.

State Resources Overwhelmed

As part of the request, the Governor must take appropriate action under State law and direct execution of the State's emergency plan. The Governor shall furnish information on the nature and amount of State and local resources that have been or will be committed to alleviating the results of the disaster, provide an estimate of the amount and severity of damage and the impact on the private and public sectors, and provide an estimate of the type and amount of assistance needed under the Stafford Act.

In addition, the Governor must certify that, for the current, or impending disaster, State and local government obligations and expenditures (of which State commitments must be a significant proportion) will comply with all applicable cost-sharing requirements.

Disaster Declaration Types

There are two types of disaster declarations provided for in the Stafford Act: Emergency Declarations and Major Disaster Declarations (Table 1-9-1 below). Both declaration types authorize the President to provide supplemental Federal disaster assistance.

Table 1-9-1: Types of Disaster Declarations	
Emergency Declaration	• Declared for any occasion or instance when the President determines Federal assistance is required. • Supplement State and local efforts in providing emergency services, such as the protection of lives, property, public health, and safety, or to lessen or avert the threat of a catastrophe in any part of the United States. • The total amount of assistance provided for a single emergency may not exceed $5 million. If this amount is exceeded, the President shall report to Congress.
Major Declaration	• The President can declare a Major Disaster Declaration for any natural event, including any hurricane, tornado, storm, high water, wind-driven water, tidal wave, tsunami, earthquake, volcanic eruption, landslide, mudslide, snowstorm, or drought, or, regardless of cause, fire, flood, or explosion, that the President believes has caused damage of such severity that it is beyond the combined capabilities of state and local governments to respond. • A major disaster declaration provides a wide range of federal assistance programs for individuals and public infrastructure, including funds for both emergency and permanent work.

Not all programs, however, are activated for every disaster. The determination of which programs are authorized is based on the types of assistance specified in the governor's request and on the needs identified during joint PDA and any subsequent PDAs.

President Directed Emergency Assistance

The President may direct emergency assistance without a Governor's request if an incident occurs that involves a subject area exclusively or preeminently the responsibility of the U.S. The President will consult the Governor of any affected State, if practicable.

In addition, FEMA may provide accelerated Federal assistance and support where necessary to save lives, prevent human suffering, or mitigate severe damage, even in the absence of a specific request. In these instances, the Governor of the affected State will be consulted if practicable, but this consultation shall not delay or impede the provision of such rapid assistance.

This page intentionally left blank.

Section 1-10: Tribes

Introduction

Tribal Overview

Tribal Relations Support Annex (TRSA)

    TRSA Policies

Tribal Liaison

Tribal Assistance Coordination Group (TAC-G)

Tribal Expectations of Federal Responders

Introduction

The U.S. has a unique legal and political relationship with Indian Tribes, established through and confirmed by the U.S. Constitution, treaties, statutes, executive orders, and judicial decisions. Federal Departments and Agencies are charged with engaging in regular and meaningful consultation and collaboration with Tribal officials in the development of Federal policies that have Tribal implications, and are responsible for strengthening the government-to-government relationship between the U.S. and Indian Tribes.

This coordination and cooperation extends to the conduct of CISAR operations during a disaster response under the Stafford Act.

Tribal Overview

The following information is provided to help CISAR planners and responders understand the status and background of Indian Tribes.

- There are over 560 federally recognized Tribes in 35 States;

- According to the U.S. Bureau of Census, the estimated population of American Indians and Alaska Natives, including those of more than one race, as of July, 1, 2007, were 4.5 million, or 1.5 percent of the total U.S. population;

- Indian lands in the continental U.S. comprise over 45 million acres of reserved lands and an additional 10 million acres in individual allotments (There are another 40 million acres of traditional native lands in Alaska);

- The U.S. recognizes Indian Tribes as domestic dependent nations, possess nationhood status, and retain inherent powers of self-government;

- The relationship between federally recognized Tribes and the U.S. is one between sovereigns (i.e., between a government and a government), a "government-to-government" principle that has helped shape relations between the Federal Government and Indian Tribes;

- States have no authority over Tribal Governments unless expressly authorized by Congress;

- While federally recognized Tribes generally are not subordinate to States, they can have a government-to-government relationship with the States as well;

- Tribes possess the right to form their own governments, make and enforce laws (both civil and criminal), tax, establish and determine membership (e.g., tribal citizenship), license and regulate activities within their jurisdiction, to zone, and exclude persons from Tribal lands;

- The Tribal Chief Executive:

  o May be called the President, Tribal Chief Executive, or Tribal Chief;

  o May have the authority to suspend or initiate Tribal laws and regulations to ensure safety and security of Tribal people; and

  o Has the authority to order quarantine or close the boundaries of Federal Tribal lands to ensure safety and security of Tribal people.

Tribal Relations Support Annex (TRSA)

In Stafford Act disaster response operations, Tribal relations are coordinated under ESF #15 (External Affairs) and are represented in the Tribal Relations Group in the Joint Field Office.

In addition to ESF #15, the Tribal Relations Support Annex (TRSA) describes the policies, responsibilities, and concept of operations for effective coordination and interaction of Federal incident management activities with those of Tribal governments and communities during incidents requiring a coordinated Federal response.

TRSA Policies

- DHS, in cooperation with other Federal Departments and Agencies, coordinates Tribal relations functions for incidents requiring Federal coordination;

- Federal Agencies shall respect Indian Tribal self-government and sovereignty, honor tribal treaties and other rights, and strive to meet the responsibilities that arise from the unique legal relationship between the Federal Government and Indian Tribal Governments;

- State Governors must request a Presidential Disaster Declaration (PDD) on behalf of a Tribe under the Stafford Act (However, Federal Departments and Agencies can work directly with Tribes within existing Agency authorities and resources in the absence of such a declaration);

- Federal Departments and Agencies must comply with existing laws and Executive Orders mandating that the Federal Government deal with Indian Tribes on a government-to-government basis, reflecting the federally recognized Tribes' right of self-government as sovereign domestic dependent nations (A Tribe may, however, opt to deal directly with State and local officials);

- Federal Departments and Agencies involved in potential or actual incidents requiring a coordinated Federal response shall consult and collaborate with Tribal Governments on matters affecting the Tribes and must be aware of the social, political, and cultural aspects of an incident area that might affect incident management operations;

- Federal Departments and Agencies provide appropriate incident management officials with access to current databases containing information on Tribal resources, demographics, and geospatial information;

- Federal Departments and Agencies recognize the unique political and geographical issues of Tribes whose aboriginal and contemporary territory is on or near the current international borders of Canada and Mexico;

- Federal Departments and Agencies shall include Tribes in all aspects of incidents requiring a coordinated Federal response that affect Tribes and incident management operations;

- A Tribal Relations Element can be established in the JFO to provide the operational capability for collecting and sharing relevant incident information, alerting, and deploying required tribal relations staff to or near the affected area, and ensuring compliance with Federal laws relating to Tribal relations; and

- For incidents that directly impact Tribal Jurisdictions, a Tribal representative shall be included in the Unified Coordination Group, as required.

Tribal Liaison

A Tribe may appoint a Tribal member to serve as a JFO Tribal liaison. As authorized by the Tribal Government, the Tribal Liaison:

- Is responsible for coordinating Tribal resources needed to prevent, protect against, respond to, and recover from incidents of all types. This also includes preparedness and mitigation activities;

- May have powers to amend or suspend certain Tribal laws or ordinances associated with the response;

- Communicate with the Tribal community and helps people, businesses, and organizations cope with the consequences of any type of incident;

- Can negotiate mutual aid and assistance agreements with other Tribes or jurisdictions;

- Can request Federal assistance under the Stafford Act through the Governor of the State when it becomes clear that the Tribe's capabilities will be insufficient or have been exceeded;

- Can elect to deal directly with the Federal Government; and

- Although a State Governor must request a PDD on behalf of a Tribe under the Stafford Act, Federal Departments or Agencies can work directly with the Tribe within existing authorities and resources.

Tribal Assistance Coordination Group (TAC-G)

The TAC-G consists of Federal Government representatives that support ESF #15 (External Affairs) and the TRSA during disaster response operations under the Stafford Act.

The TAC-G:

- Is composed of Indian Affairs (IA), Indian Health Service (IHS), and FEMA representatives;

- May be deployed to manage major and/or complex incidents, and incidents that extend into multiple operational periods and require a written Incident Action Plan (IAP).

- Provides the strategic guidance and operational context within the NIMS interagency framework;

- Coordinates with DHS to ensure Tribal relations actions are carried out in accordance with established Federal Government policies and procedures;

- Assists in providing an efficient and reliable flow of incident-related information between the Tribes and the Federal Government;

Tribal Expectations of Federal Responders

- Normally, the organizing Federal entity will meet with the Tribal Council or Tribal Chief Executive and receive permission to conduct operations on federally recognized Tribal lands;

- CISAR planners and responders should discuss any cultural or historical considerations before initiating operations;

- Death of a Tribal member may be treated differently than protocols normally found in non-Tribal populations (the Tribal Shaman or elder may be needed before removing a body); and

- Animals are held in high regard in most Tribal societies.

Overview

Territories are one of the four types of political division of the U.S., overseen directly by the Federal Government and not any part of a U.S. State. Territories can be classified by whether they are *incorporated* (part of the U.S. proper) and whether they have an *organized* government (through an Organic Act passed by the U.S. Congress) or a territorial constitution (and functioning legislature).

Under the Stafford Act, the following five U.S. Territories are able to receive disaster assistance and are included in the definition of a "State:"

- Puerto Rico;

- U.S. Virgin Islands;

- Territory of Guam;

- Territory of American Samoa; and the

- Commonwealth of the Northern Mariana Islands (CNMI).

Additionally, Stafford Act assistance is available to two sovereign nations under the compact of free association with the U.S.:

- Federated States of Micronesia (FSM); and the

- Republic of the Marshall Islands (RMI).

Table 1-11-1 on the next page provides a general overview of the U.S. Territories supported by the Stafford Act.

Table 1-11-1: U.S. Territories and Commonwealths				
Stafford Act Support Location	Capital	Total Area (Square Miles)	Population[1]	FEMA Region
Puerto Rico	San Juan	3,151	3,808,610	II
U.S. Virgin Islands	Charlotte Amalie	135	109,825	II
Territory of Guam	Agana	212	178,430	IX
American Samoa	Pago Pago	77	57,291	IX
Northern Mariana Islands	Saipan	176	69,221	IX
Federated States of Micronesia	Pohnpei	271	110,728 (2009)[2]	USAID[4]
Marshall Islands	Majuro	69.8	61,300 (2009)[3]	USAID[4]

[1] Unless otherwise stated, U.S. Census (As of 01 Apr 2010).
[2] World Bank, *2011 World Development Indicators* (Washington D.C., World Bank: April, 2011).
[3] 2009 estimate by the Marshall Islands Government health ministry.
[4] Under the service agreements of the amended Compact of Free Association with the Federated States of Micronesia and Marshall Islands, the U.S. Agency for International Development (USAID) is responsible for coordinating disaster response.

CISAR Challenges

CISAR responders will have significant challenges when deploying to disaster response operations in the U.S. Territories and Commonwealths.

Pacific Ocean

For U.S. Territories, FSM, and RMI in the Pacific Ocean, the great distances, limited airport facilities, low elevation of many of the populated islands and limited SAR resources can cause significant challenges in conducting disaster CISAR operations.

The U.S. Territories, FSM, and RMI are at risk for typhoons, as well as tsunamis caused by earthquakes. While there are USCG and military resources in Guam, as well as NPS and other Federal resources in the other Territories, any organic response to a disaster will in all likelihood be very limited.

For operations in the Pacific U.S. Territories, CISAR responders should have the appropriate U.S. passport.

Atlantic Ocean

Puerto Rico and U.S. Virgin Islands have more robust, organic SAR capabilities with USCG, NPS, and National Guard resources available on both islands.

As in the Pacific Ocean, Puerto Rico and the U.S. Virgin Islands are also at risk for major disasters caused by hurricanes and tsunamis caused by earthquakes.

Puerto Rico

Puerto Rico is an organized, unincorporated U.S. Territory with commonwealth status. Policy relations between Puerto Rico and the U.S. are conducted under the jurisdiction of the Office of the President.

Puerto Rico is an island situated between the Caribbean Sea and the Atlantic Ocean, just east of the Dominican Republic, with an area of 3,515 square miles (slightly less than 3 times the size Rhode Island). San Juan, the capital, is located on the northeastern shore of the main island; there are also 3 small islands included in the Commonwealth: Vieques and Culebra to the east and Mona to the west. San Juan's location makes it one of the Caribbean Sea's most valuable ports. The Mona Passage, off Puerto Rico's west shore, is also a crucial shipping route to the Panama Canal (Figure 1-11-1 below).

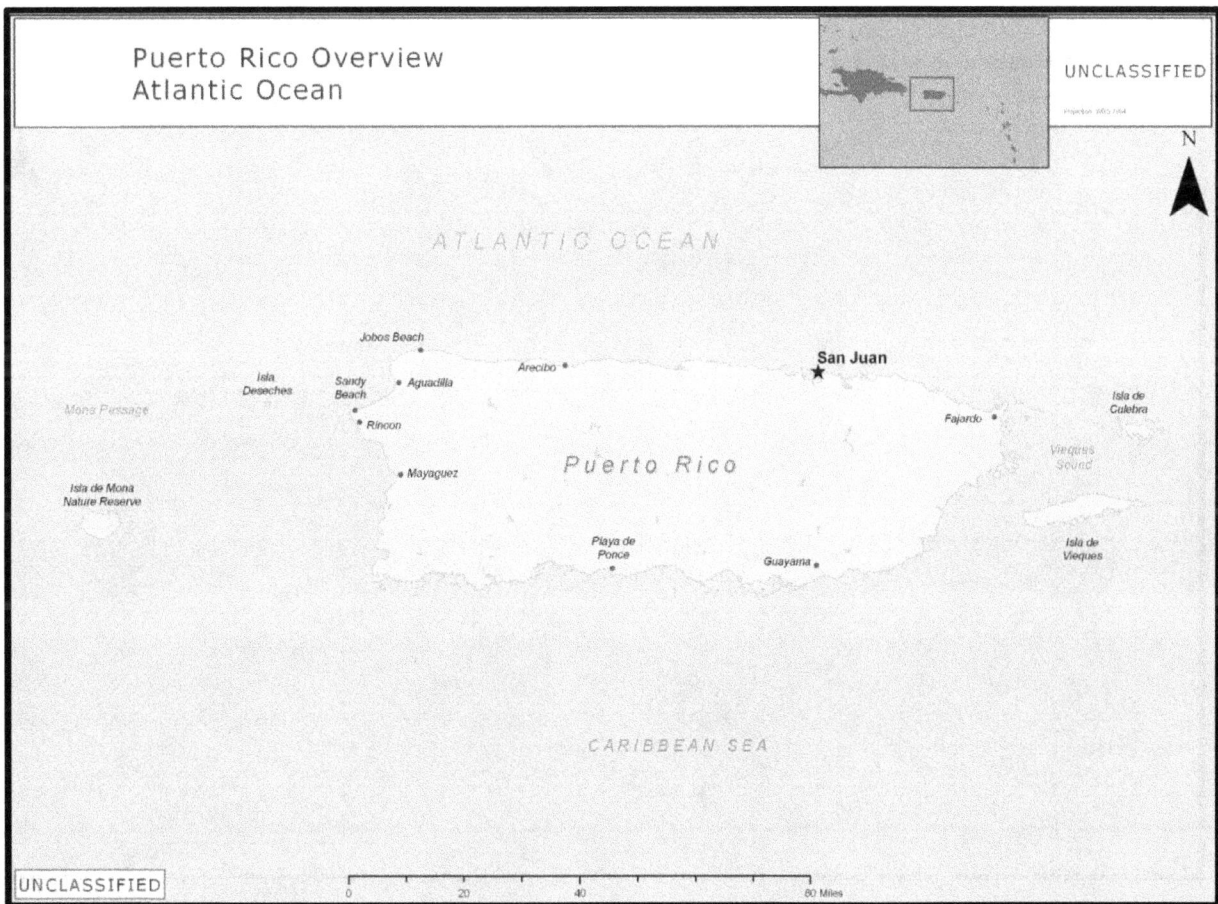

Figure 1-11-1: Puerto Rico

U.S. Virgin Islands

The U.S. Virgin Islands is an organized, unincorporated U.S. Territory located immediately east of Puerto Rico (Figures 1-11-2, 1-11-3, and 1-11-4). Although more than 50 separate islands and cays constitute this westernmost of the Lesser Antilles, only three have a size and population of any significance: St. Thomas, St. Croix, and St. John. The U.S. Virgin Islands have an area of approximately 135 square miles (twice the size of Washington D.C.). Almost all the other islets are both uninhabited and uninhabitable. Most of the population is shared equally by St. Croix and St. Thomas, although St. Croix is considerably larger in area. The capital is located in Charlotte Amalie on St. Thomas.

St. Thomas has one of the best deepwater ports in the Caribbean.

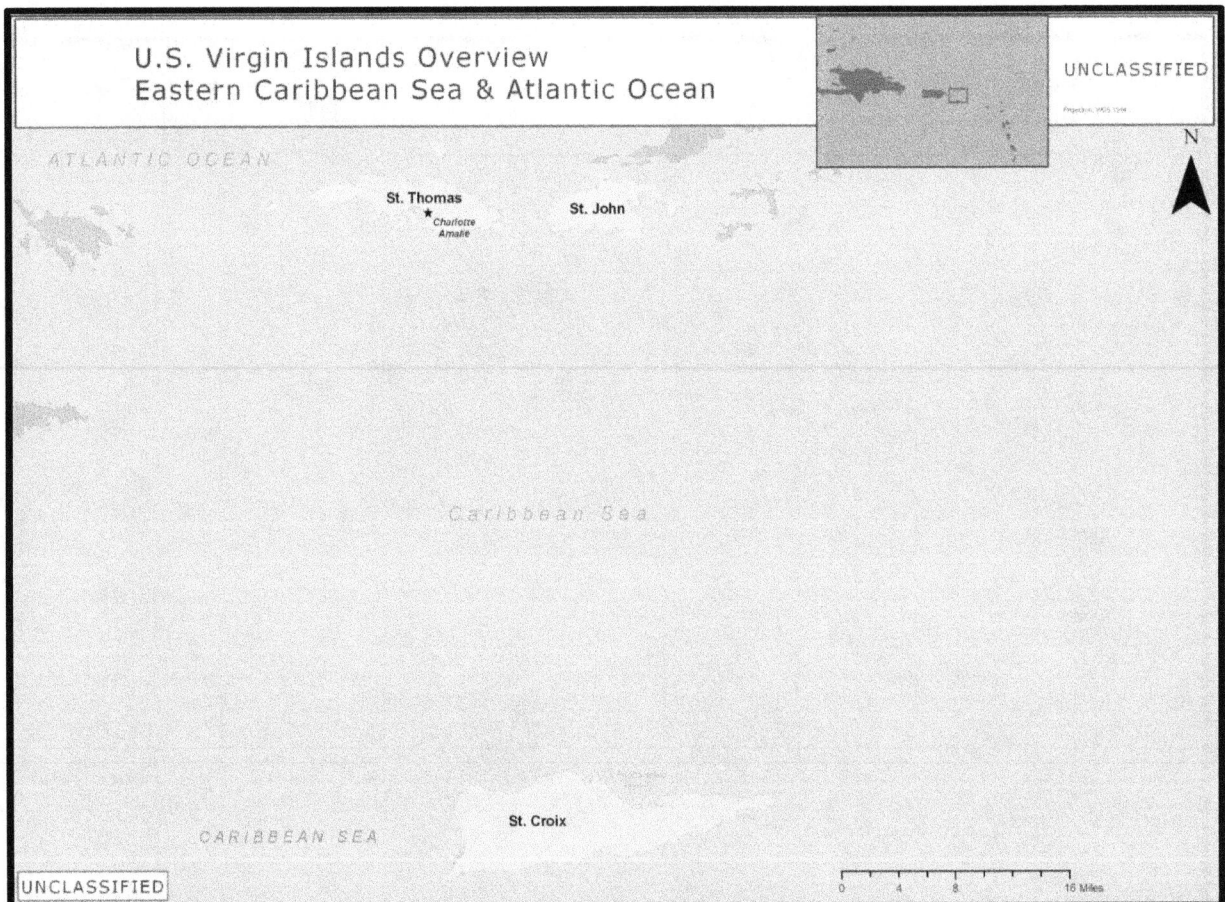

Figure 1-11-2: U.S. Virgin Islands (Overview)

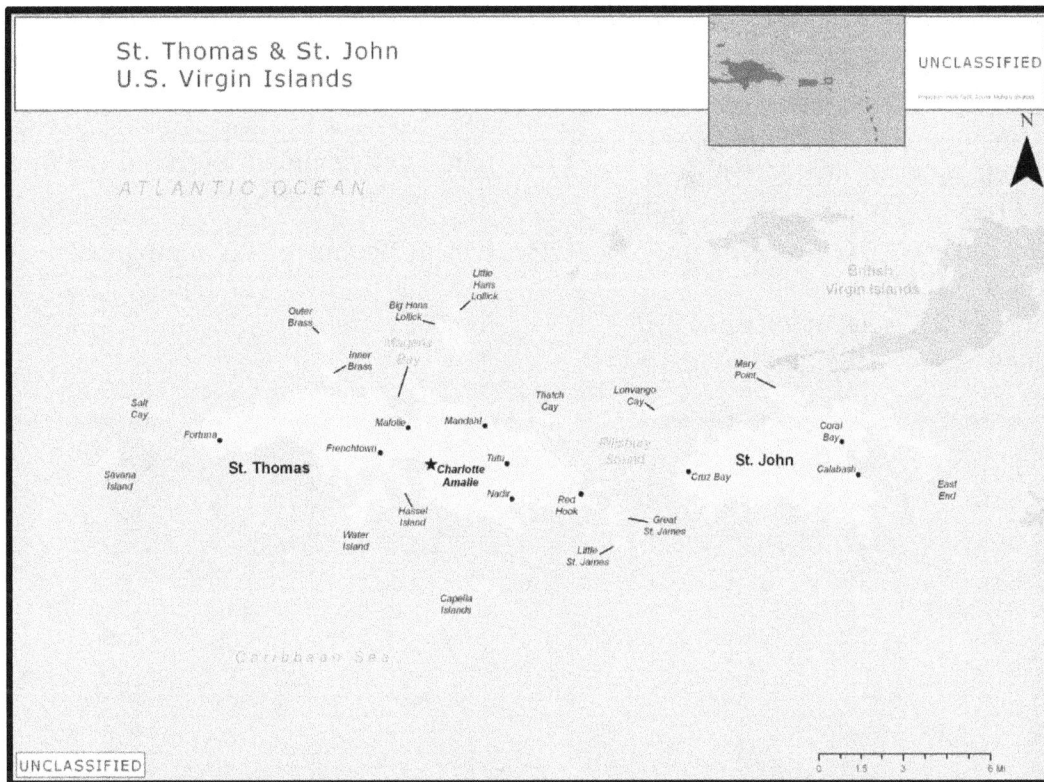

Figure 1-11-3 St. Thomas and St. John, U.S. Virgin Islands

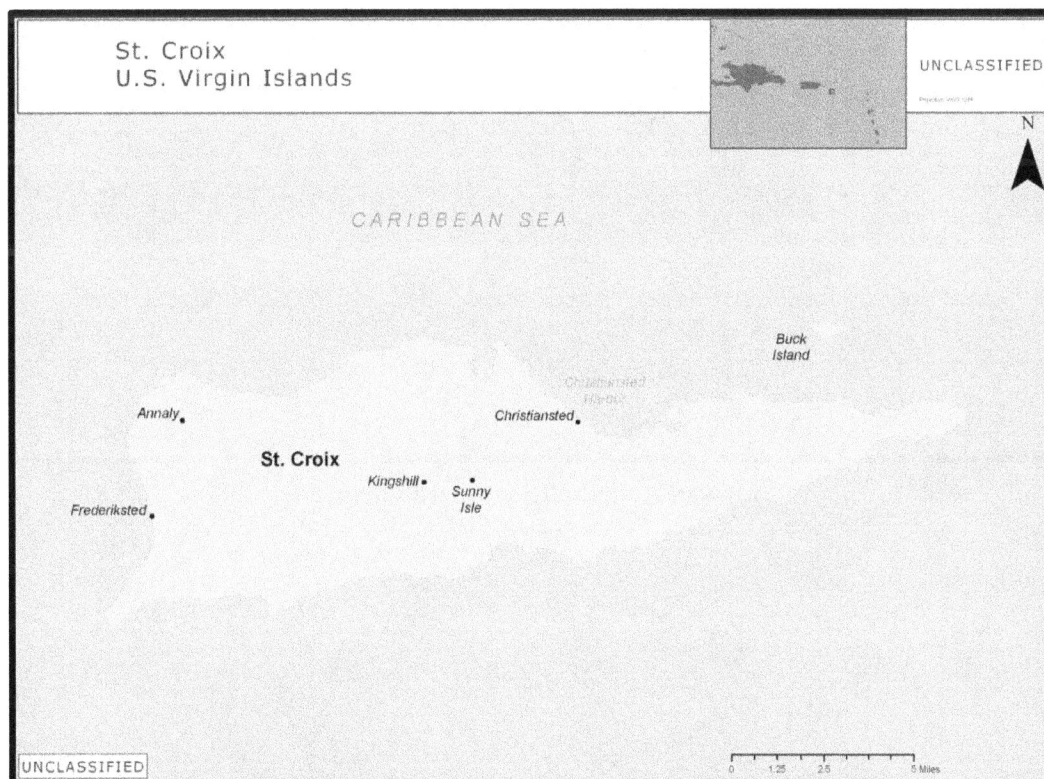

Figure 1-11-4 St. Croix, U.S. Virgin Islands

Territory of Guam

Guam is an organized, unincorporated U.S. Territory administered by the Office of Insular Affairs, DOI. Guam is 30 miles long and 9 miles wide (3 times the size of Washington D.C.), with an area of approximately 212 square miles, located 6,000 miles west of San Francisco and 3,700 miles west-southwest of Honolulu, Hawaii (Figure 1-11-5 below). Guam is the largest and southernmost island in the Mariana Islands archipelago.

Guam's population is approximately 180,000, of which 24,000 are U.S. military related.

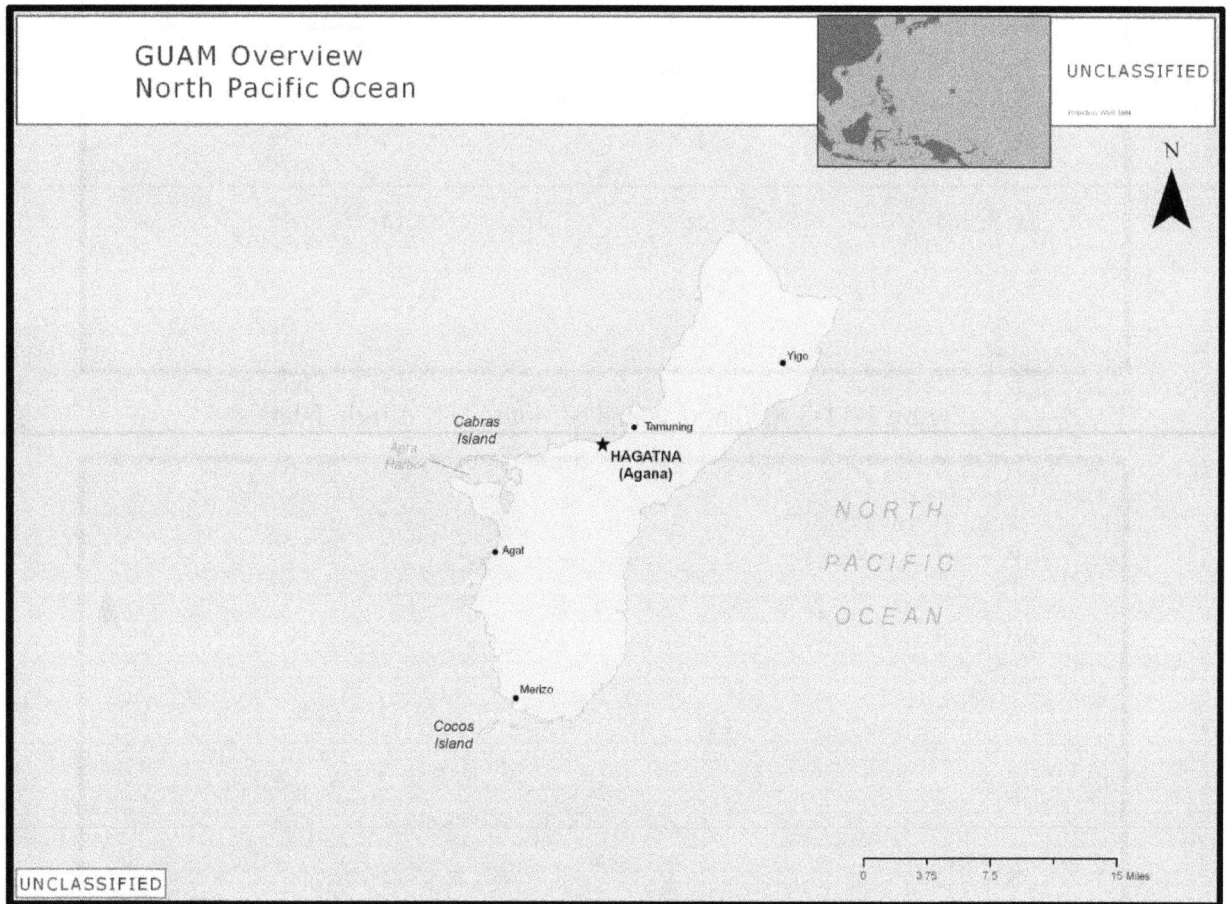

Figure 1-11-5: Territory of Guam

Territory of American Samoa

American Samoa is an unincorporated unorganized U.S. Territory administered by the Office of Insular Affairs, DOI. It consists principally of five volcanic islands and two coral atolls, for a total area of approximately 77 square miles (slightly larger than Washington D.C.). American Samoa is located approximately 2,300 miles southwest of Hawaii (about half way between Hawaii and New Zealand). The largest and most populated island is Tutuila, which has the capital city (Pago Pago).

In addition to Tutuila, the principal islands are Aunu'u and the Manu'a islands (a cluster of three islands), Ta'u, Ofu, and Olosega, located approximately 65 miles east of Tutuila.

Swains Island, a small island with a population of less than 25 and Rose Atoll, an uninhabited atoll approximately 120 miles east of Tutuila, make up the remainder of the Territory (Figure 1-11-6 below).

95 percent of the American Samoa population lives on Tutuila.

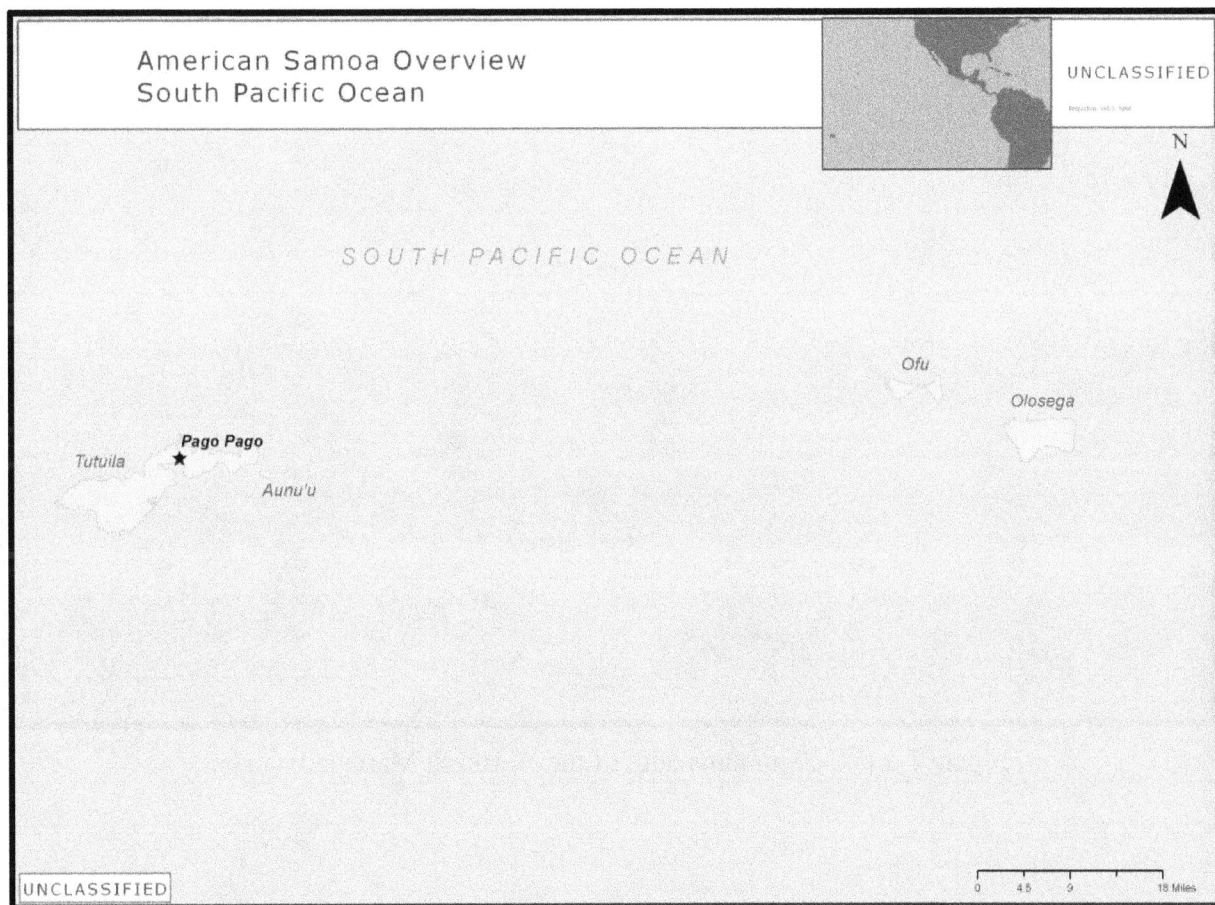

Figure 1-11-6: Territory of American Samoa

Commonwealth of the Northern Mariana Islands

CNMI is a commonwealth in political union with the U.S. under the jurisdiction of the Office of Insular Affairs, DOI. CNMI consists of 14 islands extending in a generally north-south direction for 338 miles for a land area of 176 square miles (2.5 times the size of Washington D.C.). With almost 570,000 people, Saipan is the largest island (46 square miles) and is the government center (Figure 1-11-7 below). Thousands of other people live on the islands of Rota (32 square miles) and Tinian (39 square miles), which are largely rural.

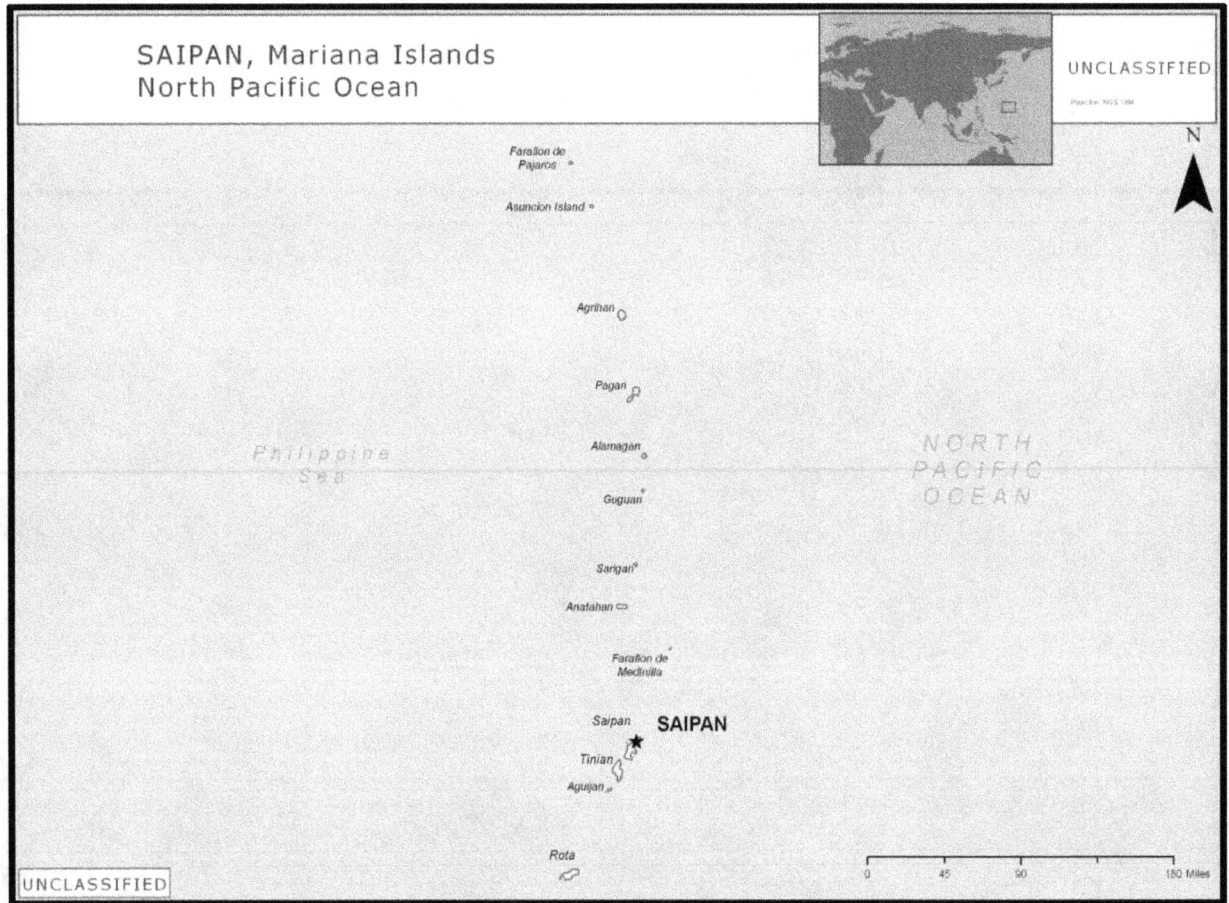

Figure 1-11-7: Commonwealth of the Northern Mariana Islands

Federated States of Micronesia

FSM (with Palau) forms the archipelago of the Caroline Islands, and lies approximately 497 miles east of the Philippines. FSM consists of 607 islands and includes (from west to east) the States of Yap, Chuuk (formerly Truk), Pohnpei (formerly Ponape), and Kosrae. FSM covers approximately 271 square miles (4 times the size of Washington D.C.) and is scattered over more than 1 million square miles of ocean (Figure 1-11-8 below). FSM's largest island cluster is Pohnpei (163 islands), with an area of 133 square miles, while the smallest cluster is Kosrae (5 islands), spanning 42.5 square miles. The islands include a variety of terrains, ranging from mountainous to low, coral atolls and volcanic outcrops.

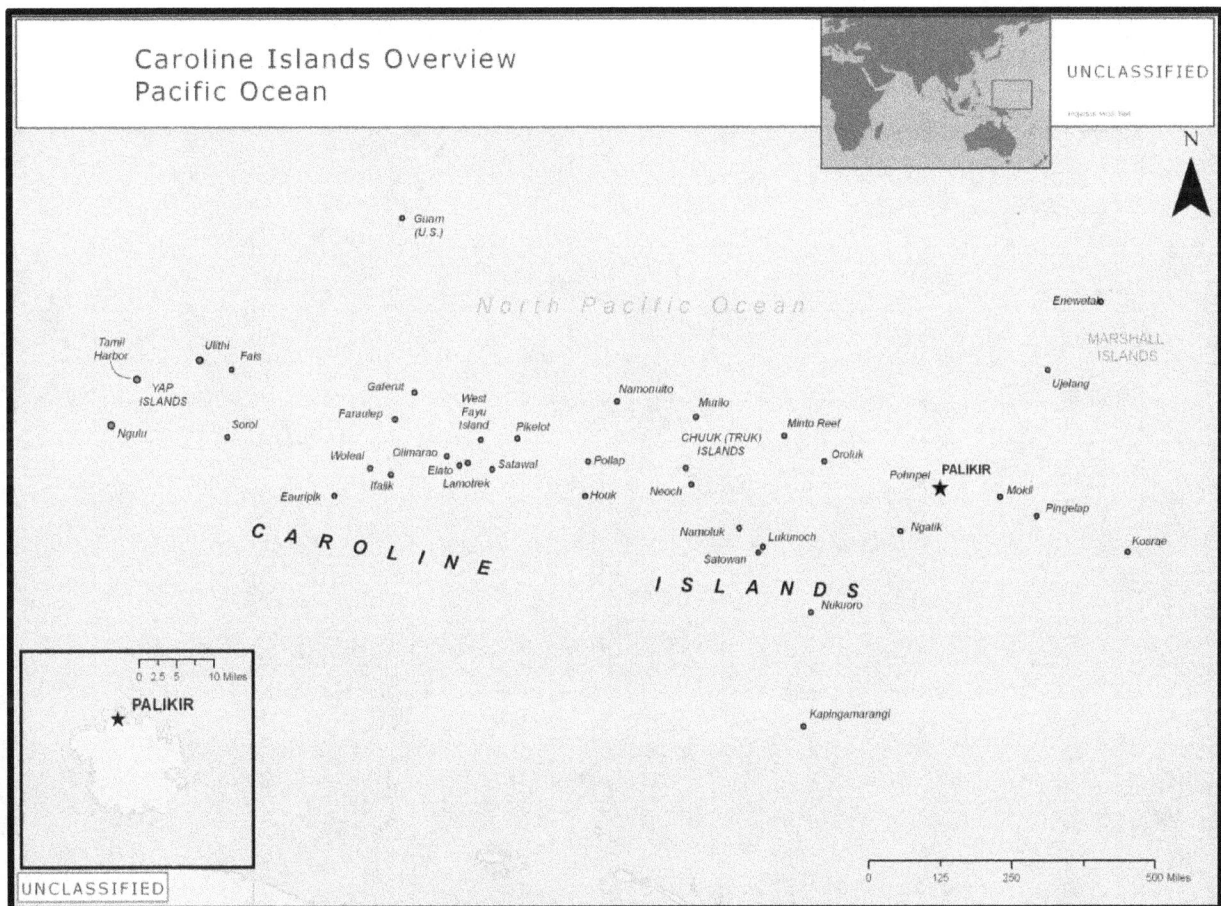

Figure 1-11-8: Federated States of Micronesia

Republic of the Marshall Islands

RMI is located in the North Pacific Ocean approximately 2,486 miles northeast of Australia. RMI consists of 2 groups of small islands, atolls (coral islands), and reefs running from northwest to southeast (Figure 1-11-9 below). The more easterly of these is the Ratak Chain, the more westerly, the Ralik Chain. It is estimated that there are 1,152 islands and 30 atolls, but only 4 islands and 19 atolls are inhabited (approximately the size of Washington D.C.).

With terrains of coral, limestone, and sand, none of the islands have any high ground, and the most elevated location of the islands is 33 feet. The total land area is approximately 70 square miles, and about 60 percent is taken up by crops. RMI's capital is Majuro, which is located on an atoll of the same name.

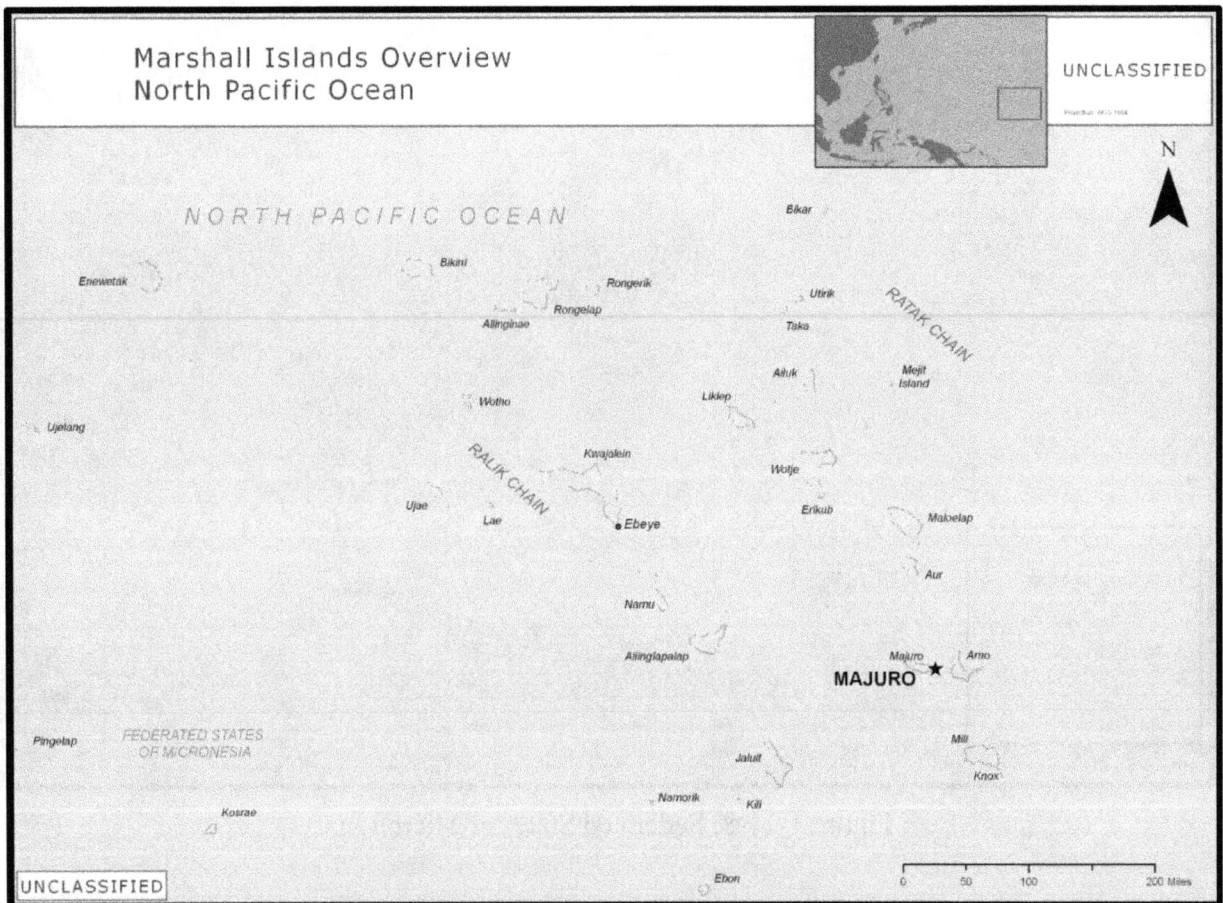

Figure 1-11-9: Republic of the Marshall Islands

Section 1-12: Nongovernmental Organizations

Overview

Functions

Working Relationships

Private Sector

Overview

In the U.S., all emergencies regardless of size or type are local events. However, when a community's resources are insufficient to respond to an incident, local government may call on nongovernmental organizations for assistance.

Nongovernmental organizations are non-profit entities with an association that is based on interests of its members, individuals, or institutions and that is not created by government, but may work cooperatively with government. Such organizations serve a public purpose, not a private benefit. Examples of NGOs include faith-based charity organizations and the American Red Cross.

For example, hundreds of NGOs responded to the January, 2010, earthquake in Haiti, ranging from very small faith-based and secular community organizations to large international organizations, such as Doctors without Borders.

As required under Homeland Security Presidential Directive (HSPD) 5, the NIMS enables responders from different communities with a variety of job responsibilities to better work together. Everyone has a role to play in NIMS implementation – including search and rescue.

Functions

Nongovernmental Organizations (NGOs) render assistance through existing EOCs and other structures. NGOs can provide invaluable assistance to CISAR operations, including shelter, emergency food supplies, assistance with animals, and other vital CISAR support services. These groups often provide specialized help for individuals with special needs, including those with disabilities.

Working Relationships

Effective interagency coordination with NGOs requires pre-event planning in order to leverage the capabilities of these cost effective resources. These organizations are capable of providing operational and logistical expertise and can also help manage volunteer services and donated goods. Voluntary Organizations Active in Disaster (VOADs) and faith based organizations will contribute services that will enhance the CISAR response. Pre-event planning is the key to leveraging NGO capabilities.

Private Sector

The private sector is responsible for most of the critical infrastructure and key national resources and thus may require assistance in the wake of a disaster or emergency. They also provide goods and services critical to CISAR response efforts, either on a paid basis or through donations.

Part 2: CISAR Management

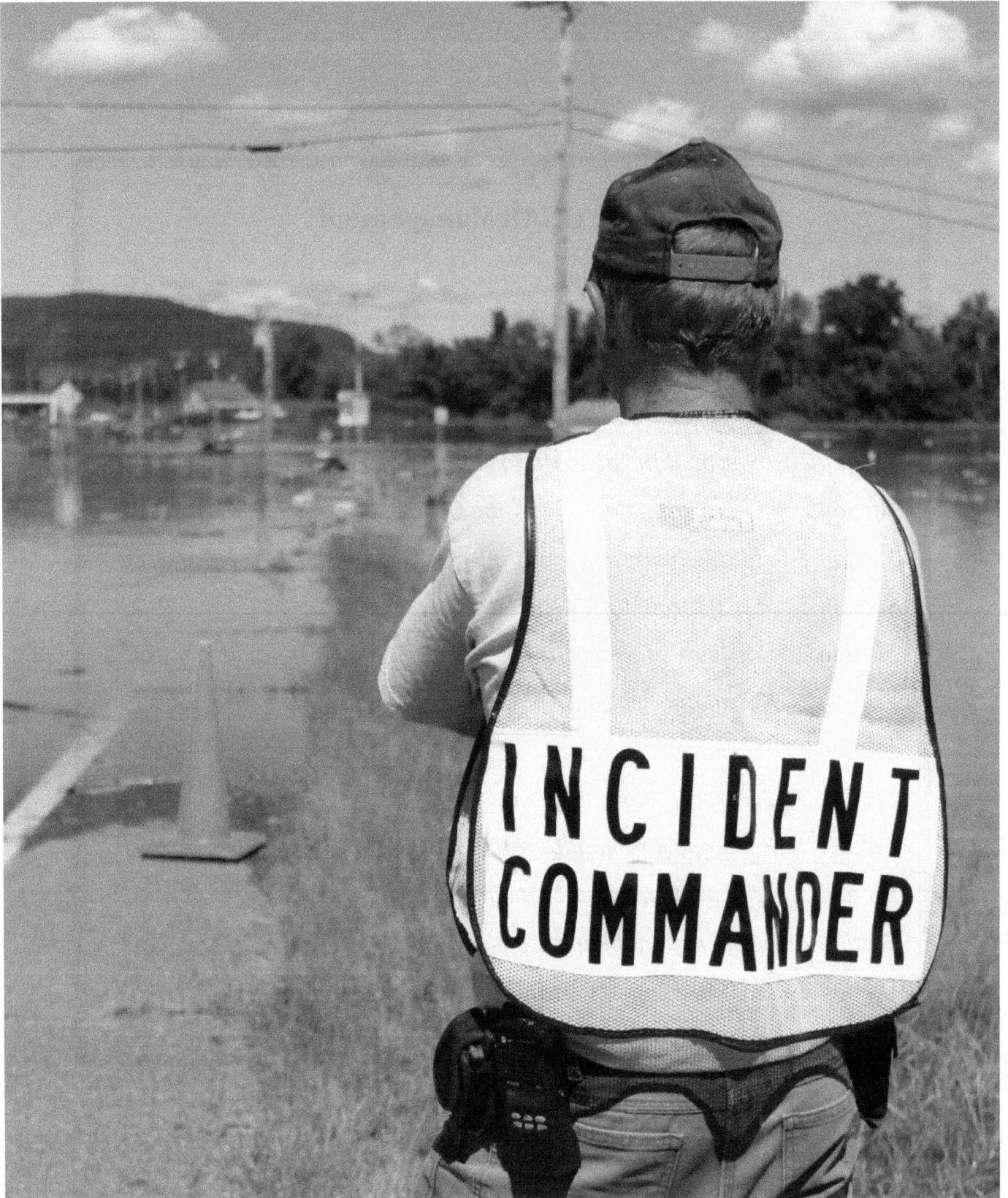

Section 2-1: CISAR Management

Introduction

Incident Commander (IC)

SAR Mission Coordinator (SMC)

On Scene Coordinator (OSC)

Transition from CISAR Operations

Introduction

The NSP and NSS are largely based on international conventions that U.S. SAR services are obligated to follow. These references are written to maximize the effectiveness of SAR operations, particularly when working with military services, SAR authorities of other nations, or with ships, or aircraft at sea.

In addition to following relevant provisions of these references for CISAR, the NIMS organizational structure will normally be implemented for overall response management. Use of NIMS is particularly important when non-SAR operations are being conducted in conjunction with a large CISAR operation.

For large incidents that may involve both SAR and non-SAR activities, the SMC will initiate action and coordinate the overall CISAR response in accordance with the references mentioned above and NIMS.

(Note: The USCG Incident Management Handbook is widely used by Federal agencies (available at www.uscg.mil/nsarc) and addresses SAR within the context of a major incident.)

Incident Commander (IC)

The IC may be designated as the SMC. However, separate individuals should carry out the IC and SMC functions if the operational tempo and/or span of control requires separate positions.

For large CISAR operations that may include other non-SAR activities, IC tasks may include:

- Standing up the IC and NIMS organization;

- Mobilize additional appropriate resources as soon as possible to stabilize the situation or assist in the non-SAR operations;

- In consultation with FAA, provide advisory air traffic service to aid pilots in maintaining safe operations;

- Assign or request a Public Information Officer (PIO) to provide initial information to the media and establish a 24 hour Joint Information Center (JIC) to provide timely information and updates on progress of SAR efforts and outline of future actions;

- Be available to provide press briefings;

- Use appropriate assistance to notify Next-of-Kin (NOK) as soon as possible.

- Maintain daily contact with NOK to provide progress of SAR efforts and outline future actions; and

SAR Mission Coordinator (SMC)

The IC will designate an individual to serve as SMC. The SMC serves as the link between the SAR system and the NIMS organization and is best placed at the Branch Director or Group Supervisor level. Under the IC's general direction, the SMC coordinates the CISAR response in accordance with the NSP and associated documents.

(Note: For further information on SMC, refer to Section 2-6: Search and Rescue Mission Coordinator (SMC)).

On Scene Coordinator (OSC)

The OSC coordinates the SAR mission on scene using available resources. The OSC may also serve as a Branch Director or Group Supervisor to manage on scene operations other than SAR, particularly after the CISAR operation is concluded and other missions take precedence, such as search and recovery. OSC duties may include the following:

- Implement the SAR Action Plan;

- Establish and maintain communications with the SMC;

- Operational control and coordination of assigned CISAR aircraft, boats, and response teams;

- Establish and maintain communications with CISAR aircraft, boats, and response teams using assigned on scene channels;

- For arriving CISAR aircraft, boats, and response teams, provide initial briefing and search instructions;

- Provide advisory air traffic service to aid pilots in maintaining aircraft separation;

- Carry out SAR action plans, and modify plans to cope with changing on scene conditions advising the SMC of all major changes;

- Receive and evaluate survivor sighting reports, and divert CISAR responders to investigate sightings;

- Obtain search results from departing CISAR aircraft, boats, and response teams; and

- Submit sequentially numbered situation reports (SITREPs) to the SMC at regular intervals.

- Establishment of a common altimeter setting for all on scene aircraft (this may be done by the ACO or senior pilot if the OSC is a surface unit;

- Require aircraft to make "operations normal" reports to the OSC. It is recommended that:

  o For helicopters: every 15 minutes;

  o For multi-engine fixed-wing aircraft: every 30 minutes.

(Note: Operations normal reports may also be instituted for non-aviation SAR aircraft at suitable intervals to monitor status, personnel safety, fatigue, operators with limited training or experience, area hazards, density and diversity of aircraft being used.)

Transitioning from CISAR Operations

For some incidents, the CISAR response may be completed or operations suspended by the time the IC is fully operational. As the CISAR operation winds down, the IC may designate the OSC in the SAR response to also serve as a Branch Director or Group Supervisor to manage on scene operations other than SAR. Likewise, CISAR responders may also be reassigned to other groups in the NIMS structure once the CISAR operation is concluded.

CISAR responders should receive sufficient NIMS training to carry out their respective duties within the ICS.

Overview

Federal SAR support in a disaster must be provided in a timely manner to save lives, prevent human suffering, and mitigate severe damage. This may require mobilizing and deploying assets before they are requested via normal NRF protocols.

The nature of a catastrophic incident may immediately overwhelm State, Tribal, Territorial, and local response capabilities and require immediate Federal support. Federal support can be provided immediately under the authority of the NSP, and also in the form of initial assistance that FEMA is able to provide under the NRF.

Lifesaving Priority

A CISAR operation may involve very large numbers of persons needing assistance, and priority must be given to human lifesaving.

Lifesaving efforts must be immediate to be effective.

Scope of Operations

The nature and scope of a catastrophic incident may result in large numbers of persons in distress and include chemical, biological, radiological, nuclear or high-yield explosive attacks, disease epidemics, and major natural or manmade hazards.

Multiple incidents may occur simultaneously or sequentially, in contiguous or noncontiguous areas.

The incident may cause significant disruption of the area's critical infrastructure, such as energy, transportation, telecommunications, and public health and medical systems.

Local CISAR response capabilities and resources (to include mutual aid from surrounding jurisdictions and support from the State, Tribe, or Territory) may be insufficient and quickly overwhelmed. State, Tribal, Territorial, and local emergency personnel who normally respond to incidents may be among those affected and unable to perform their duties.

Immediate Response

When appropriate for lifesaving, Federal Departments and Agencies that conduct SAR operations generally have authority to respond immediately (includes DoD component commands).

Normally, awareness of the need for immediate response becomes known due to direct receipt of calls for help from persons in distress (such as distress alerts to a Rescue Coordination Center - RCC), or requests for assistance from some recognized civil authority.

(Note: Nothing in any Federal plan is ever intended to preclude prompt assistance to persons in distress when it can be reasonably be provided.)

With rare exceptions (e.g., critical national security situation, CBRNE environment, interference with critical military duties), jurisdictional, legal, or financial considerations should not preclude prompt response to save lives.

Typically, a unified response builds upon the work of those providing immediate aid to those in distress.

Response Actions

Four key response actions typically occur in the conduct of CISAR operations:

- Gain and maintain situational awareness;

- Activate, pre-position, and deploy SAR resources and capabilities;

- Efficiently and effectively coordinate and conduct lifesaving response actions among agencies; and

- As the situation permits, demobilize Federal SAR resources.

Key Principles

Key principles that must be applied for successful CISAR operations are:

- Engaged partnership between Federal, State, Tribal, Territorial, and local SAR authorities;

- A tiered CISAR response;

- Scalable, flexible, and adaptable CISAR capabilities;

- Unity of effort; and

- Readiness.

In addition, the principle of using all available resources to save lives is especially pertinent in demanding CISAR operations.

CISAR Responder: Balance of Risk

In CISAR operations, lifesaving must be weighed against the risks taken by CISAR responders (see *Section 3-1: Risk Assessment*). To save lives and protect property, decisive action on scene is often required of CISAR responders. Although some risk may be unavoidable, CISAR responders can effectively anticipate and manage risk through proper training, planning, and situational awareness.

Tiered Approach

CISAR operations must be implemented through a tiered approach (See Figure 2-2-1 on the next page):

- If required, State assistance will supplement local efforts; and

- When requested, Federal assistance will supplement State, Tribal, Territorial, and local CISAR efforts.

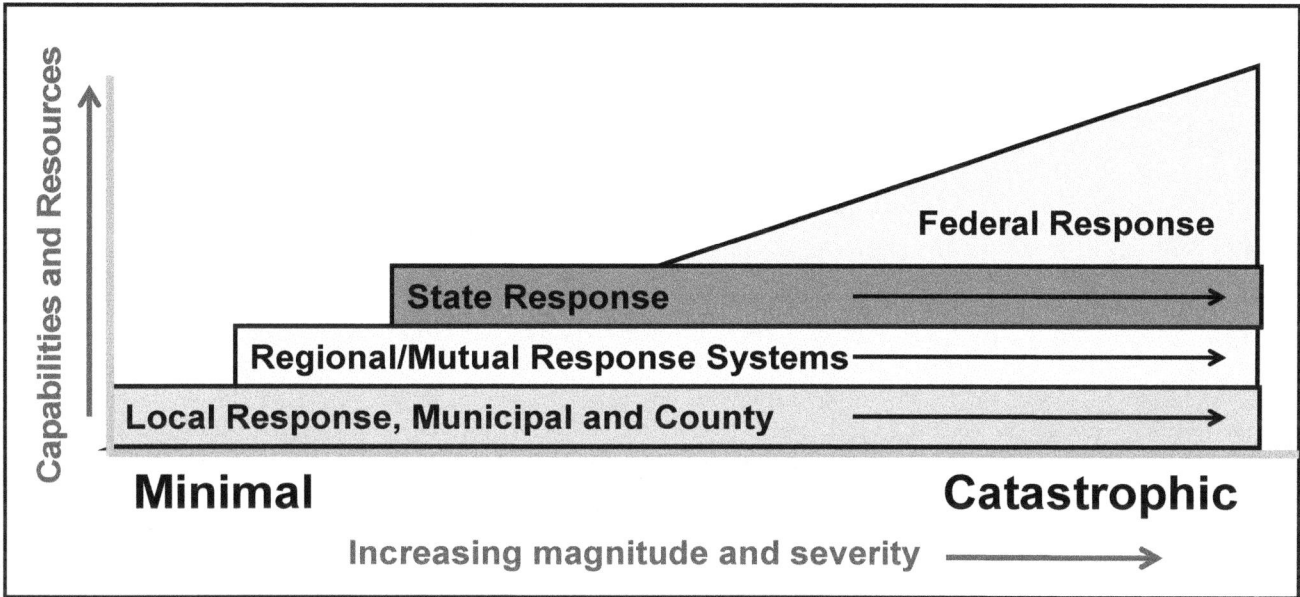

Figure 2-2-1: Tiered Approach to CISAR Operations

This page intentionally left blank.

Section 2-3: CISAR Planning Considerations

CISAR Planning Considerations

Lessons Learned: Hurricane Katrina

National Preparedness

Federal, State, Local, Tribal, and Territorial Planning Relationships

CISAR Planning Fundamentals

Time Driven Objective and Resources to Support

CISAR Chain of Events

Resource Efficiency (Capability vs. Time)

Search Theory and Search Effectiveness

Lily Pad and Transportation Planning Considerations

SAR Action Plans

Unified SAR Branch/CISAR Coordination Cell

Planning Cycle

CISAR Planning Considerations

Interoperability, synchronization, and efficient unity of effort among multiple agencies are fundamental imperatives to life-saving during overwhelming Mass Rescue Operations (MRO) within a CISAR environment.

Coordinated and integrated operational efficiency in support of State, Tribal, Territorial, and local SAR operations can only be achieved through integrated, standardized planning. This Section highlights new, emerging concepts, procedures, and planning considerations to assist the CISAR planner at all levels of government and planning architectures (strategic, operational, and tactical).

Goal of CISAR Planning

To enable a collaborative, interagency planning process and achieve standardized interoperability among all agencies and resources across all domains; to locate, rescue, and safely transport persons in distress within an unbroken operational chain of events in the timeliest manner possible.

Planning in normal day-to-day SAR is to find survivors of a distress incident as quickly as possible, subject to resources available. Subsequently, the actual rescue and transport to safety of survivors during normal SAR seldom requires extensive planning and is normally accomplished uneventfully through agency specific practices and standard operating procedures.

However, timely rescue and transport to safety in response to the immediate need of extremely large numbers of persons in distress following a catastrophic incident will likely be insufficient and hampered by damage to critical infrastructure and key resources.

During an MRO, the overwhelming magnitude of individuals needing rescue will likely dominate the immediate operational requirement. As such, the "search" portion of SAR and the scientific mathematics and algorithms used to develop optimal search plans may be of limited value in CISAR planning and operations. Therefore, the development of a thorough, all domain (e.g., land, waterborne, and air), multi-agency integrated plan to efficiently affect recovery, provide medical treatment, and transport survivors to places of safety demands the full attention of SAR planners and decision makers at all levels of government.

The prospect of overwhelming numbers of people requiring immediate life-saving assistance poses the greatest challenge to CISAR planning and effective operations. This prospect will likely require timely implementation of resources from multiple Federal, State, and local agencies.

Lessons Learned: Hurricane Katrina

In the aftermath of Hurricane Katrina (2005), the extensive flooding requiring the rescue of thousands of distressed persons overwhelmed first responders. During the response, transporting small numbers of rescued persons many miles to places of safety or placing rescued persons on higher ground in the vicinity (in order to speed further rescues) with no follow-on assistance proved grossly inefficient and left the survivors at continued risk.

Out of necessity, an all domain integrated operational chain of events was eventually developed and successfully implemented by first responders. Lily Pads (see *Section 2-9: Delivery of Survivors*) were identified for intermediate drop-off of distressed persons providing medical treatment and transport to places of safety allowing rescue resources to quickly return to scene. To be effective for future, similar operations, putting this unbroken chain of events into timely practice demands the utmost of integrated collaborative planning.

Planning is foundational for CISAR preparedness and response. CISAR planning will:

- Allow jurisdictions to influence the course of events in a CISAR operation by determining in advance the actions, policies, and processes to be followed;

- Guide other preparedness activities;

- Enable awareness of capabilities across the response community; and

- Contribute to unity of effort by providing a common response blueprint.

National Preparedness

On March 30, 2011, Presidential Policy Directive 8 (PPD-8) was published. This directive aims at strengthening the security and resilience of the U.S. through systematic preparation for the threats that pose the greatest risk to the security of the Nation, including acts of terrorism, cyber-attacks, pandemics, and catastrophic natural disasters. PPD-8 replaced Homeland Security Presidential Directive (HSPD)-8 (National Preparedness), issued December 17, 2003, and HSPD-8 Annex I (National Planning), issued December 4, 2007, except for paragraph 44 of HSPD-8 Annex I.

Individual plans developed under HSPD-8 and Annex I remain in effect until rescinded or otherwise replaced.

Our national preparedness is the shared responsibility of all levels of government, the private and nonprofit sectors, and individual citizens. Everyone can contribute to safeguarding the Nation from harm. As such, while PPD-8 is intended to galvanize action by the Federal Government, it is also aimed at facilitating an integrated, all-of-Nation, capabilities-based approach to preparedness.

The national preparedness system is being designed to help guide the domestic efforts of all levels of government, the private and nonprofit sectors, and the public to build and sustain the capabilities outlined in the national preparedness goal. The national preparedness system will include guidance for planning, organization, equipment, training, and exercises to build and maintain domestic capabilities. It provides an all-of-Nation approach for building and sustaining a cycle of preparedness activities over time.

The national preparedness system will include a series of integrated national planning frameworks, covering prevention, protection, mitigation, response, and recovery. The frameworks are built upon scalable, flexible, and adaptable coordinating structures to align key roles and responsibilities to deliver the necessary capabilities. The frameworks are coordinated under a unified system with a common terminology and approach, built around basic plans that support the all-hazards approach to preparedness and functional or incident annexes to describe any unique requirements for particular threats or scenarios, as needed. Each framework describe how actions taken in the framework are coordinated with relevant actions described in the other frameworks across the preparedness spectrum.

The national preparedness system includes an interagency operational plan to support each national planning framework that will provide a more detailed concept of operations description of critical tasks and responsibilities; detailed resource, personnel, and sourcing requirements; and specific provisions for the rapid integration of resources and personnel.

Detailed SAR Action Plans should be utilized as CISAR Tactical Plans. Standardized SAR Action Plans will help minimize the planning cycle time, assist users in finding pertinent information, and help lessen the need to provide amplification of details.

Federal, State, Tribal, Territorial, and Local Planning Relationships

Federal, State, Tribal, Territorial, and local plans describe each respective government's approach to SAR operations. Because these levels of government all provide support to operations conducted at the local level, plans have similar and overlapping functions.

Planning must be coordinated among all levels of government to ensure a singular operational focus. The goal is to ensure the effectiveness of combined Federal, State, Tribal, Territorial, and local operations through integration and synchronization.

An integrated, all-of-Nation, capabilities-based approach to planning helps define how Federal Departments and Agencies add the right resources at the right time to support State and local CISAR operations. From the States' perspective, integrated planning provides answers to questions about working with other organizations and obtaining resources.

The national frameworks are scalable, flexible, and adaptable in order to accommodate the many State, Tribal, Territorial, and local government planning formats, styles, and processes. They lay the initial foundations and provide a means for synchronizing operations across the spectrum of response operations and thus enable integrating national CISAR planning efforts both horizontally across the Federal Government and vertically among Federal, State, Tribal, Territorial, and local entities.

State, Tribal, Territorial, and local governments are encouraged to review, provide feedback, and utilize the five National Planning Frameworks (available at: www.fema.gov/ppd8).

The frameworks:

- Provide general guidelines on developing Interagency Operational Plans (IOPs);

- Promotes a common understanding of the fundamentals of planning and decision making; and

- Can help emergency planners produce integrated, coordinated, and synchronized SAR plans.

PPD-8 supports national vertical integration by clearly articulating Federal planning procedures to State, Tribal, Territorial, and local governments and establishes a consistent SAR planning process across all levels of government.

CISAR Planning Fundamentals

The challenge of planning for saving lives and property is made easier if planners consider the following common fundamentals during the planning process (Figure 2-3-1 below):

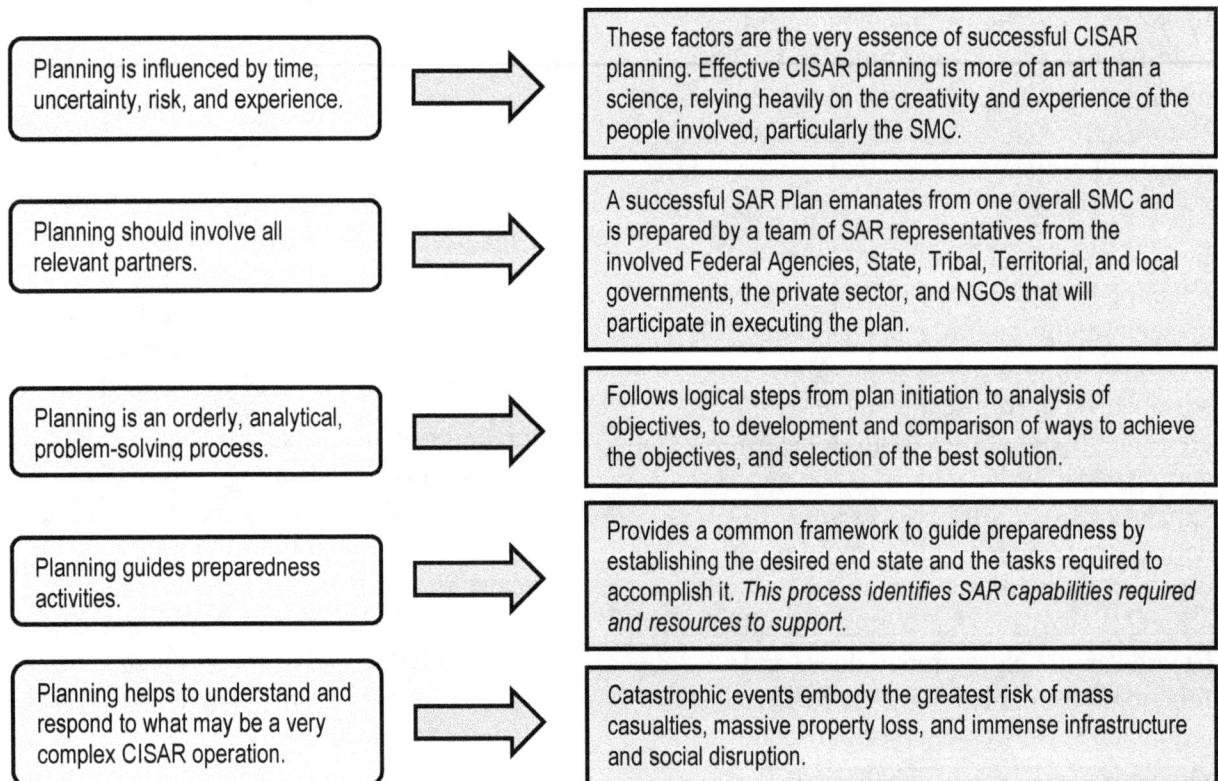

Planning is influenced by time, uncertainty, risk, and experience.	⇒	These factors are the very essence of successful CISAR planning. Effective CISAR planning is more of an art than a science, relying heavily on the creativity and experience of the people involved, particularly the SMC.
Planning should involve all relevant partners.	⇒	A successful SAR Plan emanates from one overall SMC and is prepared by a team of SAR representatives from the involved Federal Agencies, State, Tribal, Territorial, and local governments, the private sector, and NGOs that will participate in executing the plan.
Planning is an orderly, analytical, problem-solving process.	⇒	Follows logical steps from plan initiation to analysis of objectives, to development and comparison of ways to achieve the objectives, and selection of the best solution.
Planning guides preparedness activities.	⇒	Provides a common framework to guide preparedness by establishing the desired end state and the tasks required to accomplish it. *This process identifies SAR capabilities required and resources to support.*
Planning helps to understand and respond to what may be a very complex CISAR operation.	⇒	Catastrophic events embody the greatest risk of mass casualties, massive property loss, and immense infrastructure and social disruption.

Figure 2-3-1: CISAR Planning Factors

Figure 2-3-1 is continued on the next page.

Planning should address functions common to all hazards.	The causes of catastrophic incidents can vary greatly; the effects and the CISAR response normally does not.
Planning should be based on existing plans and procedures.	Planners should capitalize on plans, procedures, and lessons learned from other incidents. The State is a valuable resource for the local jurisdiction, just as the Federal Government is a valuable resource for the State.
Planning depicts the anticipated environment for action.	Early understanding and agreement on planning assumptions among all CISAR stakeholders provides the context for interaction.
Planning assigns tasks, allocates resources, and establishes accountability for CISAR ops.	Decision makers must ensure planners have the resources needed to accomplish the planning requirements.
Planning includes senior officials throughout the process to ensure both understanding and buy-in.	Planning helps decision makers not otherwise knowledgeable with CISAR to anticipate and think critically, reducing time between decisions and actions.
Planning identifies the task and purpose of the operation and facilitates cooperation and communication.	One of the benefits of CISAR planning is that important constraints and restraints that affect freedom of action and expectations of all CISAR responders are identified.
Planning is fundamentally a risk management tool.	Uncertainty and risk are inherent to response planning and CISAR operations (see Section 3-1: CISAR Risk Assessment).

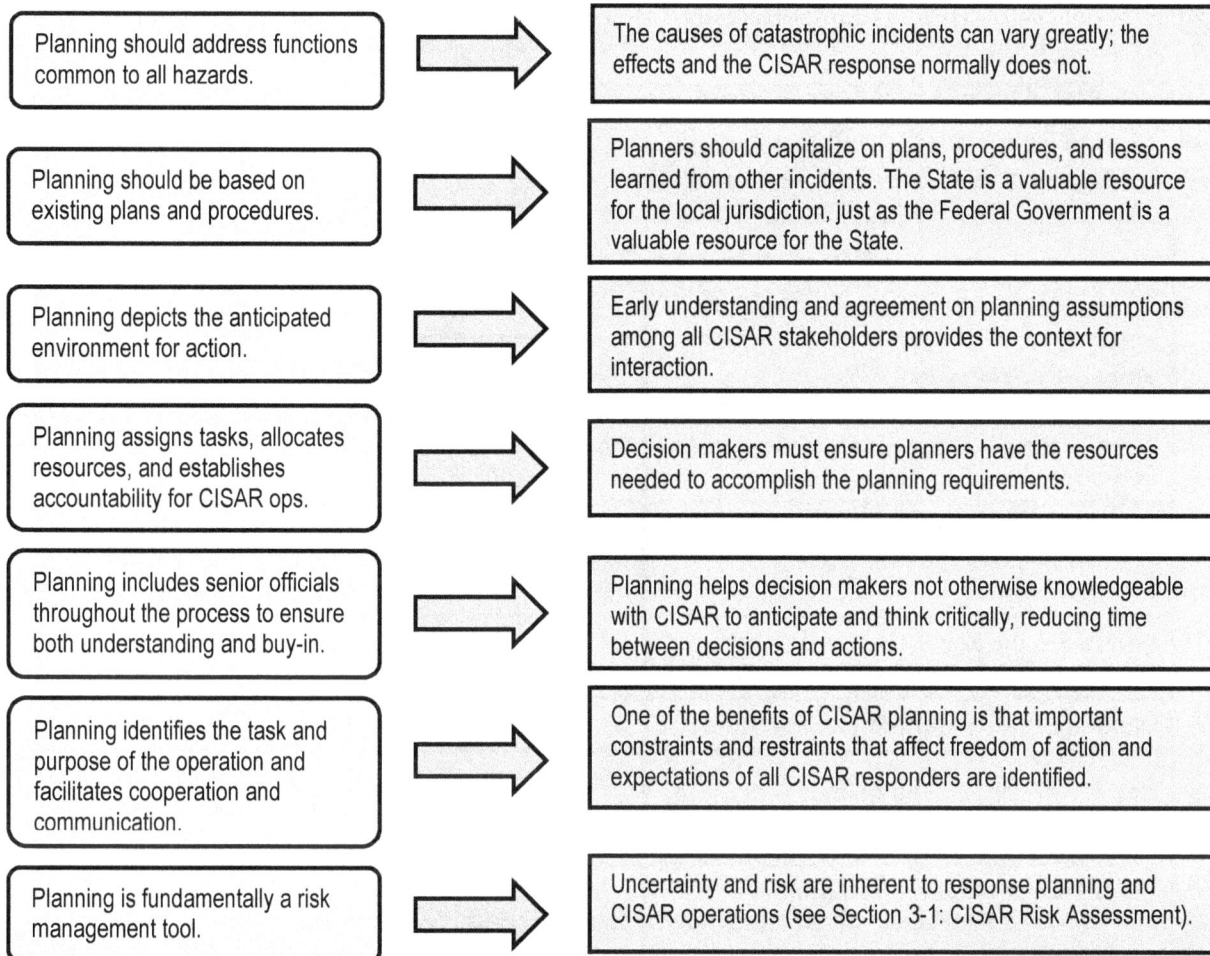

Figure 2-3-1: CISAR Planning Factors (continued)

Time Driven Objective and Resources to Support

The primary objective of CISAR is two-fold and time driven:

- Special Response Teams (SRT) and hasty searches completed within 24 hours; and

- Primary searches completed within the next 48 hours. (See *Section 2-7: CISAR Searches*.)

With the possibility of overwhelming MRO within a CISAR environment, time is the enemy. Knowledge of the SAR resources required to achieve these objectives within the specified time is paramount. Capability

gaps to achieve time driven objectives must be identified at the local level. Any additional resource requirements must be requested as early in the planning process as possible.

Planners at all levels of government should use a capabilities based approach to determine CISAR resource requirements. An incident specific gap analysis should be the basis for State, Tribal, Territorial, and local SAR planners in their initial determination of the necessary resources required to achieve CISAR objectives. This approach will also provide better clarity and justification for external resource requests (e.g., EMAC, Pre-scripted Mission

Assignment (PSMA), and Federal Request for Assistance (RFA)).

Questions may need to be posed, variables will need to be considered, and factors will need to be determined in order to better ascertain CISAR resource requirements. Table 2-3-1 below asks several questions that should be considered when conducting CISAR planning.

Table 2-3-1: CISAR Planning Questions	
Question	**Follow On Considerations**
What areas are at risk?	What is the size and the predominant domain of the area(s) that will require search and rescue?
What is the potential number of persons in distress?	What is the Population Density (Urban) or Domiciles per square mile (Rural).
What is the *(supported)* State/local SAR Plan and who is the (supported) State SAR Coordinator?	Establish immediate dialogue. Where is the SAR planning "Center of Gravity"? Will the State stand-up a Unified SAR Cell? Can you offer/provide timely SAR planning assistance?
Where are the key SAR coordination cells and which should be manned?	NRCC, FSARCG, RRCC, JFO/IMAT, JPRC, State EOC, Nat'l Guard, State SC/SMC
What are the environmental concerns?	Geographic; Forecasted weather; and Night operations.
Can CISAR objectives be met with State, Tribal, Territorial, and local resources?	Special Response Teams (SRT), Hasty Searches within 24 hours; Primary Searches within next 48 hours; Desired search effectiveness (Probability of Success (POS)).
What are the resource requirements to achieve the CISAR Objectives and what are the capability gaps?	Enable an efficient SAR Chain: Search, Rescue, Transport
What are some enabling capabilities?	Incident Awareness and Assessment imagery; Transport Routes; Lily Pad sourcing and resources; Pre-Hospital Med Teams; Air Helibase/Hot refueling; and contingency air traffic and airspace management measures.
What additional capabilities should be considered?	FAA assistance/LNOs to facilitate Airspace Mgmt Plan/TFRs

CISAR Chain of Events

Table 2-3-2 below is the general chain of events that occur during CISAR operations.

Table 2-3-2: Standard CISAR Chain of Events	
Event	**Actions**
1. Search	Locate (Special Response Teams (SRT); hasty and primary searches).
2. Rescue	Via air, land, and/or waterborne.
3. Transport	Lily Pads, Places of Safety, and/or out of area transport.

Resource Efficiency (Capability vs. Time)

As previously stated, planners should use a capabilities based approach to determine CISAR resource requirements. However, a capabilities based assessment cannot be accomplished without reasonable knowledge of the resource efficiencies of available or potential requested assets. Many key State EPLOs and State SCs have communicated a need for this knowledge and have requested resource efficiency guidance to further assist in CISAR planning.

Table 2-3-3 below provides general aviation resource and FEMA US&R Task Force efficiency rates as a guide for SAR planners to consider when determining CISAR resource requirements.

Table 2-3-3: General Resource Rates		
Helicopters	**Fixed-Wing Aircraft**	**FEMA US&R Task Force**
Search	**Search**	**Search**
• 17 - 25 square miles per hour per helicopter; • Based on 60 – 90 mph, ¼ mile track spacing • Area searched per hour may vary depending on actual search speed and track space flown as search area conditions (Urban, Suburban, Rural) and environmental factors dictate. **Rescue/Transport to Lily Pad** Assume non standard load of persons for transport (passengers), 15 minutes on scene and 5 minutes or less to/from Lily Pad: • Light Helicopter (HH65, UH-1): 15 passengers per hour • Medium Helicopter (H60, CH46): 50 passengers per hour • Heavy Helicopter (CH-53; CH-47) Ground pick-only: 150 passengers per hour	• 50 – 100 square miles per hour per airframe; • Based on 90 – 180 mph, ½ mi Track Spacing; • Light (Civil Air Patrol (CAP); Unmanned Aerial Vehicle (UAV)): 50 square miles per hour; • Medium (HU25, HC130): 100 square miles per hour; and • Area searched per hour may vary depending on actual search speed and track space flown as search area conditions (Urban, Suburban, Rural) and environmental conditions dictate.	• There is no specific US&R Task Force search rate. The number of structures (or size of area) that one US&R Task Force can search during a 12-hour operational period is incident specific. Factors such as nature of the event, type of construction, occupancy type, degree of damage, void conditions, complexity of operations, time of day, environment, and logistical challenges all impact the calculation of a general resource rate. An estimated rate will be indentified during the response, based on incident specific information. • Search efficiency can be force multiplied up to a factor of 4 if military support to US&R is requested and provided.
(Note: Extended CISAR operations may required additional aircraft.)		

Search Theory and Search Effectiveness

The science of CISAR is not nearly as developed as SAR planning and operational disciplines that have matured over many years within the maritime and isolated inland domains. However, truly effective CISAR planning is more of an art than a science and relies heavily on the creativity and experience of the people involved, particularly the SMC (see *Section 2-6: Search and Rescue Mission Coordinator (SMC)*). The CISAR planner should utilize and adapt traditional search planning methods to the CISAR environment.

Uncertainties are enormously magnified during CISAR planning. What normally is a simple determination of what SAR resources are required for day-to-day SAR becomes a challenge to even the most experienced planner in a CISAR environment.

Despite the lack of CISAR specific search theory applications, search planners should utilize existing search theory and modeling applications and mold them to the CISAR environment to determine search effectiveness.

Search effectiveness is the probability that a given search will succeed in locating the search object and is measured and defined within traditional search theory in terms of a percentage of Probability of Success (POS). POS is the resultant of Probability of Detection (POD) and Probability of Containment (POC).

The CISAR planner must have adequate resources available in order to achieve CISAR time driven objectives. In attempting to assess required resources, the planner must weigh uncertain variables against known resource efficiency and factor these assessments against the desired search effectiveness (POS).

Lily Pad and Transportation Planning Considerations

Advanced Lily Pad planning and survivor transportation capabilities are critical to enabling the efficient unbroken chain of rescuing large numbers of distressed persons in a CISAR environment.

With the primary function of enabling the swift drop off and on site care of distressed persons, Lily Pads can be as austere as an isolated open field or riverbank, or be equipped with a spectrum of services for both survivors and responders.

For potential area(s) at risk (e.g., flooding), Lily Pads, and routes to Places of Safety can be identified, named, and included to existing plans in advance. For natural

disasters such a hurricanes and other flooding scenarios, planned Lily Pads can be individually activated within a SAR Action Plan following an incident/storm passage when damage/post storm assessment dictates the most operationally efficient location.

(Note: Section 2-9: Delivery of Survivors, further addresses Lily Pads and services that can be provided to survivors awaiting further transport to Places of Safety.)

SAR Action Plans

For a CISAR incident, the State SC will normally be the overall SMC for the incident, supported by other assisting agencies and resources. CISAR operations are too large, may involve a significant number of agencies, and are too dynamic and time sensitive to expect acceptable results from anything other than management from a single SMC. Accordingly, the SMC is responsible for developing a well integrated SAR Action Plan and the employment of the most efficient and effective use of available SAR resources within all domains.

This singularly managed, all-domain approach has been proven during the conduct of other MROs for decades.

Unified SAR Branch/CISAR Coordination Cell

Before and during CISAR operations, the SMC should implement a CISAR Coordination Cell. Under NIMS organization, this cell may likely be a branch, i.e. a Unified SAR Branch under the Operations Section well-resourced with the Federal SAR Coordination Group, State, and local experienced SAR SMEs. Figure 2-3-2 on the next page depicts a generic Unified SAR Branch configuration to include ESF#9 Support Agencies such as the FAA.

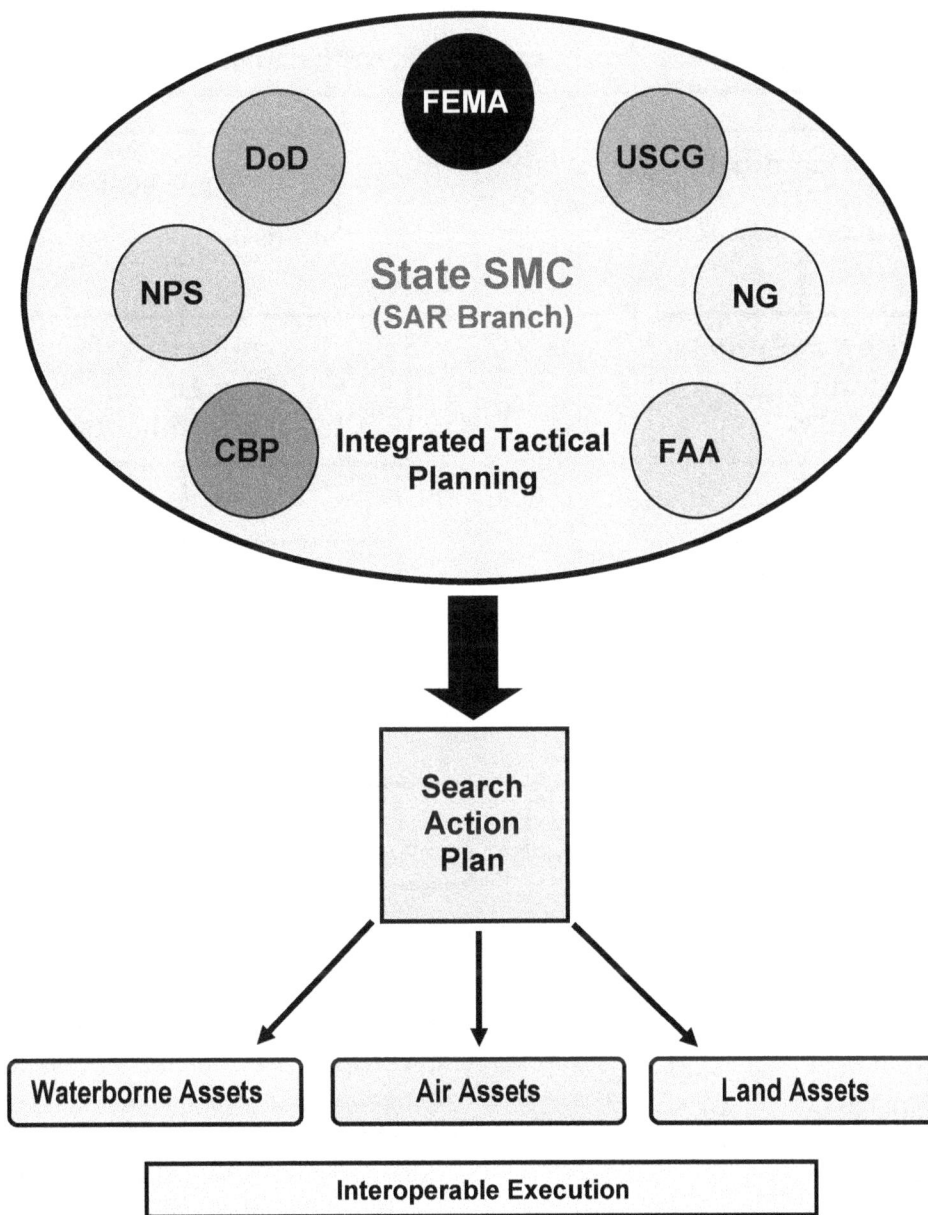

Figure 2-3-2: Generic Unified SAR Branch Configuration

Working together, the SAR Coordination Cell will develop an integrated and efficient SAR Action Plan.

The CISAR Coordination Cell is a very simple (and relatively small) group where SAR Action Plans are discussed, coordinated, developed, assigned,

disseminated for execution, and managed through the SMC.

Figure 2-3-3 on the next page compares the CISAR Coordination Cell concept in an Incident Command with the SAR chain of command organized under the international SAR system as described in the IAMSAR Manual.

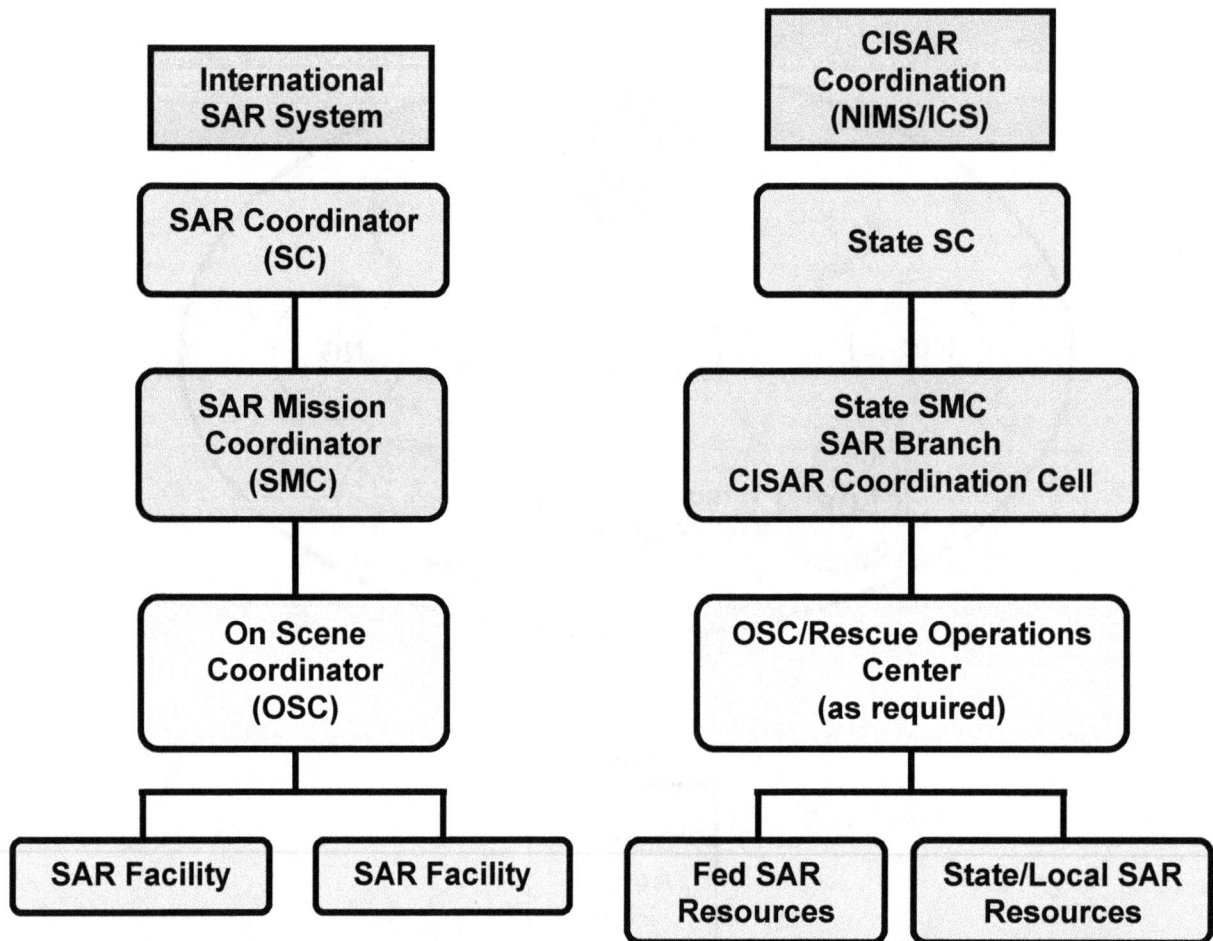

Figure 2-3-3: Comparison – International SAR System and NIMS/ICS CISAR Coordination

Planning Cycle

CISAR operations typically require more than one operational period. A regular planning cycle should provide for establishing objectives, and deploying SAR resources. Normally, the Planning "P" Operations Planning Cycle will be used to organize CISAR operations (Figure 2-3-4 on the next page).

(Note: The U.S. Coast Guard Incident Management Handbook (COMDTPUB P3120.17A) is an excellent resource for how to conduct CISAR incident planning as part of the overall response, and is located on the NSARC website at: http://www.uscg.mil/hq/cg5/cg534/nsarc/US CG%20IncidentMngmntHandbookAUG200 6.pdf.)

Figure 2-3-5 on page 2-20 is an example of a typical 24-hour planning cycle.

During this time period:
- Agree on who will present UC's response emphasis and motivation remarks
- Review task assignments, objectives, decisions & directions
- Receive operations briefing

Provide opening remarks
Review response plan as presented to ensure that Command's directions and objectives have been properly addressed
Provide further guidance and resolve issues
Give tacit approval of the proposed Plan
Agree when written plan will be ready for review & approval

During this time period:
- Meet one-on-one with Command & General Staff members for follow up on assignments.
- Prepare further guidance and clarification as needed
- Receive operations briefing

Review IAP for completion and make changes as necessary
Approve Plan

Meet and brief Command & General Staff on IC/UC direction, objectives & priorities
Assign work tasks
Resolve problems & clarify staff roles and responsibilities

Establish priorities
Identify constraints & limitations
Develop incident objectives
Identify necessary SOP's
Agree on operating policy, procedures and guidelines
Identify staff assignments
Agree on division of UC workload

Finalize UC structure
Determine overall response organization
Identify and select support facilities
Clarify UC roles and responsibilities
Determine Operational period
Select OSC & Deputy OSC
Make key decisions

Determine ICS-201 briefing timeframe & receive briefing
Clarify/request additional information
Determine incident complexity
Provide interim direction
Initiate change of command
Determine UC players
Ensure interagency notifications
Brief superiors

Provide overall guidance and clarification
Provide leadership presence and motivational remarks
Emphasize response philosophy

Monitor on-going operations
Review progress of assigned tasks
Receive periodic situation briefings
Review work progress
Identify changes that need to be made during current and future operations
Prepare for UC Update Objectives Meeting

Ensure that an appropriate initial response is deployed
Provide direction as needed
Monitor initial response operations

Cycle diagram (center):
Tactics Meeting
Preparing for the Planning Meeting
Planning Meeting
Preparing for the Tactics Meeting
IAP Prep & Approval
Command & General Staff Meeting / Briefing
Operations Briefing
IC / UC Develop/ Update Objectives Meeting
Execute Plan & Assess Progress
New Ops Period
Initial UC Meeting
Incident Brief ICS-201
Initial Response
Notification
Incident/Event

Initial Response

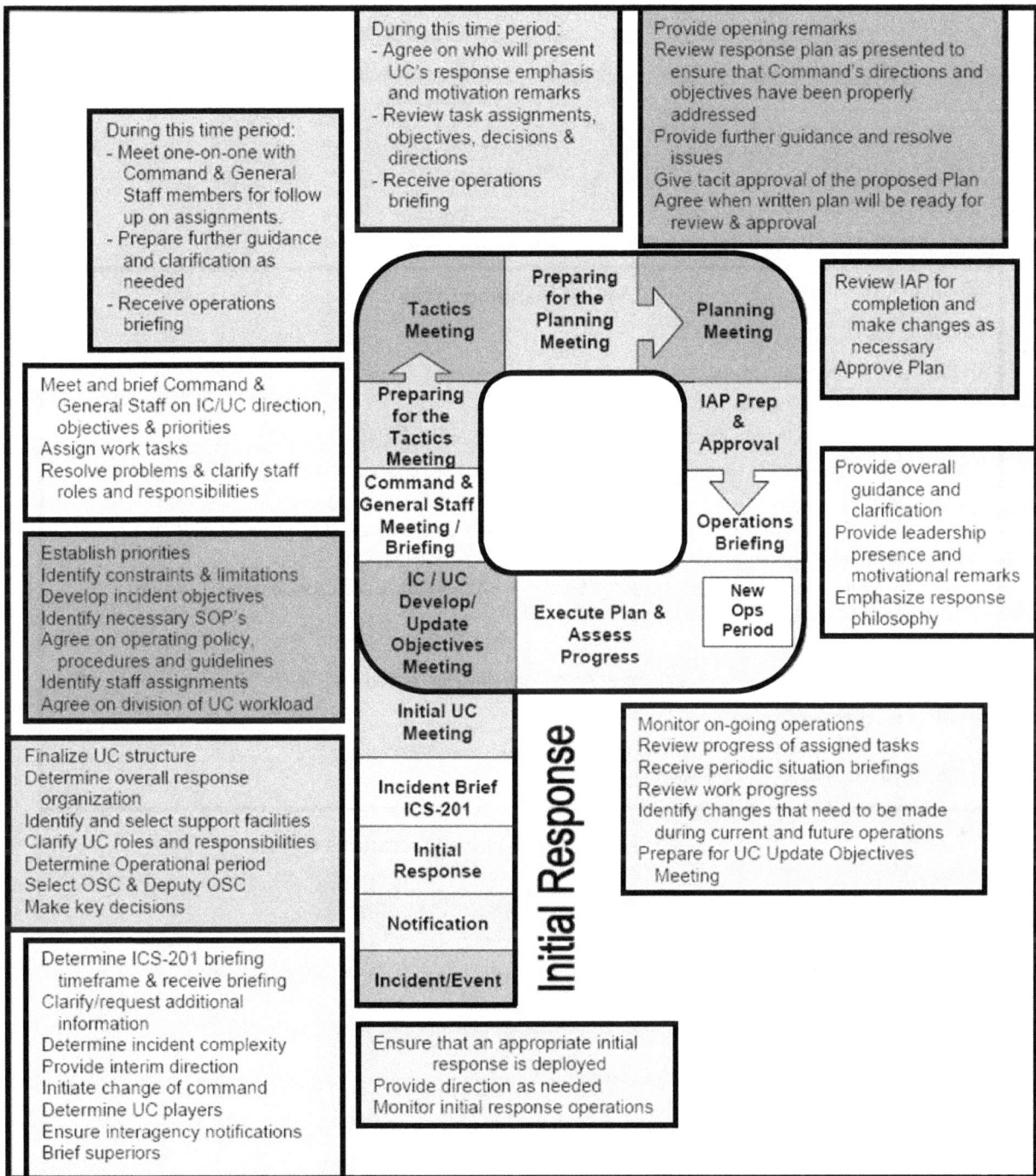

Figure 2-3-4: Planning "P" Operations Planning Cycle

2-19

24-Hour Planning Cycle

06:00	Operations brief.
07:00	Begin operational period.
09:00	Command & General Staff meeting.
13:00	Branch level Unified Command meeting (State, USCG, FEMA/US&R, DOI, DoD).
15:30	Field units report to Group/Division Supervisors on progress and operational needs for next operational period.
16:00	Group/Division Supervisors report to Branch Director(s) on progress and operational needs for next operational period.
16:30	Branch Director(s) report to Operations Section Chief on progress and operational needs for next operational period.
17:00	Tactical meeting.
19:00	Planning meeting to establish Incident Action Plan (IAP) for next operational period.
20:00	Approval of IAP.
00:00	SAR Situation Recon / Night SAR as necessary.
03:00	Tactical meeting for Operations brief.

Figure 2-3-5: Example 24-Hour Operations Cycle

Section 2-4: Notice and No-Notice Events

Introduction

Pre-event Opportunities

Preparations

Pre-Event Actions

Introduction

A catastrophic incident may occur with little or no warning.

Substantial information (e.g., storm forecasts) or clues (e.g., disease symptoms) may provide additional time to prepare before or in case CISAR operations are required.

Other events, such as earthquakes, may occur with no warning. Such events are referred to as "no-notice" events. A CISAR response caused by, for example, a hurricane, would be considered a "notice" event.

Pre-event Opportunities

Notice events, such as approaching hurricanes, provide CISAR responders with the opportunity to predict the response demands, pre-position CISAR aircraft, boats, and response teams, and develop specific action plans.

No-notice events like earthquakes and some man-made disasters happen unexpectedly.

Response activities to notice and no-notice events are similar. However, maximum advantage should be taken of any time available before an expected event to maximize the success of post-event efforts.

Preparations

Before a notice event actually occurs, SAR authorities should:

- Designate the SMC;
- Review applicable SAR plans;
- Update the readiness status of available SAR resources;
- Determine capability gaps and additional SAR resource requirements;
- Track the status of evacuations; and
- Monitor the storm or other threat.

Pre-Event Actions

To further prepare for a notice event, the following actions may be taken:

- Carry out CISAR operations just before and after the event, as appropriate;
- Issue standby orders;
- Keep all status reports up-to-date;
- Inform key authorities of intended SAR actions;
- Prepare the IC;
- Review and brief reporting procedures;
- Support CISAR aircraft, boats, and response team logistics requirements;
- Ensure that no SAR command and control issues are left unresolved;

- Implement plans for supplemental personnel and resources;

- Pre-position CISAR responders;

- Ensure that charts and grids of the geographic region are available for use by CISAR responders;

- Determine what Temporary Flight Restrictions (TFRs) will be required;

- If the use of CISAR aviation resources is anticipated, coordinate with the FAA on the need for enabling air traffic and airspace management measures, including TFRs, and an overarching contingency air mission and airspace management plan;

- Ensure that the evacuation of personnel and equipment is implemented that would otherwise compound subsequent CISAR operations; and

- Identify and ensure that lily pads and other places of safety are ready.

Section 2-5: National Incident Management System (NIMS)

NIMS

CISAR Unity of Effort

Incident Commander

SAR System

NIMS

"The Secretary [of Homeland Security] shall develop, submit for review to the Homeland Security Council, and administer a National Incident Management System (NIMS). This system will provide a consistent nationwide approach for Federal, State, and local governments to work effectively and efficiently together to prepare for, respond to, and recover from domestic incidents, regardless of cause, size, or complexity. To provide for interoperability and compatibility among Federal, State, and local capabilities, the NIMS will include a core set of concepts, principles, terminology, and technologies covering the incident command system; multi-agency coordination systems; unified command; training; identification and management of resources (including systems for classifying types of resources); qualifications and certification; and the collection, tracking, and reporting of incident information and incident resources."

Homeland Security Presidential Directive 5 (HSPD-5), Paragraph 15

NIMS

As indicated in HSPD-5 and the NRF, the National Incident Management System (NIMS) is to be used for ESF #9 and CISAR operations. All Federal Departments and Agencies involved in CISAR operations must adopt and implement NIMS where applicable.

The NRF discusses the NIMS core concepts, principles, terminology, and technologies.

Information and training for NIMS is available from FEMA (http://training.fema.gov/IS/NIMS.asp) and other sources.

CISAR Unity of Effort

Unity of effort during CISAR operations requires a clear understanding of the roles of participants and seamless coordination across jurisdictions while respecting the chain of command of participating organizations. Familiarity with the NIMS concepts and principles is essential to the seamless integration of the CISAR operation with the other aspects of the incident response.

NIMS emphasizes:

- A single set of objectives;

- A collective, strategic approach;

- Optimizing information flow and coordination;

- Understanding joint priorities;

- Respecting legalities; and

- Maximizing probability of success under a single plan.

Incident Commander (IC)

Under NIMS the IC has overall command and management of the incident response. This includes establishing and communicating strategic goals and operational objectives to all responding agencies and personnel.

SAR System

When the Incident Command is implemented, the SMC function will be placed under the umbrella of the NIMS organizational structure. Typically, the SAR Branch Director or SAR Group Supervisor is placed in the Operations Section, where the CISAR response system is integrated into the Incident Command (Figure 2-5-1 below). The CISAR response may also include an On Scene Coordinator (OSC) and an Aircraft Coordinator (ACO) to assist managing critical SAR resources.

In some cases, the person serving as IC may also be designated as the SMC. The terms "Incident Commander" or "Operations Section Chief" are not interchangeable with titles associated with SAR response functions.

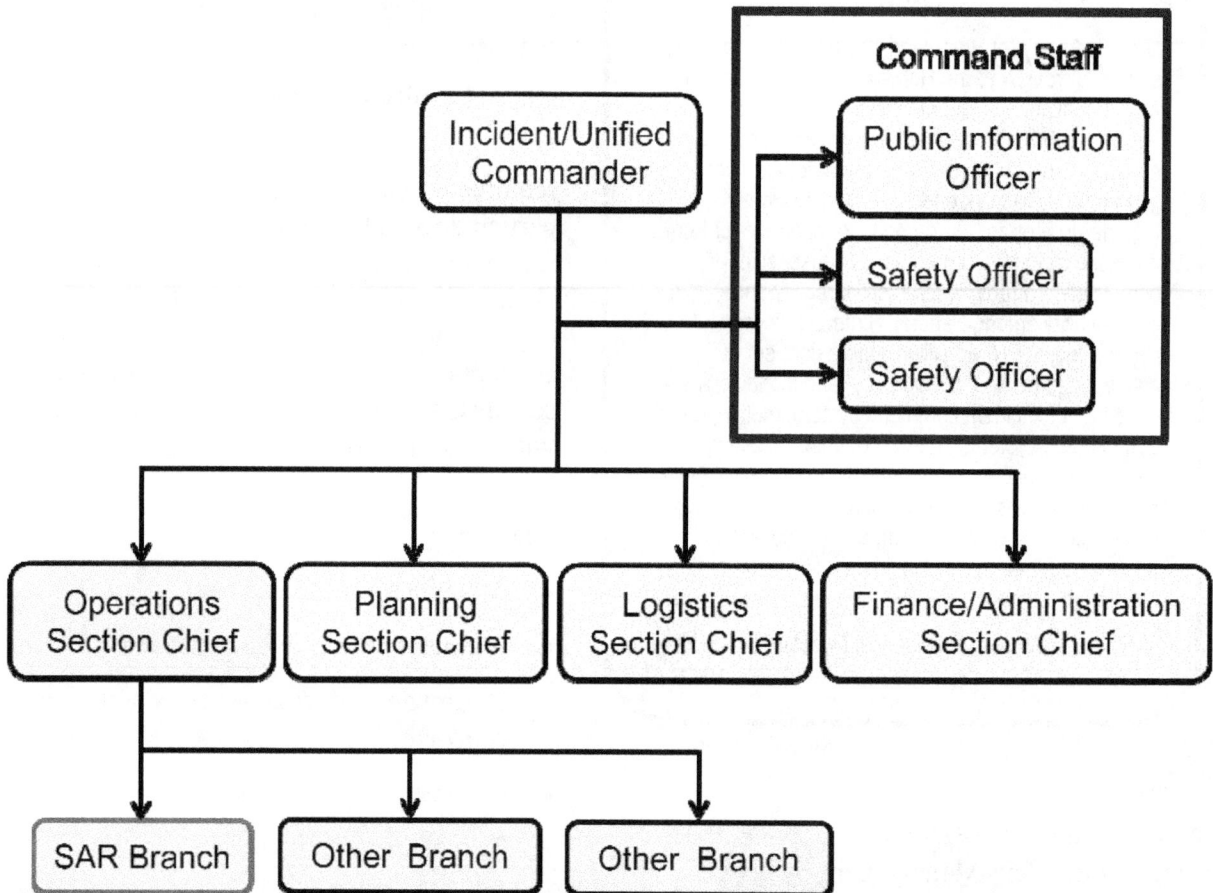

Figure 2-5-1: Incident Command with SAR Branch

Section 2-6: SAR Mission Coordinator (SMC)

Introduction

Who is SMC?

General Guidelines

SMC Duties

SMC Briefings

References

Introduction

The SMC is that person responsible in the Incident Command for coordinating, directing, and supervising the CISAR operation. The SMC function is normally located in the SAR Branch of the Operations Section.

SMC is an extremely challenging position, responsible for coordinating time critical ESF #9 aircraft, boat, and response team SAR operations throughout the affected area. It is the SMC who ensures that Federal, State, Tribal, Territorial, and local SAR assets conduct effective, efficient, and coordinated CISAR operations in a safe manner.

Who is SMC?

Federal Department and Agency CISAR responders normally conduct CISAR operations at the request of the State, Tribe or Territory. As such, the SMC responsible for overall management of CISAR operations is a representative or organization from the respective State, Tribe, or Territory that requested Federal assistance.

In exceptional circumstances, when Federal Agencies are required to conduct CISAR operations without a State, Tribal, or

Territorial request for assistance, FEMA will designate which agency will function as the ESF #9 overall Primary Agency. The overall Primary Agency will then ensure an SMC is appointed to coordinate the on scene CISAR response, with all other agencies providing support.

General Guidelines

Lifesaving CISAR operations are time critical, therefore the SMC should be identified early, as well as knowledgeable in SAR response planning and operations.

The SMC should plan for the worst when considering the complexity of CISAR operations, particularly when determining resource requirements. Then as circumstances dictate and more information is obtained, the SMC should adjust and refine resource requirements as required.

Federal CISAR responders will follow their respective agency policies concerning the conduct of CISAR operations. If a SAR resource is unable to conduct a particular SAR mission, other available resources will be considered in keeping with individual agency policy, risk assessment, and safety.

CISAR operations can be accomplished both day and night, depending on resource, type of incident, weather, circumstances, agency specific policies and procedures, etc. However, extreme care must be exercised while conducting nighttime CISAR operations.

SMCs should ensure that CISAR responders understand their respective missions, conduct proper risk assessment, and have an appropriate plan in place to conduct the operation in a safe manner. There is always risk, but knowing and mitigating the risks associated with a particular CISAR operation will help maximize the CISAR responder's safety.

SMC Success: CISAR mission planners working together

A critical aspect of SAR planning that helps mitigate the complex coordination of CISAR resources is to have each participating agency support the SMC. Only through a coordinated team effort between the SMC and the participating agencies will CISAR operations be safely managed.

The SMC is responsible for the coordinated CISAR response, but from beginning to end, it must be a team effort with all Federal, State, Tribal, Territorial, and local SAR agencies participating.

Additionally, ESF #9 operations will not happen in a vacuum. The SMC must ensure SAR mission planning is coordinated with other ESFs, as required.

CISAR operations and lifesaving must take precedence. However, logistics, media, medical services, air traffic coordination, and other critical events will also be ongoing. These other events must be considered when conducting CISAR mission planning.

SMC Duties

The following is a list of SMC duties that should be considered during CISAR operations. The list is not all-inclusive; each CISAR operation will have its own unique challenges that will need to be addressed.

Under the direction of the Incident Commander, the SMC should be:

- In charge of the CISAR operation until efforts are terminated or suspended by the Incident Commander;

- Familiar with NIMS/ICS and be experienced in coordinating large scale SAR operations;

- Well-trained in SAR planning and execution procedures;

- Thoroughly familiar with the respective State, Tribal, or Territorial SAR plan;

- Responsive to safety or capability concerns raised by CISAR responders (aviation, boat, response teams) and modify SAR mission plans as appropriate; and

- Familiar with the geo-referencing matrix (see *Section 2-11: Geo-referencing*) to ensure effective communication of position information between CISAR responders and the Incident Command.

Additionally, the SMC should:

- Develop Search Action Plans and Rescue Action Plans;

- Dispatch CISAR responders;

- Assign one or more On Scene Coordinators (OSCs) and Aircraft Coordinators (ACOs), as required;

- Not hesitate to ask for any additional SAR resources required to accomplish CISAR operations;

- Optimize the use of available CISAR resources and coordinate the provision of necessary supplies and other support equipment;

- Obtain and evaluate all information concerning the CISAR response;

- Re-evaluate any new information and modify the search plan, reassigning CISAR responders as appropriate;

- Assess CISAR operational risk and continue to do so throughout the operation (see *Section 3-1: Risk Assessment*);

- Remain informed on prevailing environmental conditions;

- Identify each area to be searched, decide on methods and SAR facilities to be used;

- Identify communication frequencies to be used by CISAR responders and develop an all-domain compatible communications plan (ICS 205);

- Ensure effective communications procedures are in place to coordinate CISAR responders;

- Ensure CISAR responders are aware of other ongoing response efforts and can coordinate operations between themselves;

- Arrange for briefing/debriefing of CISAR personnel;

- Evaluate reports from any source and modify SAR plans as necessary;

- Ensure fueling of aircraft is arranged;

- Ensure that aircraft have a proper and safe search altitude;

- Ensure aircraft can coordinate operations amongst themselves for safety of flight and with response team/boat SAR facilities;

- Ensure the care, logistics, and medical support of survivors is arranged;

- Arrange for and coordinate the use of lily pads and places of safety with appropriate authorities;

- Account for all rescued survivors until delivered to a place of safety, and for all passengers and crew if the event is a transportation incident;

- Ensure the IC remains informed on the status of ongoing CISAR operations;

- Ensure CISAR mission progress is provided to public affairs personnel; and

- Terminate or suspend search operations if further efforts are unlikely to be successful (see *Section 2-16: Conclusion of CISAR Operations*).

SMC Briefings

The SMC should conduct briefings prior to launching or diverting resources for a particular CISAR mission. CISAR personnel should be given relevant details of the mission and any instructions for mission coordination. The briefing, at a minimum, should discuss the mission objective and all foreseeable hazards that might be encountered by the responding units. Known risks may include, but are not limited to:

- Hazardous weather;

- Poor visibility;

- Hazardous conditions for CISAR responders (aviation, boat, team, etc.); and

- Any problems that may be encountered for a particular mission.

References

The information in this Section was obtained from the following sources:

- *Coast Guard Addendum to the U.S. Search and Rescue Manual to the IAMSAR Manual*; and

- *International Aeronautical and Maritime Search and Rescue Manual.*

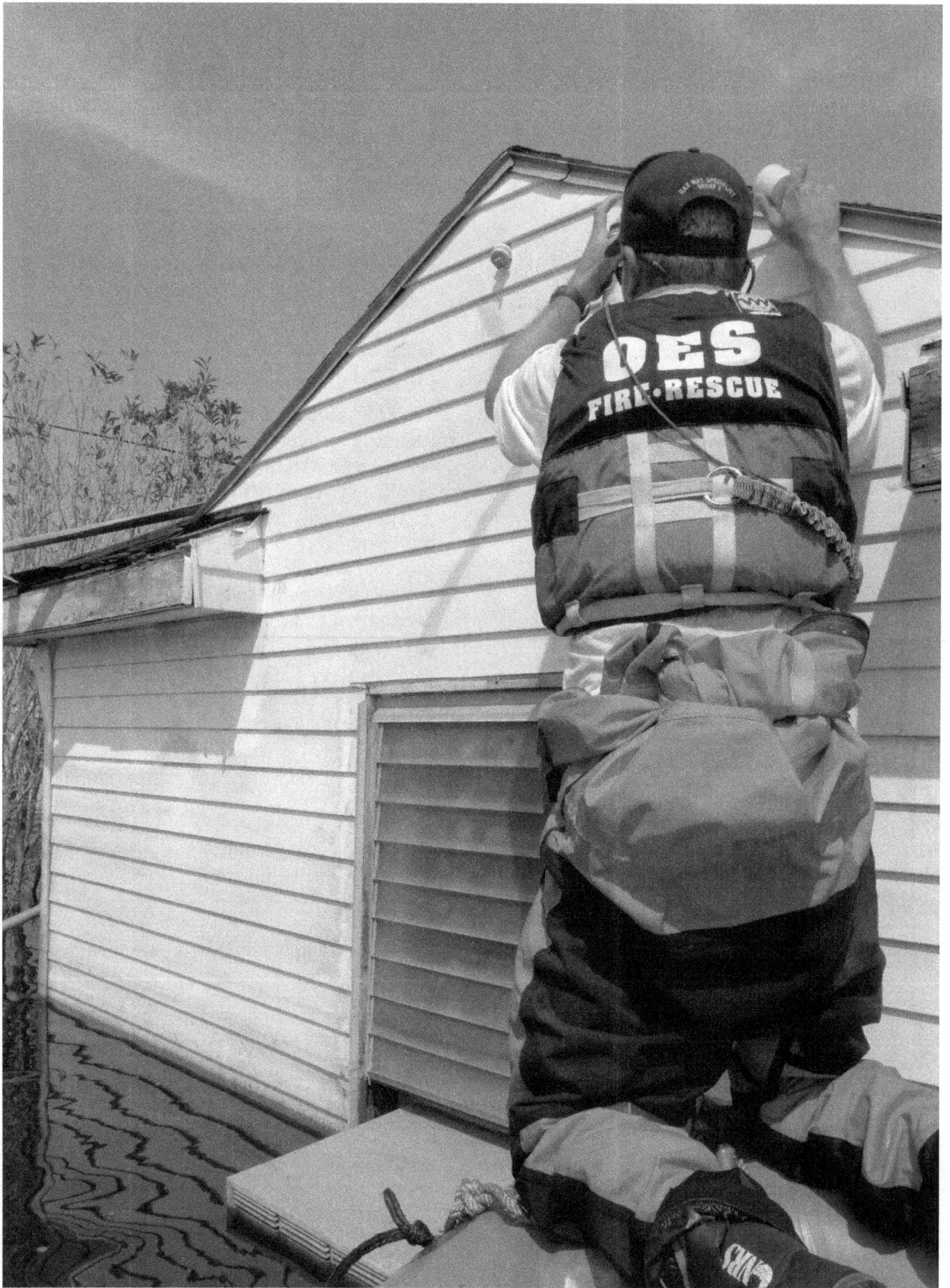

Section 2-7: CISAR Searches

Overview

Hasty Search

*   Reconnaissance*

Primary Search

Secondary Search

Probability of Detection

Special Response Teams (SRT)

Human Remains

Overview

The search terms discussed in the following paragraphs are particularly useful during CISAR operations involving populated land or flooded areas and are familiar to most State, Tribal, Territorial, and local CISAR responders. Federal CISAR responders should be familiar with and use these terms and concepts as well.[1]

SAR responders attempt to conduct as many rescues as possible while the search operations continue. CISAR aircraft, boats and response teams may need to continually switch between conducting search operations and rescue operations, especially during the hasty search when many persons are in distress.

Most CISAR search operations generally progress through a sequence of hasty, primary, and secondary searching. These search phases may overlap or vary within

different portions of the overall search area. Increasing levels of search planning are customarily involved as searching progresses to the primary and then to the secondary phases.

Hasty Search

Hasty search is an umbrella term that covers a variety of search activities. Other terms used for hasty search include "rapid" search or "initial reflex" search, depending on the background and training of the CISAR responders involved. Additional terms are also used by some CISAR responders to describe certain aspects of hasty searches, such as "spot" searches and "smart" searches. Emergency plans may provide additional guidance on what terms to use when CISAR responders from various organizations and levels of government are working together for a particular incident.

A hasty search is conducted as soon as possible to:

- Save as many lives as quickly as possible;

---

[1] These search terms are not normally used for national or international civil SAR; nor for CISAR operations involving aeronautical or maritime distress situations.

- Target high probability locations for searching;
- Quickly sweep targeted areas to find survivors; and
- Help determine the scope, nature, and magnitude of CISAR operations.

The time required for a hasty search will depend on factors such as available search resources, challenges inherent in the search area, and the number and needs of survivors. The SAR objective for a hasty search: completion within 24 hours.

The hasty search typically involves a fast-paced visual inspection of the area accompanied by vocal or audio hailing. This may be an air effort with surface units called in as necessary, or vice versa. The search can be compared to medical triage when it helps prioritize initial efforts in an expansive situation.

Search tactics may include:

- *Trolling*: zigzag area search with additional aircraft in trail;
- *Spot Search*: visual scan for survivors starting at a point followed by expanding squares;
- *Sound Sweep*: loud hailers and sound of aircraft bring people out; and
- *Quick scan*s: conducted around structures and in selected voids.

Reconnaissance

Particularly if no area assessments were conducted before CISAR aircraft, boats, and response teams arrive on scene, CISAR responders should conduct reconnaissance and assessments throughout the hasty search. Reconnaissance is used to influence immediate and future SAR efforts and provide information to the IC that may be pertinent to non-SAR operations, including body recovery. The SMC may designate selected personnel or response teams solely to conduct reconnaissance efforts. The information gained will be used to help prioritize search areas and optimize resource allocation during the more systematic primary and secondary searches.

Primary Search

Following the hasty search, a primary search is conducted (SAR objective for a primary search: completion within 48 hours). The primary search is typically conducted by surface SAR responders supported by aircraft. Primary searches involve complete circumnavigation of buildings and other structures, looking in doors and windows while hailing for survivors and entering buildings where evidence of life and the risk to rescuers is acceptable.

Normally, personnel involved in primary searches have received some prior, or on-the-spot training and instructions, use standard procedures to mark structures searched (see *Section 2-8: Structural Marking Systems*), document the results, and can call for aircraft, boats, or ground support to either facilitate their own efforts or remove survivors.

Secondary Search

Following the primary search, a secondary search is used to systematically search enclosed areas, such as each room in a particular building. Forced entry, debris removal, or authorization for entry may be required. Searchers will need instructions on whether they should conduct rescues when they find survivors or provide information to enable others to perform the rescue. In many instances, this will depend on the condition and situation of the survivor.

Probability of Detection (POD)

The probability of detecting survivors is affected by factors such as the complexity of the location to be searched, search pace, search team size and capabilities, environmental factors, and available sensors. These variables must be taken into account when determining the probability that survivors had been in a searched area, they would have been found. This information is critical for trained search planners to optimize use of available search resources during current and subsequent area searches.

Special Response Teams (SRTs)

SRTs may be used, especially during the hasty search, to respond to large concentrations of persons in distress, or persons with special needs in known locations (see *Section 3-4: Persons with Special Needs*). These teams may have specialized training or equipment to deal with the anticipated situation to which they are responding, such as medical or law enforcement capabilities.

SRTs should target locations already identified in advance in relevant emergency plans, such as:

- Child care and school facilities;

- Hospitals, nursing homes, and mental institutions;

- Shelters and marshalling points;

- Prisons and jails; and

- Areas of last refuge (possibly where local first responders will be found).

Human Remains

Discovered human remains are typically documented and bypassed during hasty and primary searches. Remains recovery should commence concurrently with CISAR secondary searches.

CISAR personnel should be instructed on how to arrange for human remains recovery as efforts to locate and assist survivors continues (see *Section 3-7: Handling of Human Remains*).

Reluctant Survivors

Response personnel should be aware of the possibility that some survivors will, to various degrees, resist rescue or evacuation efforts, and may pose a threat to rescuers if they think they will be forced to leave.

Notify the Incident Command if this situation occurs.

Section 2-8: Structural Marking Systems

Objective

FEMA Building Marking System

Building Marking Template

United Nations Marking System

Objective

Federal, State, Tribal, Territorial, and local, CISAR responders must have a uniform, standardized system for marking buildings or other structures to indicate the search status. Having a common search marking system reduces the possibility of redundant searches.

FEMA Building Marking System

In the United States, the FEMA building marking system is used. Markings are placed on the front of searched structures, to the right, as high as possible from the main entrance. Orange paint or duct tape with construction grade crayons may be used. Figures 2-8-1, 2-8-2, and 2-8-3 below and on the next page detail the marking system.

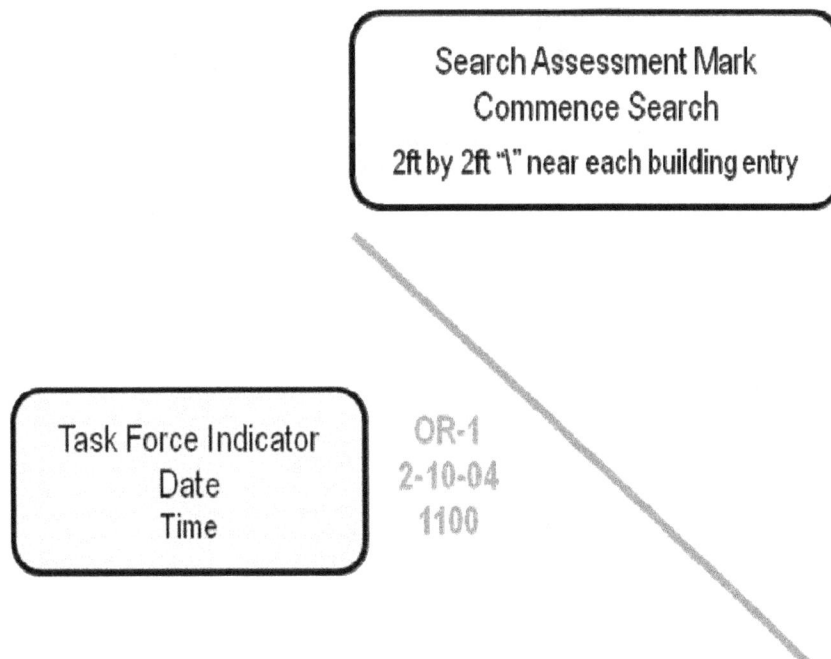

Search Assessment Mark
Commence Search
2ft by 2ft "\" near each building entry

Task Force Indicator
Date
Time

OR-1
2-10-04
1100

Figure 2-8-1: FEMA Building Marking System (Commence Search)

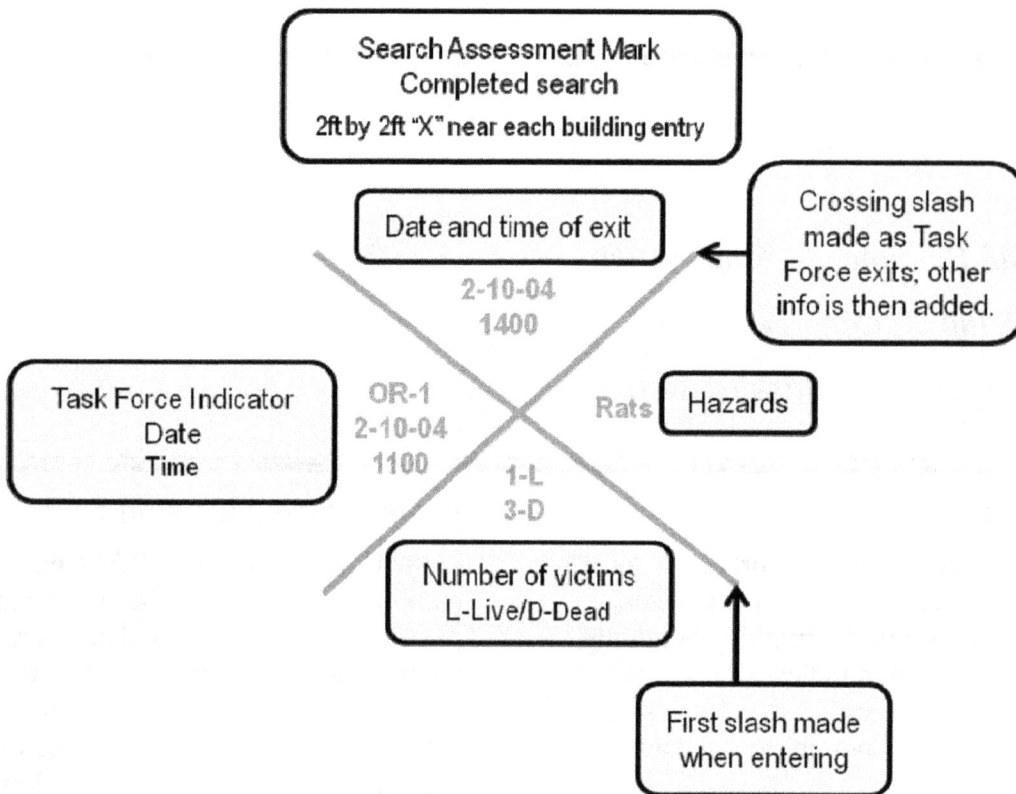

Figure 2-8-2: FEMA Building Marking System (Completed Search)

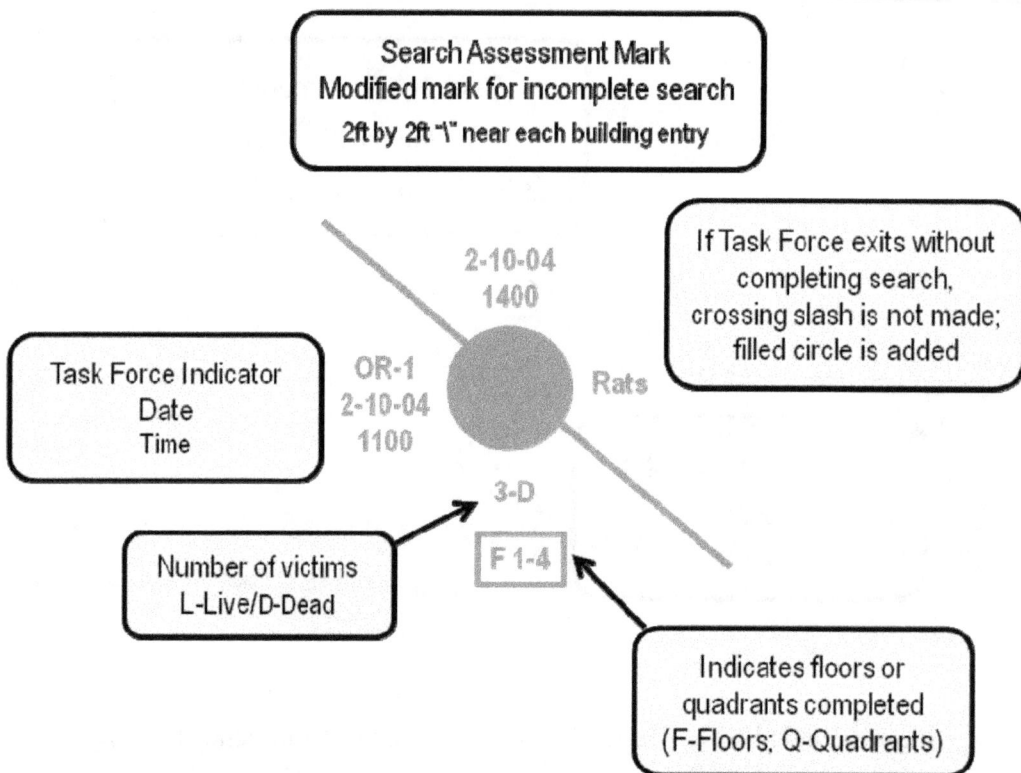

Figure 2-8-3: FEMA Building Marking System (Modified Mark-Incomplete Search)

Building Marking Template

To assist in maintaining standardized marking and recognized materials during multi-agency wide area searches, a peel and stick laminated placard, marked with an indelible marker, can be used in place of paint, tape, or crayons (Figure 2-8-4 below). These placards can greatly speed the marking process, are weather resistant, and have been designed to be comfortably carried by responders (5.5" x 4.25").

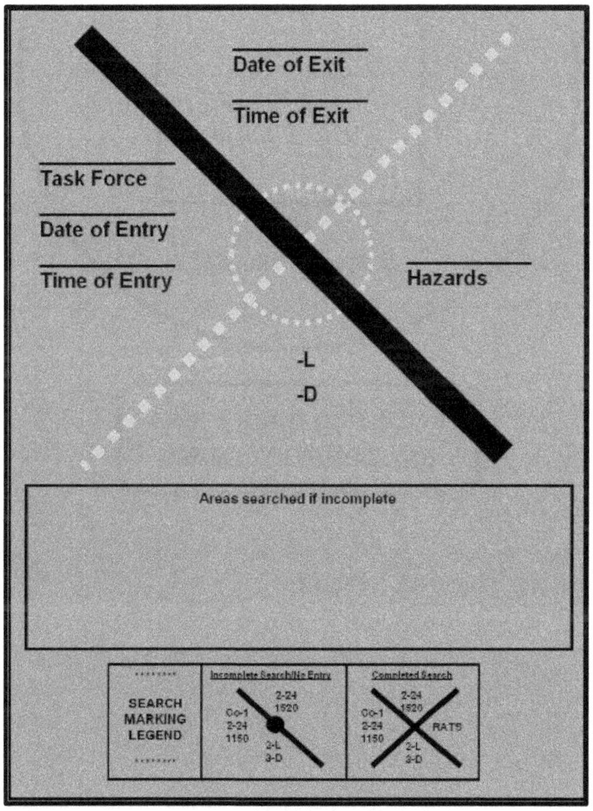

Figure 2-8-4: Building Marking Template

Placards can be quickly manufactured by local commercial office supply companies. A suggested building marking template for the placards of various colors can be found at the National SAR Committee website (www.uscg.mil/nsarc; Land SAR Section).

United Nations Marking System

Outside of the U.S., the United Nations International Search and Rescue Advisory Group (INSARAG) searched building marking system is used. The INSARAG system is explained below and depicted in Figure 2-8-5 on the next page.

- A one meter by one meter square with G ("go") or N ("no-go"), the team conducting the search, the date and time of the start of the search, and the date and time of the completion of the search written inside.

  (Note: "G" and "N" describe whether the building is safe to enter for SAR responders.)

- The number of live victims removed is written to the left of the square. The number of dead victims removed is written to the right of the square. Persons unaccounted for and/or location of other victims is written below the square.

- Additional information on hazards pertaining to the structure is written above the square.

- Any reference to building floor numbers use ground as G, 1 as the first floor above G, B1 as the first floor below G, and so forth. This is contrasted with US floor numbering that starts with 1 as the ground level.

- INSARAG marking squares are usually written in day-glow orange.

- The circle around the mark is made at complete of the building search.

INSARAG Structure
Assessment/Search Mark
(Start Search)

INSARAG Structure
Assessment/Search Mark
(Completed Search)

Hazard info INFO

```
                Go or No Go
# Live                                # Dead
Removed         TEAM                  Removed
            Date/Time Start
             Date/Time End
```

Persons Unaccounted for
Location of other Victims

1m X 1m box near entry.
Made when entering

Chem-Gases
Need Shoring
Rats

```
              G
2                              7
        Austrian Team
        20/11 0730 HR
        21/11 1730 HR
```

12?
2 dead in elevator

Encircling mark made
as operations cease

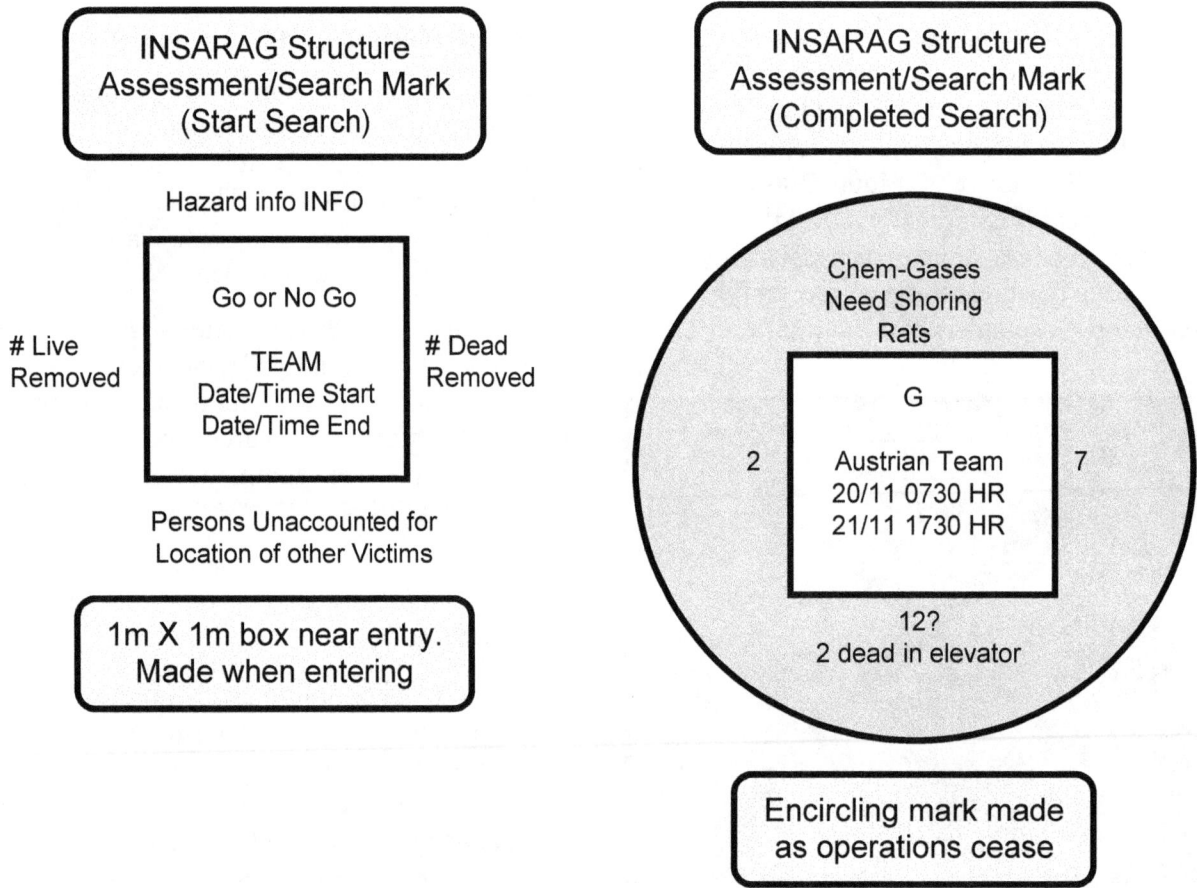

Figure 2-8-5: INSARAG Building Marking System

Section 2-9: Delivery of Survivors

Places of Safety

Lily Pads

Lily Pad Services

Lily Pads: State, Tribal, Territorial, and Local Government Responsibility

Other Federal Government Support Responsibilities

Places of Safety

The IAMSAR Manual describes a place of safety as a location where:

- Rescue operations are considered to terminate;

- The lives of survivors are no longer threatened;

- Basic human needs (such as food, shelter and medical needs) can be met; and

- From which transportation arrangements can be made for the survivors' next or final destination.

Lily Pads

A lily pad is an interim stopping point during rescue operations where survivors can be accounted for, possibly have some initial basic needs cared for, and from which they can be transported to a place of safety.

For large numbers of persons in distress, it may be necessary to establish a temporary safe delivery point for intermediate handling of survivors. In major aircraft or marine disasters a short distance offshore, survivors might be transported to a suitable nearby landing area where a temporary emergency care center could be established. The survivors should be processed, provided with emergency care, and transported to a

permanently established emergency care center or a place of safety.

By using a temporary delivery point, a large number of survivors can be evacuated quickly. Secondary CISAR responders can then transfer survivors to medical care centers.

Use of lily pads can help CISAR responders remain focused more on rescue operations and less on transportation.

Lily Pad Services

The following are typical of services that may need to be provided at lily pads:

- Helispots/landing zones;

- Medical triage and first aid;

- Food and shelter;

- Onward movement transportation;

- Law enforcement/force protection;

- Animal containment;

- On Scene Commander capability;

- Communication with the SMC, SAR facilities, and places of safety;

- Refueling arrangements for SAR facilities; and

- Arrangements for food and rest for rescue personnel, and possibly for crew changes.

For the following reasons, planners should consider identifying schools, particularly State controlled public high schools, as a primary source for Lily Pads:

- *Abundance.* Schools are proportionally dispersed among the populations, both rural and urban. One or more are likely to be close to any incident scene.

- *Landing Zones.* Even in the most congested of urban areas, schools normally afford ample landing areas for helicopters (football, baseball, soccer fields, etc.).

- *Sustenance.* Schools have facilities to prepare and/or serve food (cafeterias).

- *Onward Movement.* Schools have large parking lots and pre-existing entry and exit routing that can facilitate organized arrival and departure transport of survivors.

Lily Pads: State, Tribal, Territorial, and Local Government Responsibility

Large search areas involving large populations may require the use of multiple dispersed locations where lily pads and places of safety for CISAR operations will be established.

State, Tribal, Territorial, or local authorities are normally responsible for the establishment and support of lily pads (if required) and places of safety as well as welfare of the survivors once delivered. Depending on the extent to which a lily pad

or place of safety is to be used, a person should be designated to facilitate and oversee services at these support locations.

It is important to avoid, if possible, delivering survivors to locations where their needs for care and further transportation cannot be met.

CISAR authorities are normally responsible for transport of survivors from lily pads to places of safety. However, the IC may assign this responsibility to others. This function can often be planned for and provided by authorities responsible for ESF functions other than ESF #9. Similarly, arrangements must be made to transport survivors with critical medical or other special needs to facilities that can meet these needs.

Relevant State, Tribe, Territory, and local plans should be clear on how this is to be handled.

Other Federal Government Support Responsibilities

In addition to ESF #9, the following ESFs may be pertinent to lily pad support operations:

- ESF #1, Transportation;

- ESF #2, Communications;

- ESF #6, Mass Care, Housing, and Human Services;

- ESF #7, Logistics Management and Resource Support;

- ESF #8, Public Health and Medical Services; and

- ESF #13, Public Safety and Security.

Section 2-10: Communications

Effective CISAR Communication

Communications Plan

National Interoperability Field Operations Guide (NIFOG)

ESF #2 – Communications

Other Interoperability Standards

SAR Frequencies

NOAA Weather Radio

Effective CISAR Communication

Effective communication is required to meet the anticipated needs of CISAR responders, SAR planners, decision makers, the media, and the public. Mobilization, deployment and employment of personnel, equipment, and communications systems will require interagency coordination to ensure timely and accurate information is available to all stakeholders.

Communications Plan

Communications include all written, spoken, and electronic interaction among all audiences based upon their task-related needs. An interagency, all domain, interoperable communications plan is imperative to successful CISAR operations. A comprehensive ICS 205 (Incident Radio Communications Plan) should be developed, vetted among participating agencies, tested, and disseminated as early in the planning process as possible. The SMC should enlist the assistance of both State and Federal ESF #2 Disaster Emergency Communications (DEC) officers to collaborate in developing and managing the communications plan.

An interoperable communications plan:

- Is a necessary part of any CISAR integrated operation;

- Will help to ensure timely and effective communications resources are installed and supported with the appropriate personnel;

- Describes:

  o Who will require interoperable communications capabilities;

  o What will be done with available communications capabilities;

  o How the objectives will be accomplished; and

  o How the success of the communications plan will be measured.

- Should provide for a heavy volume of communication use, as a CISAR incident will normally involve many responding organizations that need to communicate effectively with each other; and

- Should include objectives, goals, and tools for all communications requirements, including information concerning:

- o Radio communications (terrestrial and satellite, digital and voice, frequencies);
- o Print publications;
- o Online communications;
- o Media and public relations materials; and
- o Signs.

Advance arrangements should be made to link means of interagency communications that are not inherently interoperable. Interagency communications must use standard terminology understood by all CISAR responders.

National Interoperability Field Operations Guide (NIFOG)

Responders at every level of government need a communications plan that effectively addresses interoperable communications for events of any potential scope. Of course, these plans must be supported with arrangements for the communications capabilities prescribed in the plans. The National Interoperability Field Operations Guide (NIFOG) provides a framework for interoperable communications.

The NIFOG is a pocket-sized guide of technical reference material for technicians responsible for communications used in disaster response applications. The NIFOG covers regulations on interoperability, available channels, and commonly used emergency frequencies.

The NIFOG is not a replacement for a communications plan, but provides specific guidance and frequencies that may be utilized and included in the development of an interoperable communications plans.

The NIFOG can be downloaded from the NSARC website (www.uscg.mil/nsarc; Catastrophic Incident SAR Section), or directly from the DHS website: http://www.dhs.gov/files/publications/gc_1297699887997.shtm.

ESF #2 - Communications

FEMA activates ESF #2 when a significant impact to the communications infrastructure is expected or has occurred. When activated, ESF #2 provides communications support to the impacted area, as well as internally to the Unified Coordination Group (UCG) and associated UCG teams. Under ESF #2, FEMA provides communications support to CISAR responders, as well as short-term restoration of government communications.

Other Interoperability Standards

FEMA recommends adoption of the following standards that support interoperability among communications and information management systems:

- ANSI INCITS 398-2005: Information Technology – Common Biometric Exchange Formats Framework (CBEFF);
- IEEE 1512-2006: Standard for Common Incident Management Message Sets for Use by Emergency Management Centers;
- NFPA 1221: Standard for Installation, Maintenance, and Use of Emergency Services Communications Systems;
- OASIS Common Alerting Protocol (CAP) v1.1; and
- OASIS Emergency Data Exchange Language (EDXL) Distribution Element v1.0.

SAR Frequencies

Table 2-10-1 on the following page lists NIFOG SAR frequencies.

Table 2-10-1: NIFOG Communications Frequencies	
Type of SAR	**Frequencies Available**
Land SAR	Typical Frequencies: 155.160, 155.175, 155.205, 155.220, 155.235, 155.265, 155.280 or 155.295 MHz. If Continuous Tone-Controlled Squelch Systems (CTCSS) is required, try 127.3 Hz (3A).
Water SAR	156.300 MHz (VHF Marine ch. 06) Safety and SAR; 156.450 (VHF Marine ch. 09) Non-commercial supplementary calling; 156.800 (VHF Marine ch. 16) Distress and calling; 156.850 (VHF Marine ch. 17) State control; 157.100 (VHF Marine ch. 22A) Coast Guard liaison.
Coast Guard Auxiliary	138.475, 142.825, 143.475, 149.200, 150.700 MHz (NB only).
Aeronautical SAR Coast Guard/DOD Joint SAR	3023, 5680, 8364 kHz (lifeboat/survival craft); 4125 kHz (distress/safety with ships and coast stations); 121.5 MHz emergency and distress; 122.9 MHz SAR secondary and training; and 123.1 MHz SAR primary. 243.0 MHz AM initial contact; 282.8 MHz AM working.
Military SAR	40.50 wideband FM U.S. Army/USN SAR; 138.450 AM, 138.750 AM; 121.5 MHz and 243.00 MHz AM USAF SAR
VHF Marine Channels	6, 9, 15, 16, 21A, 23A, 81A, 83A

NOAA Weather Radio

NOAA Weather Radio is a nationwide network of radio stations broadcasting continuous weather information direct from a nearby National Weather Service (NWS) Office. It is an outstanding resource for CISAR responders in obtaining information on a particular disaster.

The network:

- Has more than 650 transmitters, covering all 50 States, adjacent coastal waters, Puerto Rico, the U.S. Virgin Islands, and the U.S. Pacific Territories;

- Broadcasts NWS warnings, watches, forecasts, and other hazard information 24-hours per day;

- Has an average reception range of 40 miles from the transmitter, depending on topography;

- Can broadcast post-event information for all types of hazards – both natural (e.g., earthquakes, hurricanes, volcanoes, etc.) and environmental (e.g., chemical and oil spills);

- Require a special radio receiver or scanner capable of receiving the signal; and

- Are found in the public service band and broadcast on seven frequencies (Table 2-10-2 on the next page).

Table 2-10-2: NOAA Weather Radio Bands	
Channel	Frequencies (MHz)
1	162.550
2	162.400
3	162.475
4	162.425
5	162.450
6	162.500
7	162.525

Section 2-11: Geo-referencing

Introduction

In the aftermath of Hurricane Katrina, the review of the Federal, military, State, and local SAR response found that SAR agencies used different methods to communicate geographic information. This added confusion and complexity to an extremely large scale SAR operation.

Federal, State, Tribal, Territorial, local, and volunteer CISAR responders working together in a CISAR environment face numerous challenges, including those relating to a lack of geospatial awareness. Three issues were identified during the Hurricane Katrina response:

- How do CISAR responders navigate when landmarks such as street signs and homes are blown away?

- How do CISAR responders communicate position in a common language?

- CISAR resource de-confliction: the ability to ensure multiple assets are not inappropriately operating in the same area, which can be a significant problem for CISAR responders.

Resource de-confliction is a matter of safety, particularly with aircraft, to ensure the likelihood of a mid-air collision is minimized. Additionally, resource de-confliction is a matter of efficient and effective use of limited resources so that all areas receive appropriate, available CISAR response assets.

What is Geo-referencing?

To geo-reference is to define location in physical space and is crucial to making aerial and satellite imagery useful for mapping. Geo-referencing explains how position data (e.g., Global Positioning System (GPS) locations) relate to imagery and to a physical location.

Different maps may use different projection systems. Geo-referencing tools contain methods to combine and overlay these maps with minimum distortion.

Using geo-referencing methods, data obtained from observation or surveying may be given a point of reference from topographic maps already available.

No single map/chart projection or coordinate/grid system will be perfect for all applications. In the case of projecting the earth's curved surface on a flat surface, distortion of one or more features will occur.

The conventions for locating points on the earth's surface for purposes of nautical and aeronautical navigation (long distances on small scale charts) is generally best conducted using latitude and longitude (spherical coordinates). Locating points on large-scale maps and for ground navigation is generally best accomplished with Cartesian-style plane coordinates (i.e., USNG). Large scale-maps can treat the Earth's surface as a plane – taking advantage of that simple geometric shape and math – rather than a complex sphere. Properly constructed large-scale maps – such as topographic maps take curvature of the Earth into account. Simple linear increments (i.e., meters) of plane coordinates are significantly easier for large-scale map users to handle accurately at high precision in the field than the more complex angular increments of latitude and longitude (i.e., degrees).)

Geo-Referencing Methods

Three geo-referencing methods are to be used for CISAR operations anywhere in the U.S., as indicated in the National SAR Committee geo-referencing matrix located at the end of this Section.

(Note: State/local SAR authorities and the local IC may utilize natural landmarks in combination with geo-referencing methods to identify a position, or in some cases by by both natural and manmade landmarks (e.g. "Search Red River, from the I-94 bridge, south to latitude forty-six degrees, forty-six decimal zero minutes North"). CISAR planners and responders must adapt to the

geo-referencing method used during a CISAR response.

If a SAR responder requests that a position be converted to a particular format, every effort should be made to accommodate the request.)

U.S. National Grid (USNG)

The USNG is intended to create a more interoperable environment for developing location-based services within the U.S. and to increase the interoperability of location services appliances with printed map products by establishing a preferred nationally-consistent grid reference system.

The USNG system:

- Is the primary geo-referencing source utilized by most State/local fire/rescue and FEMA US&R teams;

- Can be extended for use world-wide as a universal grid reference system, and can be easily plotted on USGS topographic maps by using a simple "read right, then up" method; and

- May be used for area gridding, as well as for pinpoint locations.

(Note: the USNG and the Military Grid Reference System (MGRS) are functionally equivalent when referenced to NAD 83 or WGS 84 datums.)

The coordinates are easily translated to distance, as they are actually in meters. Thus the distance between two coordinates can quickly be determined in the field.

Figure 2-11-1 on the next page is an example of the USNG.

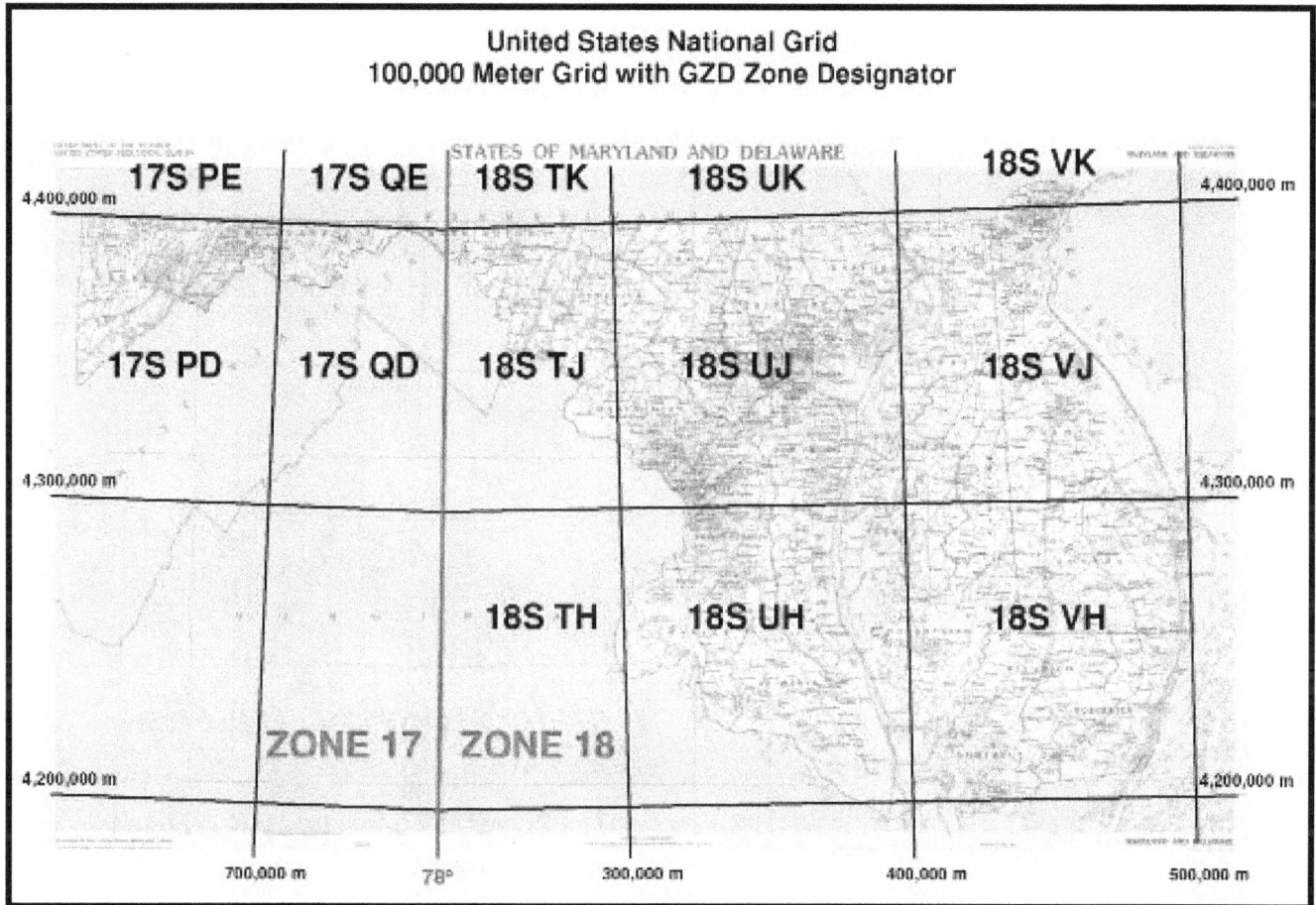

Figure 2-11-1: USNG

Figure 2-11-2 on the following two pages
explain how to find a position using USNG.

US National Grid (USNG) Coordinates: *World wide context.*

The example below locates the Jefferson Pier at USNG: 18S UJ 23371 06519.

A USNG value has three components.

U.S. National Grid
100,000-m Square ID
UJ
_____43 00
UH
Grid Zone Designation 18S

Some maps may give this leading information in a grid reference box.

Grid Zone Designation (GZD):
6° x 8° longitude zone / latitude band.
100,000-m Square Identification:

18S UJ 2337 0651 ↑

"Read right, then up."

Grid Coordinates:
Read *right*, then *up.*↑

USNG values have three components as seen above. The Grid Zone Designation gives a USNG value world-wide context with 60 longitudinal zones each 6° wide. Zones 10 - 19 cover the conterminous U.S. as seen below left. UTM zones are divided into 8° latitudinal bands. Together these 6° zones and 8° bands compose Grid Zone Designations. Example: 18S

UTM/USNG Grid Zone Designations

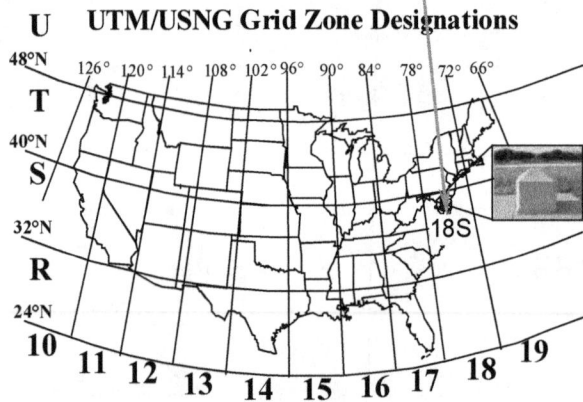

U.S. National Grid
100,000m Square Identification

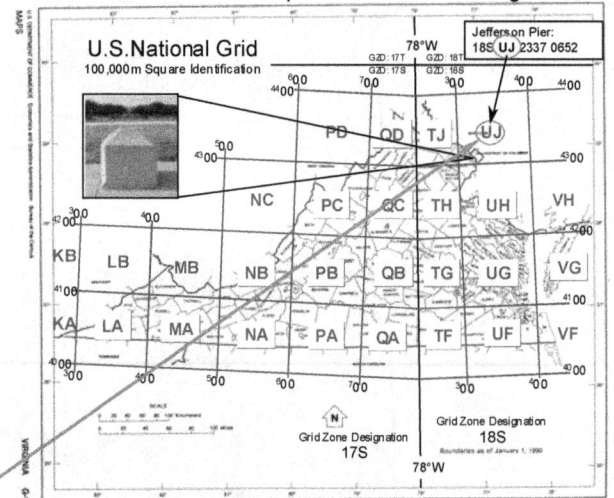

Jefferson Pier: 18S UJ 2337 0652

Grid Zone Designation 17S

Grid Zone Designation 18S

100,000-m Square Identifications
Example: UJ

GZDs are further subdivided into 100-km x 100-km squares with 100,000-m Square Identifications. In this example, the Jefferson Pier is located in UJ. These squares are organized and lettered so they do not repeat themselves but every 18°, which is approximately 1,000 miles in the mid-latitudes. The illustration at right depicts how far one must go before the letters UJ repeat. In the conterminous U.S. this ensures a given value such as UJ 2337 0651 is unique out of the entire state it is located in – as well as all surrounding states.

The Power of Truncated USNG Values

Jefferson Pier, Washington, DC
Grid: UJ23370651

+ = repeat of UJ23370651

Each 2 letter/8 digit USNG value
(10-m posting) in the outlined area is unique.

In general, people in a local community may use the grid coordinates alone – for example: 233 065. The same numbers recurs about every 60 miles but normally that will not cause a problem when the general location is understood. This is similar to the way you tell someone only the last digits of a phone number when the area code is obvious. If there is a possibility of confusion include the letter pair also – for example: UJ 233 065. A letter pair recurs about every 1000 miles so even in a disaster relief effort there should be no other point with those coordinates nearby. A complete USNG reference such as 18S UJ 233 065 is nationally and globally unique. Typically a GPS receiver or other electronic device requires a complete USNG reference since unlike a human it does not intuitively understand the general location from context. You should always give a complete USNG reference whenever abbreviated coordinates might not be clear or when listing them on letterhead, a business card or advertisement.

Ed: 20080420-USNGInstruct No2, page 1 of 2

Reading US National Grid (USNG) Coordinates: *"Read right, then up."*

The example below locates the Jefferson Pier at USNG: 18S UJ 23371 06519.

A USNG value has three components.

U.S. National Grid
100,000-m Square ID
UJ
———— ⁴³ 00
UH
Grid Zone Designation 18S

Some maps may give this leading information in a grid reference box.

Grid Zone Designation (GZD):
6° x 8° longitude zone / latitude band.
100,000-m Square Identification:

18S UJ 2337 0651

Grid Coordinates:
Read *right*, then *up*.↑

"Read right, then up."

- Grid lines are identified by Principal Digits. Ignore the small superscript numbers like those in the lower left corner of this map.

Reading USNG Grid Coordinates.

- Coordinates are always given as an even number of digits (i.e. 23370651).

- Separate coordinates in half (2337 0651) into the easting and northing components.

- [1] Read *right* to grid line 23. [2] Then measure right another 370 meters. (Think 23.37)

- [3] Read *up* to grid line 06. [4] Then measure up another 510 meters. (Think 06.51)

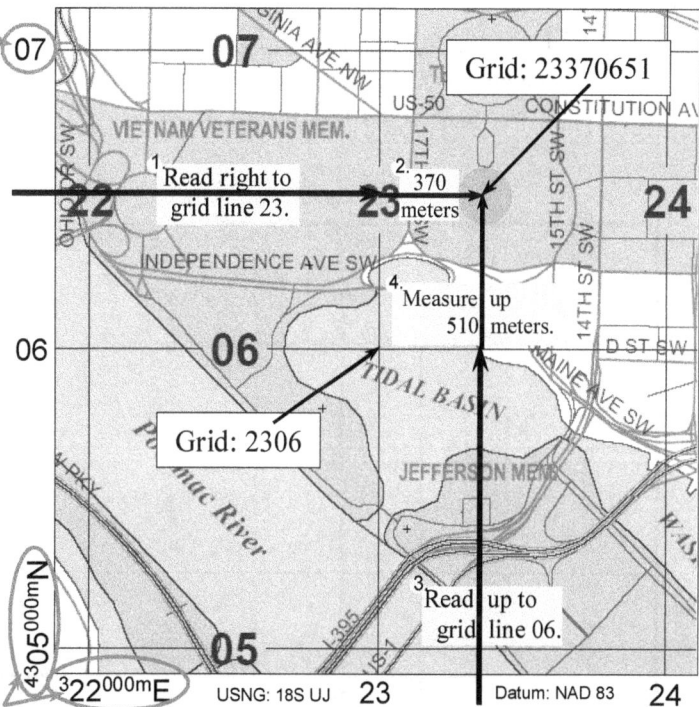

Grid:	Point of Interest:
228058	FDR Memorial:
231054	George Mason Memorial:
2338 0710	Zero Milestone:
2275 0628	DC War Memorial:
222065	Lincoln Memorial:

Ignore the small UTM superscript numbers that are provided for reference purposes. UTM numerical values are best suited for determining direction and distance as in surveying. USNG alpha-numeric values are best suited for position referencing because they can be given as only grid coordinates in a local area and with only the required precision for a particular task.

Users determine the required precision. These values represent a point position (southwest corner) for an area of refinement.

Four digits:	23 06	Locating a point within a 1,000-m square.
Six digits:	233 065	Locating a point within a 100-m square (football field size).
Eight digits:	2337 0651	Locating a point within a 10-m square (modest size home).
Ten digits:	23371 06519	Locating a point within a 1-m square (man hole size).

A modest size home can be found or identified in a local area with only an 8-digit grid. →

Complete USNG value:　18S UJ 2337 0651 - Globally unique.
Without Grid Zone Designation (GZD):　UJ 2337 0651 - Regional areas.
Without GZD and 100,000-m Square ID:　2337 0651 - Local areas.

This illustrates how nationally consistent USNG coordinates are optimized for local applications. They serve as a universal map index value in a phone or incident directory for field operation locations. Unlike classic atlas grids (i.e. B3), these can be used with any paper map or atlas depicting the national grid and in web map portals such as the Washington, DC GIS (http://dcgis.dc.gov).

They can also be used in consumer GPS receivers to directly guide you to the location. This is especially beneficial at night, in heavy traffic, or major disasters when street signs are missing.

Point of Interest	Street Address	USNG Grid: 18S UJ	Telephone: (202)
Subway Sandwich & Salads	2030 M St., NW	2256 0826	223-2587
Subway Sandwich & Salads	430 8th St., SE	2698 0567	547-8200
Subway Sandwich & Salads	3504 12th St., NE	2740 1120	526-5999
Subway Sandwich & Salads	1500 Benning Rd, NE	2815 0757	388-0421

Ed: 20080420-USNGInstruct_No2, page 2 of 2

Latitude-Longitude

Latitude and Longitude is used by aircraft and boats during CISAR operations. The Latitude-Longitude is a geographic coordinate system used for locating positions on the Earth's surface. Latitude and longitude are an angular measurement in degrees (using the symbol, " ° "), minutes (using the apostrophe symbol, " ' "), and seconds (using the quotation symbol, " " ").

Lines of Latitude are horizontal lines shown running east-to-west on maps and are known as "Parallels," due to being parallel to the equator. Latitude is measured north and south ranging from 0° at the Equator to 90° at the poles (90° N for the North Pole and 90° S for the South Pole).

Lines of Longitude are vertical lines shown running north and south on maps and are known as "Meridians," intersecting at the poles. Longitude is measured east and west ranging from 0° at the prime meridian to +180° East and -180° West (Figure 2-11-3 below).

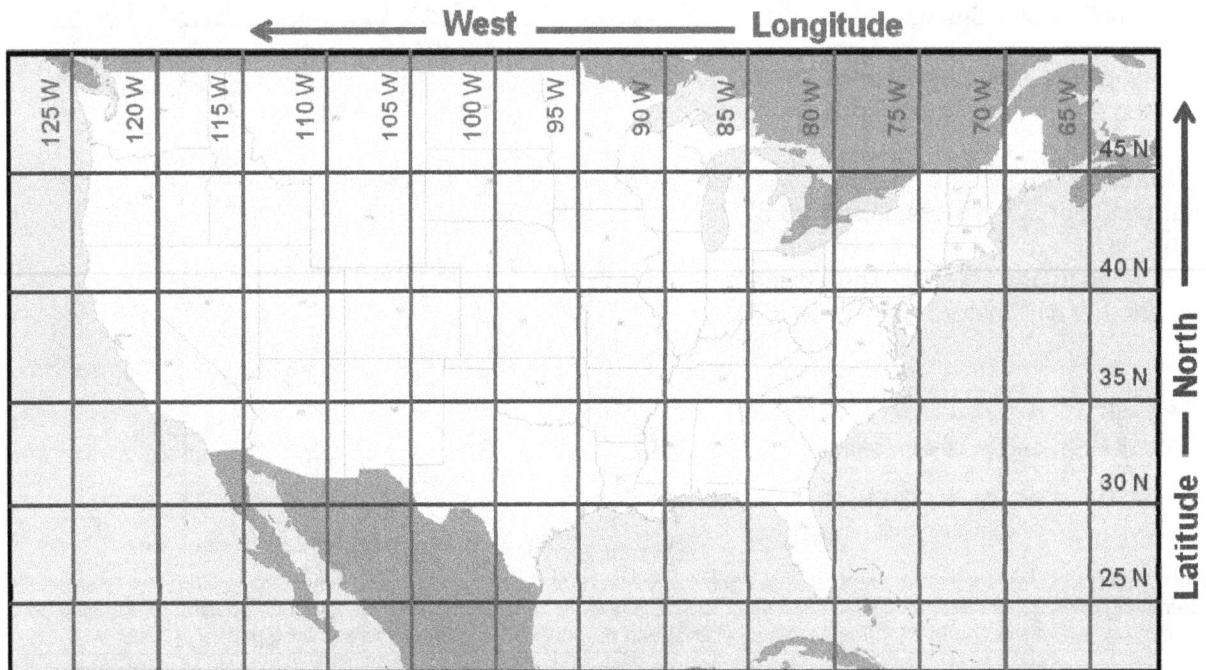

Figure 2-11-3: Latitude and Longitude

Latitude and Longitude can be read and written in three different formats:

- Degrees, Minutes, Decimal Minutes (DD° MM.mm');

- Degrees, Decimal Degrees (DD.DDDD°); and

- Degrees, Minutes, Seconds (DD° MM' SS").

When geo-referencing in Latitude and Longitude, be cautious in reading and receiving coordinates. As illustrated above, some users define location using degrees, minutes and seconds. Although the error is usually small, it can, and has meant the difference in a timely SAR response.

Confirm the coordinates if unsure of the format in which they are being communicated.

Speaking Latitude and Longitude should also be spoken in a standardized format (below).

Standard Latitude/Longitude format for CISAR operations

The standard Latitude/Longitude format for CISAR operations is Degrees, Decimal Minutes (DD° MM.mm').

Latitude is always read and written first noting "North" since the U.S. is North of the Equator. Longitude is always read and written last noting "West" since the U.S. is West of the Prime Meridian.

Speaking Latitude and Longitude

For example, 39° 36.06'N by 76° 51.42'W, should be stated as per the following:

"Three nine degrees, three six decimal zero six minutes North by seven six degrees, five one decimal four two minutes West."

The words, "degrees," "minutes," and "decimal" must be spoken.

Global Area Reference System (GARS)

GARS is a standardized geospatial area reference system for military and civil SAR application, and is based on lines of longitude and latitude. GARS provides a common language between the components and simplifies communications.

How GARS Works

- GARS is a worldwide system that divides the earth's surface into 30-minute by 30-minute cells.

- Each cell is identified by a five-character designation. (ex. 006AG).

- The first three characters designate a 30-minute wide longitudinal band. Beginning with the 180-degree meridian and proceeding eastward, the bands are numbered from 001 to 720, so that 180° E to 179° 30'W is band 001; 179° 30'W to 179 00'W is band 002; and so on.

- The fourth and fifth characters designate a 30-minute wide latitudinal band.

Beginning at the south pole and proceeding northward, the bands are lettered from AA to QZ (omitting I and O) so that 90° 00'S to 89° 30'S is band AA; 89° 30'S to 89° 00'S is band AB; and so on.

- Each 30-minute cell is divided into four 15-minute by 15-minute quadrants. The quadrants are numbered sequentially, from west to east, starting with the northernmost band. Specifically, the northwest quadrant is "1"; the northeast quadrant is "2"; the southwest quadrant is "3"; the southeast quadrant is "4".

- Each quadrant is identified by a six-character designation. (ex. 006AG3) The first five characters comprise the 30-minute cell designation. The sixth character is the quadrant number.

- Each 15-minute quadrant is divided into nine 5-minute by 5-minute areas. The areas are numbered sequentially, from west to east, starting with the northernmost band.

- The graphical representation of a 15-minute quadrant with numbered 5-minute by 5-minute areas resembles a telephone keypad.

Each 5-minute by 5-minute area or keypad "key" is identified by a seven-character designation. The first six characters comprise the 15-minute quadrant designation. The seventh character is the keypad "key" number. (ex.006AG39).

Figure 2-11-4 below graphically depicts GARS; Figure 2-11-5 on the next page is an example GARS overlay.

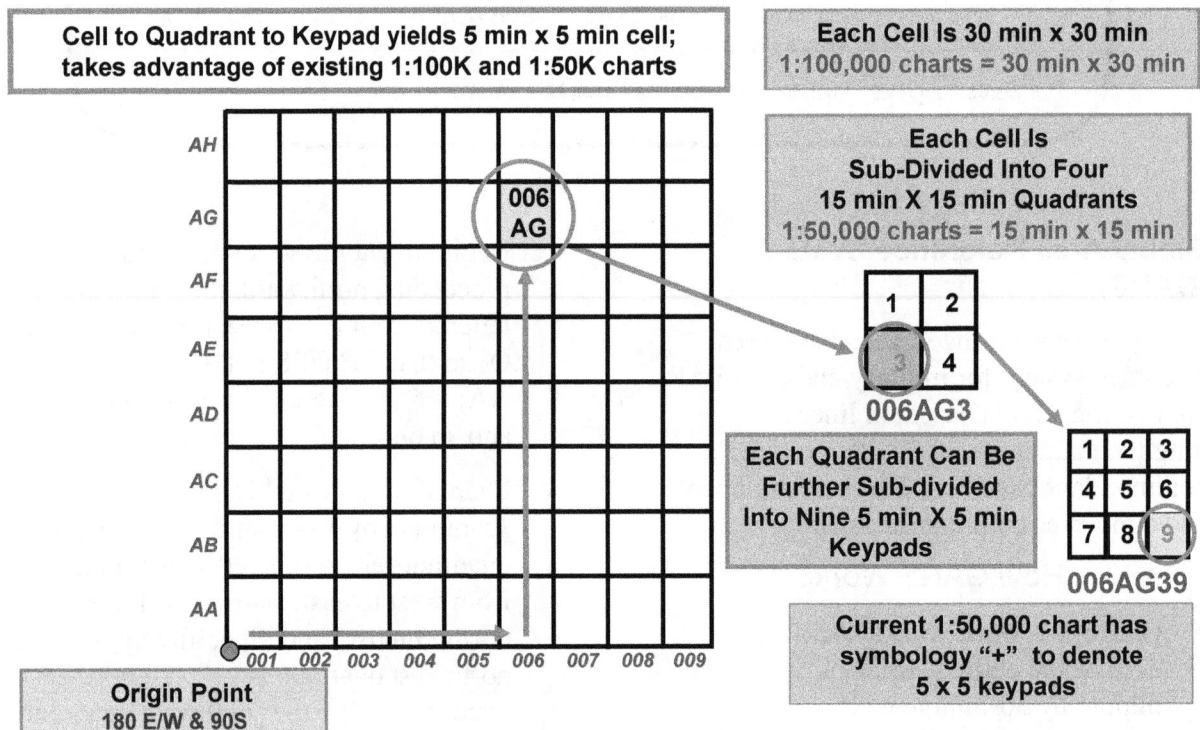

Cell to Quadrant to Keypad yields 5 min x 5 min cell; takes advantage of existing 1:100K and 1:50K charts

Each Cell Is 30 min x 30 min
1:100,000 charts = 30 min x 30 min

Each Cell Is Sub-Divided Into Four 15 min X 15 min Quadrants
1:50,000 charts = 15 min x 15 min

006AG3

Each Quadrant Can Be Further Sub-divided Into Nine 5 min X 5 min Keypads

006AG39

Current 1:50,000 chart has symbology "+" to denote 5 x 5 keypads

Origin Point
180 E/W & 90S

Figure 2-11-4: GARS

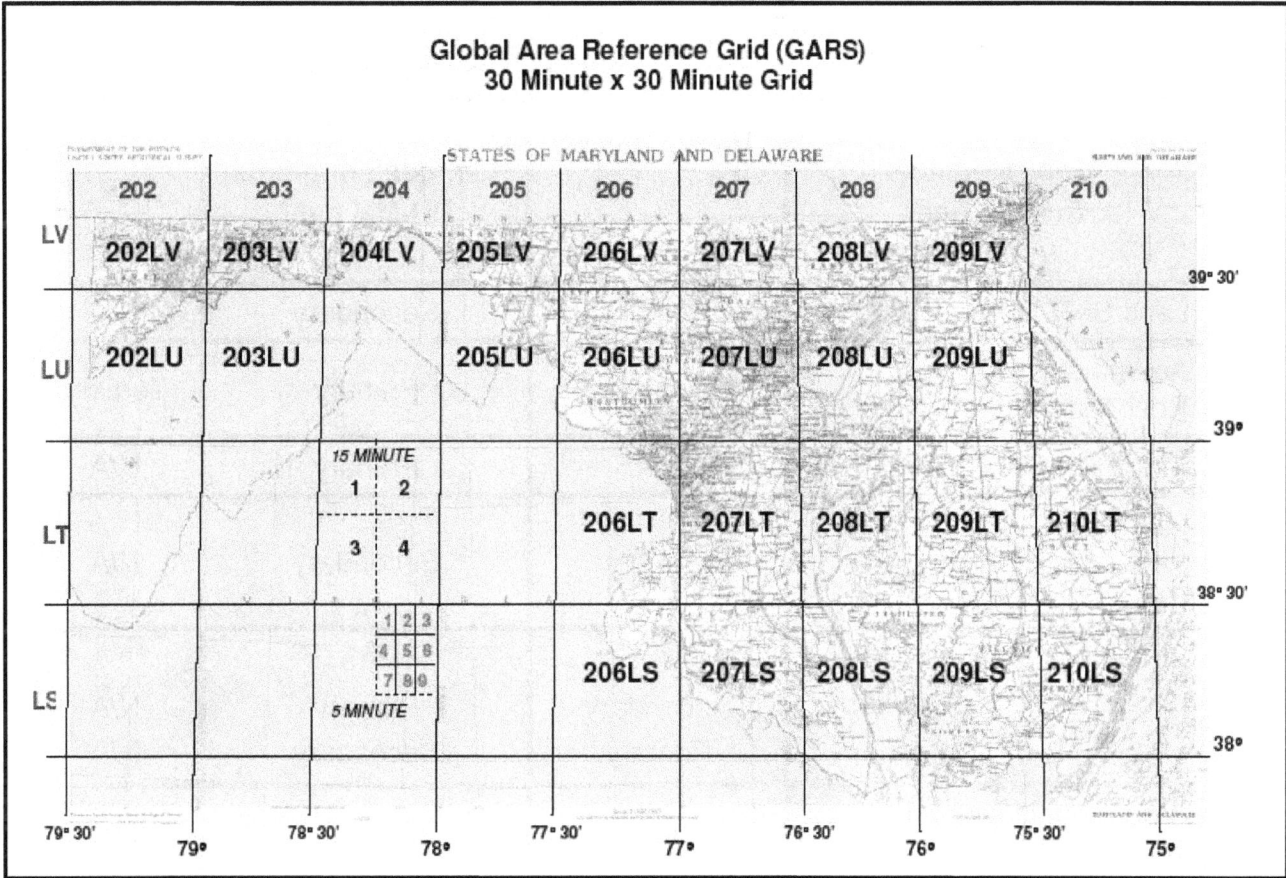

Global Area Reference Grid (GARS)
30 Minute x 30 Minute Grid

Figure 2-11-5: GARS Example

Geo-referencing Matrix

A fundamental requirement for a geo-reference system is the ability to easily interface between the Incident Command, the land CISAR responder (or maritime CISAR responder) and the aeronautical CISAR responder. Because each has unique geo-referencing requirements, effective interface between each component is vital to a successful CISAR response.

The geo-referencing matrix (Table 2-11-1 on the next page) minimizes confusion and provides guidance on what geo-referencing system each CISAR responder should be using.

Map Datum

North American Datum 1983 (NAD 83) and World Geodetic System 1984 (WGS 84) are equivalent at scales smaller than 1:5000.

Table 2-11-1: CISAR Geo-referencing Matrix			
Geo-reference System User	United States National Grid (USNG)	Latitude/Longitude DD-MM.mm[1]	GARS[2]
Land SAR Responder[3]	Primary	Secondary	N/A
Aeronautical SAR Responders[4]	Secondary	Primary	Tertiary
Air Space Deconfliction[5]	N/A	Primary	N/A
Land SAR Responder/ Aeronautical SAR Responder Interface.[6]	Primary	Secondary	N/A
Incident Command:			
Air SAR Coordination	Secondary	Primary	N/A
Land SAR Coordination	Primary	Secondary	N/A
Area organization and accountability[7]	Secondary	Tertiary	Primary

[1] During CISAR operations (and to avoid confusion) Latitude and Longitude should be in one standard format: DD-MM.mm. If required, use up to 2 digits to the right of the decimal. If required, allow 3 digits in the degrees field for longitude (i.e., DDD-MM.mm). Do not use leading zeros to the left of the decimal for degrees or minutes that require fewer than the maximum number of possible digits to express their value. The minimum number of digits is always one, even if it is a zero. (Example: Recommended: 39° 36.6'N 76° 51.42'W; Not Recommended: 39° 36.600'N 076° 51.420'W).

[2] GARS: Global Area Reference System.

[3] Land SAR responders use U.S. National Grid; however, a good familiarity with latitude and longitude is necessary to ensure effective interface between Land and Aeronautical SAR responders (Note: Land SAR includes SAR on flooded terrain).

[4] Aeronautical SAR responders will use latitude and longitude for CISAR response. However, aeronautical SAR responders that work directly with Land SAR responders should understand the U.S. National Grid system for effective Land SAR/Aeronautical SAR interface.

[5] Air space deconfliction will *only* be implemented and managed using Latitude and Longitude.

[6] Aeronautical SAR responders working with Land SAR responders have the primary responsibility of coordinating SAR using USNG. However both groups must become familiar with both georeference systems.

[7] Describes the requirement for providing situational awareness of CISAR operations geographically to Federal, military, state, local and tribal leadership.

Introduction

A key element in conducting safe CISAR operations is effective and safe management of aviation resources. The confluence of many helicopters and fixed-wing aircraft conducting multiple sorties in a complex and congested environment increases the risk of mid-air collision. CISAR planners must be able to coordinate CISAR aircraft and ground responders to ensure their effective use and safety of each.

While each disaster presents a unique set of circumstances, the ability of response agencies to conduct these contingency aviation operations safely and effectively consistently faces a number of key challenges, including the variables listed below:

- Air Traffic Management (ATM) and other Air Navigation Services (ANS) such as communications, navigational aids, and surveillance (e.g. radar) normally provided by the Federal Aviation Administration (FAA) may be disrupted or otherwise degraded;

- In addition to their ANS shortfalls, airports and airfields that need to be used as aviation enabled response and recovery nodes may have constrained capacity and/or Maximum on Ground (MOG) levels due to unavailable or degraded airport operations, including aircraft ground support services such as refueling, passenger/cargo handling, security, etc.;

- Large numbers of low altitude air missions (e.g., CISAR flights) may be needed, necessitating these aircraft to operate under Visual Flight Rules (VFR) and in Visual Meteorological Conditions (VMC);

- A diverse mix of rotary wing and fixed wing platforms, as well as Unmanned Aircraft Systems (UAS), is often used to support response efforts;

- Response air missions are frequently carried out by multiple operators from different Federal, State, Tribal, Territorial, and local agencies, using aircrews, who may be unfamiliar with the disaster area and conducting contingency flight operations with each other;

2-53

- Air mission tasking may be carried out through agency specific command-and-control channels complicating unified management of aviation operations in the disaster area;

- Air missions may be conducted using private contractor operators, which could have unclear coordination links to the aviation operations management mechanisms used by the IC; and

- Aircraft not participating in response activities (e.g. General Aviation (GA) and commercial carriers) may need and be authorized to access the airspace over the disaster area as part of the FAA's mandated effort to mitigate, recover and resume normal National Airspace System (NAS) operations.

The above constraining factors are frequently present during disasters in combination and amplified by the consistent need to conduct response flights in a shared volume of airspace.

Air Operations Branch (AOB)

NIMS provides for the optional establishment of an AOB. The Operations Section Chief of the Incident/Unified Command may establish an AOB if:

- Warranted by the nature of the incident and the availability of air assets;

- The complexity of air operations requires additional support and effort; or

- The incident requires mixing tactical and logistical use of helicopters and other aircraft.

The AOB:

- Oversees all aircraft operations related to the incident, not just operation of CISAR response aircraft;

- Should be staffed with a supervisor, a tactical group, helicopter coordinator,

fixed-wing coordinator, and other staff as required;

- Establishes and operates bases for rotary-wing air assets and maintains required liaison with off-incident fixed-wing bases;

- Is responsible for timekeeping of aviation assets assigned to the incident;

- Should coordinate with the deployed FAA operations liaison officer (LNO) to the AOB/State EOC and the FAA Airspace Access Response Cell (AARC) to request approval to enter TFR airspace and to ensure compliance with any flight restrictions and/or special instructions; and

- Should obtain from the deployed FAA LNO or the AARC a discrete beacon code for those aircraft requesting to fly into the TFR airspace.

CISAR Aircraft Coordinator (ACO)

In addition to the AOB, the SMC may assign an Aircraft Coordinator (ACO) to help maintain safe and effective use of aircraft involved in CISAR operations.

An ACO:

- Is the person, team, or facility that will coordinate multiple aircraft in CISAR operations in support of the SMC or OSC;

- Is a supporting and advisory service;

- Must be familiar with planning and conducting search operations using several aircraft at once;

- Will normally be performed by the facility with the most suitable mix of communications, radar, and plotting capability combined with trained personnel to effectively coordinate the involvement of multiple aircraft in

CISAR operations while maintaining flight safety; and

- Should perform aircraft coordination duties at the most effective location (e.g., Incident Command, EOC, fixed-wing aircraft, ship, etc.).

Depending on needs and qualifications, the ACO may be assigned duties that include:

- Coordinating the airborne CISAR resources in a defined geographical area;

- Maintaining flight safety – issue flight information;

- Ensuring flow planning (e.g., aircraft point of entry and point of exit);

- Prioritizing and allocate tasks;

- Coordinating the coverage of assigned search areas;

- Ensuring aircraft communications are maintained; and

- Making consolidated situation reports (SITREPs) to the SMC and OSC, as appropriate.

Normally, the ACO would be located with the SMC in the SAR Branch of the Operations Section, but may also work in the Air Operations Branch if one is established. The ACO's focus is on CISAR response aircraft. Obviously, ACO duties will have to be closely coordinated with the AOB if one is established.

Figure 2-12-1 on the next page shows a typical organizational structure for SAR air operations with the utilization of an Air Operations Branch.

```
                        ┌─────────────────┐
                        │   Operations    │
                        │  Section Chief  │
                        └─────────────────┘
                                 │
        ┌────────────────────────┴──────────────────┐
        │                                            │
┌───────────────────┐              ┌─────────────────────┐
│   SAR Branch/      │              │   Air Operations    │
│ SAR Mission        │              │  Branch Director    │
│ Coordinator        │              │    ("Air Boss"}     │
│    (SMC)           │              └─────────────────────┘
└───────────────────┘                        │
        ┊                                     │
┌───────────────────────────────────┐        │
│ CISAR Aircraft Coordinator (ACO)   │        │
│ (Optional. Designated by SMC to    │        │
│ manage SAR aircraft, as required.  ┊ ─ ─ ─ ─▶│
│ If established, the ACO may work in │        │
│ the Air Operations Branch)         │        │
└───────────────────────────────────┘        │
                          ┌───────────────────┴───────────────────┐
                          │                                        │
              ┌───────────────────┐                   ┌───────────────────┐
              │ Air Support Group  │                   │ Air Tactical Group │
              │    Supervisor      │                   │    Supervisor      │
              └───────────────────┘                   └───────────────────┘
                  │                                        │
         ┌────────┴────────┐                     ┌─────────┴──────────┐
┌──────────────┐  ┌──────────────┐      ┌──────────────┐  ┌──────────────┐
│ Helibase(s)  │  │ Fixed-Wing   │      │ Helicopter   │  │ Fixed-Wing   │
│  Manager     │  │  Base(s)     │      │ Coordinator  │  │ Coordinator  │
│              │  │  Manager     │      │              │  │              │
└──────────────┘  └──────────────┘      └──────────────┘  └──────────────┘
      │                                        │                  │
┌──────────────┐                      ┌──────────────┐  ┌──────────────┐
│Heliport/     │                      │ Helicopters  │  │ Fixed-Wing   │
│Helispot      │                      │              │  │  Aircraft    │
│Manager       │                      └──────────────┘  └──────────────┘
└──────────────┘
```

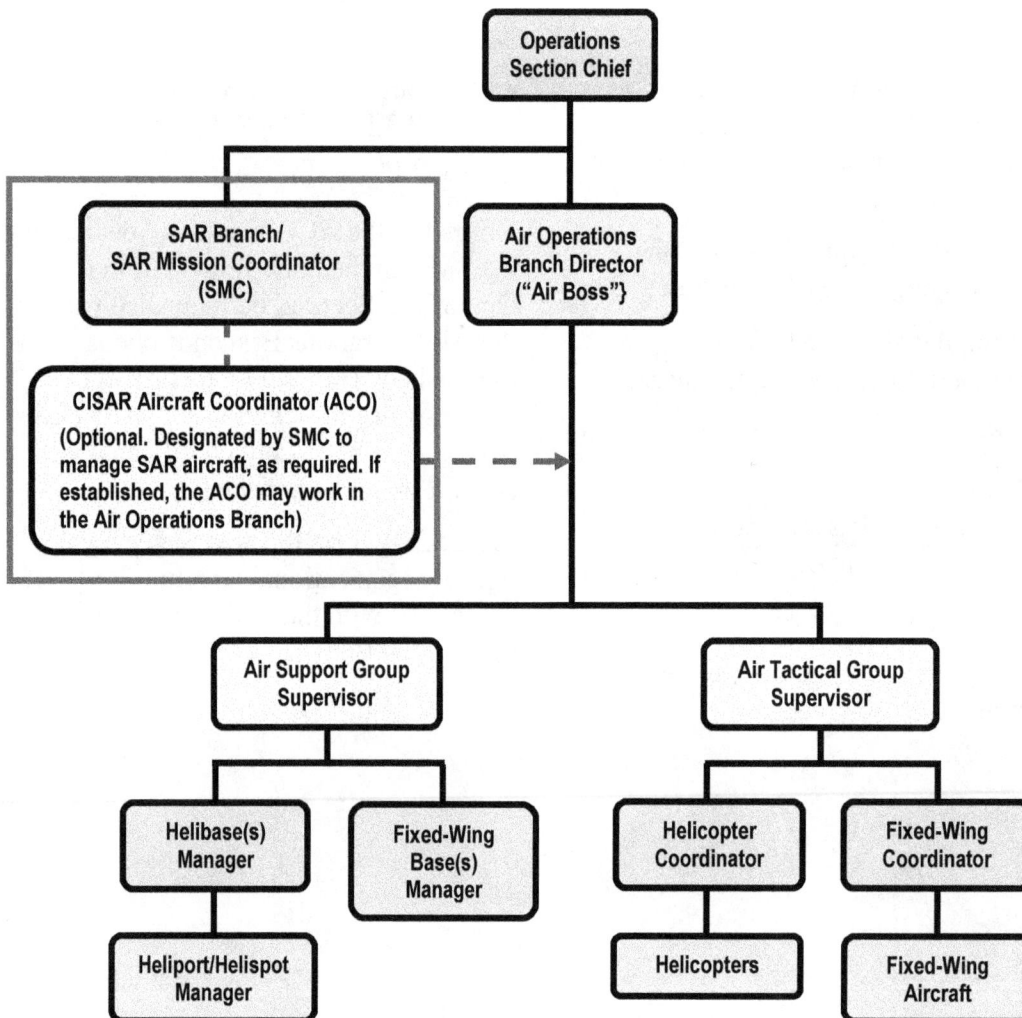

Figure 2-12-1: Typical Air Operations Organization with SAR Branch

Use of Aircraft for CISAR Operations

Lifesaving has priority over all other aircraft missions. With this in mind, delivery of vital supplies (e.g., water, medical supplies) may also become an additional priority.

CISAR response aircraft can quickly search large areas, intercept and escort aircraft or other CISAR response units, and perform aerial delivery of supplies, equipment, and personnel.

The aircraft pilot will always make the final determination as to whether the aircraft can perform the assigned mission.

The SMC/OSC/ACO should be aware of the specifications of the aircraft being used during CISAR operations in order to make informed decisions when allocating CISAR aircraft resources.

CISAR aircraft pilots should carefully evaluate the SAR action plan to ensure pattern orientation for the assigned search area and those for adjacent assignments meet safety requirements and provide the best opportunity for detecting the search object. The SMC must be notified immediately upon discovery of any safety issues and should be notified of all other apparent deficiencies as early as practicable.

When planning aircraft search areas, consider the following:

- Orientation of aircraft search patterns and CSP placement is based on safety, aircraft endurance, navigation, environmental conditions, available resources, and any other factors critical to the CISAR operation.

- If a systematic search of a particular area is conducted, the aircraft's CSP should be placed close to the point of the aircraft's departure location to facilitate the start of the search effort;

  (Note: A different CSP may be designated to take into account factors such as the aircraft's next destination, mission, fuel replenishment location, etc.)

- When planning an aircraft search pattern, consider the location of the sun, especially early and late in the day. Looking into the sun can make detection of people in distress difficult.

Risk Management

The IC, SMC, ACO, OSC and aircraft commanders continuously make operational decisions during CISAR operations. As missions progress, each must weigh and continually reassess the urgency of the mission and the benefits to be gained versus the risks involved. The safety of the aircrew and aircraft must always be one of the primary considerations for planning and conducting aviation missions.

For CISAR operations, potential risks to the aircraft and crew should be weighed against risks to the personnel and/or property in distress if the mission is not undertaken. Probable loss of the aircrew is not an acceptable risk.

Aircrew Fatigue

Physical factors impact the ability of flight crews to exercise good judgment. Chief among those factors is fatigue.

Stressors such as constant vibration, loud noises from machinery and radios, illness, poor physical conditioning, improper diet, and irregular or insufficient sleep patterns can create both acute and chronic fatigue.

Most aviation units and organizations have developed crew standards that protect personnel while maximizing support for the CISAR operation. Agency specific fatigue standards must be adhered to.

Flight Safety

Flight safety is of paramount concern in complex CISAR operations and must be considered in mission planning.

For aviation safety to be truly effective, safety must be a pervasive notion supported at all levels of the Incident Command.

Most aviation mishaps are preventable and usually the result of human error, mechanical failure, or combination of both. Most mechanical failures may be attributed to a human error at some point, either in the design, maintenance, or operation of equipment.

If mishaps are to be prevented, it is necessary to detect and guard against human error at every stage of an air operation. This requires a continuous review and communication between all activities affecting aviation operations and maintenance so that mistakes or potential mistakes can be identified, evaluated, and corrected.

During CISAR operations, where hazards to aircraft safety can and will occur, these hazards must be identified and effectively reduced or eliminated, to minimize the potential for a mishap and ensuring the aircraft's continued operational availability.

Each individual connected with air operations, whether in an operational or supporting role (e.g., aircrew, scheduling, maintenance), contributes directly to the effectiveness of aviation safety. Effective safety is a team effort and requires the active participation of "all hands."

Aircraft Prioritization

Aircraft mission prioritization, a responsibility of the IC, is critical to an effective incident response. Normally, CISAR and other lifesaving missions should be considered the first priority.

An example prioritization scheme is in Table 2-12-1 below.

| Table 2-12-1: Example Aircraft Prioritization Based on Mission ||
Priority	Missions
Priority 1: Direct lifesaving operations	• CISAR; • Incident Awareness and Assessment; • Evacuation; • Medical Response (Pre-hospital Medical Response); • Patient Movement; • Aero-Medical Evacuation; and • Air Ambulance. *(Note: In Priority 1, the rescue of persons in distress must be the highest priority)*
Priority 2: Direct life sustaining, damage prevention, support and enabling tasks	• Command and Control/Incident Management; • Security; • Medical Response (e.g., DMAT - Disaster Medical Assistance Teams); • Life sustaining commodities distribution; • Firefighting; • Airfield opening; • Levy and infrastructure repair; and • Mission critical personnel movement.
Priority 3: Other response-related missions	• Non-mission critical personnel movement; • Non-time critical logistics distribution; and • Direct response support missions.
Priority 4: Other mission on non-interference basis	• News; • Other non-critical movements; and • Recovery Missions.

Helicopters

For planning purposes, helicopters are generally excellent SAR platforms capable of recovering persons from a wide variety of distress situations on land and water.

Helicopters involved in CISAR operations should generally have some or all of the following capabilities:

- Wide area and hover search;

- Hoisting/winching;

- Marking (e.g., smoke, sea dye, flares, etc.);

- EMT/rescue swimmer/pararescuemen delivery;

- Delivery of survivors/equipment;

- Confined area landing;

- Direction finding;

- Night illumination;

- Search sensors; and

- Command and control/incident awareness and assessment.

This page intentionally left blank.

Section 2-13: Temporary Flight Restrictions (TFRs)

Definitions and Types

Requests and Issuance

TFR Proportions

Mission Type Altitude Stratification

Additional Notes Regarding Altitude Stratification Schema (Figure 3-2-2)

Definition and Types

A Temporary Flight Restriction (TFR) is a regulatory action issued via the U.S. Notice to Airmen (NOTAM) system to restrict certain aircraft from operating within a defined area, on a temporary basis, to protect persons or property in the air or on the ground.

The Code of Federal Regulations, Title 14 (14 CFR) identifies TFRs for a variety of situations, including disaster response which is covered in 14 CFR 91.137.

14 CFR 91.137 TFRs may be issued to:

- Protect personnel and property on the surface or in the air from a hazard associated with an incident on the surface (14 CFR 91.137 (a) (1));

- Provide a safe environment for the operation of disaster relief aircraft (14 CFR 91.137 (a) (2)); or to

- Prevent an unsafe congestion of sightseeing and other aircraft above an incident or event which may generate a high degree of public interest (91.137 (a) (3)).

(Note: Properly accredited news representatives are allowed into 91.137 (a)

(2) and 91.137 (a) (3) TFRs if they file a flight plan with the appropriate FAA ATC facility specified in the Notice to Airmen and the operation is conducted above the altitude used by the disaster relief aircraft unless otherwise authorized.)

The second type of TFR that could be issued for a catastrophic incident or national security event is provided for in 14 CFR 99.7, Special Security Instructions. The FAA, in consultation with DoD and/or Federal security and intelligence agencies, may issue special security instructions to address situations determined to be detrimental to the interests of national defense.

For example:

- The FAA regularly uses 99.7 TFRs to protect National Special Security Events (NSSE) and other security sensitive events such as the Presidential inauguration; and

- When President Bush visited New Orleans and toured the Gulf Coast, a 14 CFR 91.141 TFR was implemented.

For CISAR operations, a 14 CFR 91.137 TFR will normally be used.

Requests and Issuance

Only the FAA may issue a TFR. However, the following entities may request the FAA issue a TFR for a specific disaster area:

- Military commands;

- Federal security and intelligence agencies; regional directors of the Office of Emergency Planning;

- State civil defense directors;

- Authorities directing or coordinating organized relief or response air operations;

- State Governors;

- FAA Flight Standards District Offices; and

- Aviation and sporting event officials.

Non-FAA authorities should contact the nearest air traffic control facility to request a TFR (If possible, have the location latitude/longitude in degrees/minutes/seconds format.)

FAA authorities should contact their respective service area representative for non-emergency requests, the cognizant air route traffic control center for emergency requests and the System Operations Support Center (SOSC) for VIP or security TFRs.

(Note: TFR information or assistance can be obtained from the SOSC by calling 202-267-8276; for any other security related question or concerns relating to aviation contact the Domestic Events Network (DEN) at 866-598-9522. Also note that the subject contingency TFRs and other related air traffic and airspace management measures used by the FAA to support CISAR and other response aviation operations will be planned, coordinated, and implemented in accordance with the FAA's Airspace Management Plan for Disasters (AMP).)

TFR Proportions

TFRs can be issued as a cylinder based on a point, a polygon, or other shapes. Latitude and longitude, and/or a fixed radial distance from a navigational aid identify the point or corners of the TFR. A TFR always includes a range of altitudes. Rules are enforced about who can enter or leave a TFR and what can be done within the TFR airspace.

TFRs are sized to minimize disruption of surrounding airspace while meeting the needs of the requestor. Figure 2-13-1 below shows two types of TFR shapes.

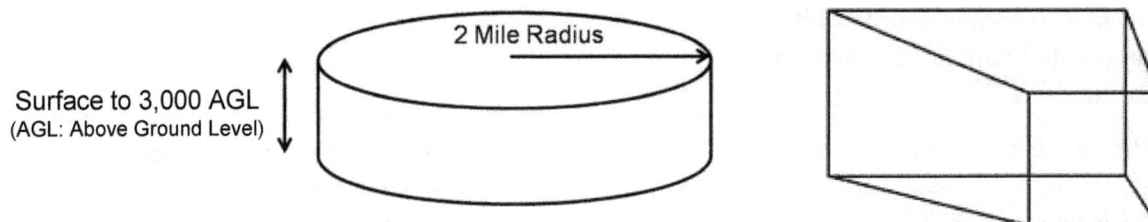

Figure 2-13-1: Types of TFR Shapes

Mission Type Altitude Stratification

In the event there is a need for relatively high density, low altitude VFR response air missions (e.g., SAR and sling loads), the FAA may segregate select operations by mission type using altitude blocks within a TFR (see Figure 2-13-2 on page 2-68).

Figure 2-13-2 is a template developed by the FAA as a guide for AOB Directors (AOBD) and SMCs to utilize in the development of an air operations plan. The FAA (specifically the FAA liaison to the AOB) can be of great assistance to the AOBD/SMC by developing and managing TFR/altitude stratification customized to the operational needs and local requirements of the response.

Mission type based altitude stratification is designed to be used for VFR operations (i.e., "see and avoid" based flight).

Air missions operating within designated altitude blocks must remain in VFR operation under VMC. If a pilot is unable to do so because of expected or encountered Instrument Meteorological Conditions (IMC), the flight must depart the defined TFR stratified altitude structure by the safest route possible. The flight in question may resume its response air mission if and when it is able to resume VFR operations under VMC within the designated altitude block. If air traffic control (ATC) is available, the subject Pilot-in-Command (PIC) must also advise ATC and comply with any consequent ATC instructions.

The FAA will adjust this altitude stratification schema to meet the specific conditions of each disaster and the incident mission requirements.

Also note that response air missions, both those identified in the altitude stratification diagram and others, may be authorized to operate within altitude blocks not normally designated for their use. This can be done to address overriding safety concerns and/or mission needs (e.g., time critical life-saving activities). The FAA will also make location and disaster specific modifications to this altitude stratification scheme to accommodate any overlap with Class B or other controlled airspace, which may still be provided service by an FAA facility.

Additional Notes Regarding Altitude Stratification Schema (Figure 3-2-2)

As indicated in Figure 2-13-2 on the next page, participating aircraft flying within this altitude stratification may be further separated through the use of flight direction altitude blocks in accordance with the following:

- NORTH/EAST – from 360° to 179°

- SOUTH/WEST – from 180° thru 359°

All aircraft will use the local altimeter setting as directed by the appropriate ATC facility or nearest aeronautical weather reporting facility.

In addition, these altitude blocks do not constitute or supersede any ATC instructions.

VFR OPERATIONS (SEE AND AVOID)

Safety Note: This altitude segregation structure may be altered to support local mission needs and operating environment – refer to latest NOTAMs. Participating aircraft in this TFR altitude structure must operate VFR and stay under VMC within the designated mission type altitude blocks. If unable, participating aircraft must suspend mission and exit TFR via safest route. This requirement may be overridden by ATC or other safety of life critical considerations. Also note that disaster response mission types including those identified on this diagram and others (e.g., CIKR IAA missions, both government and private sector) may be authorized to enter the designated altitude blocks on a case-by-case basis as long as deconfliction procedures are implemented.

OTHER PARTICIPATING LOW LEVEL FIXED WING AND APPROVED MEDIA

SOUTH / WEST
5,000' – 5,999' AGL

NORTH / EAST
4,000' – 4,999' AGL

ROTARY WING NON-SAR (E.G., LE)

SOUTH / WEST
3,500' – 3,999' AGL

NORTH / EAST
3,000' – 3,400' AGL

HELICOPTER REFUELING OPS
(To extent practicable refueling operations will be established outside of the TFR)

2,500' AGL

ROTARY / FIXED WING SAR TRANSITION

SOUTH / WEST
1,500' – 1,999' AGL

NORTH / EAST
1,000' – 1,400' AGL

FIXED WING – ACTIVE SAR

ROTARY WING – ACTIVE SAR AND SLING LOAD

COORDINATION DRAFT

6,000' AGL
(up to but not including)

4,000' AGL

4,000' AGL
(up to but not including)

3,000' AGL

3,000' AGL
(up to but not including)

2,000' AGL

2,000' AGL
(up to but not including)

1,000' AGL

1,000' AGL
(up to but not including)

500' AGL

500' AGL
(up to but not including)

SFC

Figure 2-13-2: Mission Type Altitude Stratification within a TFR

Section 2-14: Boat Operations Management

Introduction

Boat Operations Branch (BOB)

Health and Safety

Boat Operations

Boat Crews

CISAR Boat Preparation

Points to Consider in Flooding Scenarios

Introduction

In many instances, boat operations are a vital part of CISAR operations. Responses from the water presents unique challenges and major safety concerns for boat crews. CISAR boat operations are usually required for flooding incidents or large passenger ship disasters. Boat crews must be familiar with best practices and lessons learned from past incidents to ensure safe and effective boat operations.

Boat Operations Branch (BOB)

NIMS outlines the optional establishment of a Boat Operations Branch (BOB). The Operations Section Chief of the Incident/Unified Command may establish a BOB as warranted by the nature of the incident and availability of assets and personnel.

The BOB would oversee boat operations related to the overall incident, not just CISAR operations.

To enhance SAR operational efficiency and effectiveness (the right resource, to the right place, at the right time), the IC/SMC may consider consolidating boat, land, and aviation operations within a State/Federal Unified Air Marine Operations Branch" as an all-domain integrated SAR Branch to the Operations Section.

Health and Safety

CISAR responders on the water may face serious health concerns. Awareness and preparation can help to avoid serious injury or illness. As the response progresses, keep in mind the following:

- Two boat teams should be assigned to work together to support mutual safety requirements;

- In the event of a flood, sewage and waste flow freely into the waterways exposing boat crews to possible disease and contamination;

- Boat crews must be cognizant of unsanitary vessel spray and water splash;

- Boat crewmembers must be dressed in appropriate protective equipment and remain aware of potential unsanitary conditions when maneuvering or choosing a safe operating speed;

- Special consideration concerning vessel spray should be taken when operating airboats in an unsanitary environment;

- For missions where contaminants or waste are known, crews should be made aware of associated or prevalent diseases and be able to recognize symptoms;

- Responding to incidents that involve victims who have been exposed to hazardous materials, boat crews should be properly protected and exercise their duties with caution;

- To help boat crews avoid heat exhaustion, dehydration, and exhaustion, boats should be stocked with plenty of fresh water and food;

- Because boats may operate in difficult environments for extended periods of time, a duty rotation should be implemented to ensure crews do not become overly fatigued; and

- Response personnel should be aware that deceased persons may pose potential health risks.

Boat Crew Decontamination

In flooding scenarios, boat crews will inevitably get wet despite precautions. Decontamination of personnel and equipment is essential for the CISAR responder's health and safety. Showers, change of clothes/boots, bleach, sanitized wipes, laundry, etc., will be essential to ensure personnel and equipment are decontaminated at the end of each mission.

If boats are deployed via trailers from a base camp, CISAR responders should set up a decontamination area away from living quarters to ensure other responders are not contaminated.

(Note: Freshwater wash down may be limited due to contaminated local water.)

Boat Operations

In certain situations, such as flooding or extreme weather conditions, the altered environment will present unique challenges:

- Flooding can adversely impact land transportation by destroying roads and bridges, thus hindering the response;

- Local infrastructure may be unable to support CISAR responders (e.g., logistics, housing, food, etc.);

- Launch ramps may be unusable;

- Large areas of land covered in water during a flood can become uncharted waterways in which crews will be required to operate and could potentially conceal numerous hazards (e.g., cars, trees, signposts, etc.);

- Waterway aids to navigation may be moved off location or damaged, becoming useless or dangerous to CISAR responders and the boating public if the off station condition is not recognized;

- Obstacles under water and large floating debris may damage boat propellers, jet drives, and hulls;

- Operating boundaries will need to be set or adjusted for increased and uncharted waterways after a flood to ensure full coverage;

- Boat crews should have extra fuel, knowledge of the nearest operating fueling station, or a plan for fuel delivery if local fueling locations are damaged or unusable; and

- Searching for survivors should be conducted as per *Section 2-7: CISAR Searches* and *Section 2-8: Structural Marking Systems*.

Because of these challenges, keeping a good lookout and monitoring boat speed will help keep crews safe and boats free from damage.

Boat Crews

Leadership and the need for decisive action are a necessity in the event of a CISAR response. Boat crews need clearly defined roles, responsibilities, and lines of authority.

Responders can be called on to conduct operations where immediate decisions will be required to save lives and accomplish the mission. Boat operators and boat crews need to be made aware of their scope of authority in order to effectively act with limited guidance from the IC. Enabling boat operators to act quickly and decisively will reduce the reliance on communications systems that may sometimes fail.

In all boat operations, the final decision authority for the safety of the mission rests with the boat's operator (Coxswain). If the Coxswain believes an evolution is unsafe or would cause undue harm to the crew, then they should have the authority to cease operations. Risk evaluation is a healthy and necessary part of every mission. If the risk outweighs the probable benefit, a mission should be reevaluated.

CISAR Boat Preparation

Situations cannot be predicted which is why training and maintenance are so vital to preparing for CISAR boat operations. Boat crews must train and remain ready to respond.

Likewise, the maintenance of resources is equally important. Daily boat checks and regularly scheduled maintenance must be performed to ensure boats are ready to respond in the event of a catastrophe.

Points to Consider in Flooding Scenarios

The following points are provided for consideration when using boats in CISAR flooding scenarios:

- Flat bottom jon boats may be more effective than Rigid Hull Inflatable Boats (RHIBs) or other types of inflatable boat due to chain link fences, barbed wire, and other debris that can cause damage to inflatable cells;

- Air boats can be effective in a flood scenario (headsets are required for maintaining communication), but use caution: air boats can spray toxic water into the air and be ingested by SAR responders and flood victims;

- Poles, oars, and paddles should be carried in all boats because motors and propellers take a beating on debris and unseen obstacles;

- Every boat should carry extra boat propellers (propellers can be damaged be debris);

- Personal Flotation Devices (PFDs) should be carried in every boat and worn by all rescue personnel and survivors (include extra large and small size PFDs)

- Consider carrying a lightweight ladder for extricating persons from roof tops, waders, chain saws, and dead man sticks to lift power lines;

- Every boat should be equipped with at least a first aid kit (a better equipped medical kit is preferred);

- Ideally, every boat crew should have an Emergency Medical Technician (EMT);

- In flooding scenarios, animals and insects may pose additional challenges to boat crews;

- Boats should be equipped with sturdy, water resistant radios or use radio bags to limit damage due to submersion;

- Boats should be equipped with lights and chemical sticks;

- Boat managers should implement routine boat reporting requirements to ensure boat crew safety while deployed;

- Responders should be current on all vaccines and if necessary, be provided broad-spectrum antibiotics to combat the effects of contaminated water;

- Boat crews should have a minimum of three days food and supplies (anticipate limited logistical support for the first 72 hours; and

- Logistical support (e.g., berthing, laundry, portable toilets, portable shower trailers, fuel trucks, communications equipment, etc.) must be implemented as soon as possible to ensure the long term success of any CISAR operation.

CISAR operations messages include Situation Reports (SITREPs), Search Action Plans (SAPs), and Rescue Action Plans (RAPs). These messages should be unclassified, in plain language, and require no key to interpret. A standard sample message file, or computer templates and programs, should be established to aid in quickly drafting and releasing the types of messages used regularly.

Situation Reports (SITREPS)

SITREPs are reports from the OSC to the SMC, or SMC to interested agencies that provide information on current conditions and mission progress. The OSC sends SITREPs to the SMC (unless otherwise directed) to keep the SMC informed of on scene mission progress and conditions. The SMC uses SITREPs to keep superiors, the IC, and any other interested agencies informed of mission progress. For cases where other threats or non-SAR operations exist on scene, other appropriate agencies should be information addressees on all SITREPs.

Often a short SITREP is used to provide prompt information and urgent details. A more complete SITREP is used to pass amplifying information.

Initial SITREPs should be transmitted as soon as details of an incident become clear enough to confirm the need for SAR involvement, and should not be delayed unnecessarily for confirmation of all details. Further SITREPs should be issued as soon as other relevant information is obtained. Information already passed should not be repeated. During prolonged operations, the IC may decide to require "no change" SITREPs be submitted at intervals of about three hours to reassure recipients that nothing has been missed and that the unit remains operational. When the on scene response has been completed, a final SITREP should be issued.

More information on SITREPs and their contents, including sample formats, is available in the IAMSAR Manual (Volume 2); however, SAR agencies may have their own formats.

Search Action Plans (SAPs)

The SMC should develop a SAP and a Rescue Action Plan (RAP) as appropriate. In some situations these plans may be combined and promulgated in one message.

After a SAP is developed, it is provided to the OSC and CISAR responders on scene in a search action message (or other applicable format). The message should include:

- A summary of the on scene situation, including the nature of the emergency;

- Areas to be searched;

- What to search for (search object(s));

- Detailed search instructions;

- Weather information;

- Summary of SAR facilities on scene;

- A listing of the appropriately-described search area(s) and sub-areas that can be searched by the SAR facilities in the allotted time;

- On land or flooded areas, geo-referencing grids may be added to designate positions; and

- The assignment of communications frequencies to include primary and secondary control channels, on scene, monitor and press channels, and special radio procedures, schedules, or relevant communication factors.

It is better to release the SAP early. If a "first light" search is being planned; parent agencies providing CISAR responders should typically receive the SAP at least six hours before departure time if practicable, bearing in mind that the SAP can be expanded or amended later.

Rescue Action Plans (RAPs)

In conjunction with the SAP and if required, the SMC may then develop a RAP. It is provided to the OSC and CISAR responders on scene. The RAP is similar to the SAP, but with search instructions replaced with rescue directions.

Five-Line Brief

For large-scale CISAR operations, air rescue support can be reported by the OSC or the OSC's supporting ACO using a succinct five-line briefing. The five line briefing (Table 2-15-1 below) is a voice data message that contains the following information:

Table 2-15-1: Five Line Brief
Number of persons for rescue
Conditions of persons (ambulatory, non-ambulatory, etc.)
Location: Latitude/Longitude (DD MM.MM/DDD MM.MM)
Type of pickup and device to be used (hoist, land, etc.)
Hazards identified in area/amplifying information

Example 5-line radio call:

Air Blue 8, Rover 3
Line 1: 6 passengers
Line 2: non-ambulatory
Line 3: 36°01.33'N / 114°43.45'W

(State as: "Three six degrees, zero one decimal three three minutes North / one one four degrees, four three decimal four five minutes West")
Line 4: hoist with litter
Line 5: large 150 foot tower to the south

Section 2-16: Conclusion of CISAR Operations

Introduction

Termination

Suspension

Reopening Suspended CISAR Operations

Introduction

As per the NSP, CISAR operations shall normally continue until all reasonable hope of rescuing survivors has passed.

The SMC will recommend to the IC when to discontinue CISAR operations. This person should have the training and experience to make and defend this difficult decision, which should be objectively based on the facts of the operation. The IC will make the final decision.

If no SMC is assigned, the IC should normally make this decision.

Often external pressure (i.e. political, families of unaccounted for victims, etc.) may insist that CISAR operations continue beyond the time when there is any reasonable hope of rescuing survivors. Normally this can be minimized by keeping the public well-informed of the CISAR operation's progress and ensure those not directly involved understand the level of effort expended, and that the probability of success of further operations is negligible.

As per Table 2-16-1 below, two terms are used to indicate the conclusion of CISAR operations:

Table 2-16-1: Conclusion of CISAR Operations	
Termination	**Active Search Suspended (ACTSUS) Pending Further Developments**
All known person(s) are located and accounted for.	When the CISAR operation cannot be terminated because person(s) remain missing and further search efforts appear futile, the CISAR operation may be suspended.
No other CISAR issues arise.	
If person(s) remain missing at the conclusion of the CISAR effort, the CISAR operation should not be formally closed, but suspended.	

Termination

The decision as to whether to terminate CISAR operations should consider:

- The probability that survivors are still alive accounting for prevalent environmental factors since the incident;

- The cumulative probability that survivors would have been found; and

- The availability of CISAR responders to continue the search.

Termination of CISAR Operations

In CISAR operations, when all persons in need of rescue are accounted for and no other CISAR issues arise, then the CISAR operation can be terminated.

The IC, in consultation with the SMC, should consider on the basis of reliable information, that a rescue operation has been successful, or that the emergency no longer exists.

The Incident Commander shall promptly inform Federal, State, Tribal, Territorial, and local CISAR responders, authorities, or services which have been activated or notified.

Suspension

In making the decision to suspend a CISAR operation:

- Care should be taken not to end the search prematurely; and

- The humanitarian significance of the effort must be considered, but also understanding there must be a limit to the time and effort that can be devoted to CISAR as dictated by the circumstances.

Prior to suspending CISAR operations, a thorough review should be made. The decision to suspend operations should be based on an evaluation of the probability that there were survivors from the initial incident, the probability of survival after the incident, the probability that any survivors were within the search area, and the effectiveness of the search effort. The reasons for suspending CISAR operations should be clearly recorded.

The review should also examine:

- Whether search decisions were based on proper assumptions and reasonable planning;

- Certainty of the location of victims and other factors used in determining the search area;

- Whether significant clues and leads should be re-evaluated;

- The search plan to ensure that:
  - o All assigned areas were searched;
  - o There was effective use of air, boat, and ground CISAR responders to provide the best probability that victims would be located; and
  - o Compensation was made for search degradation caused by the environment, location (urban, rural, mountain, maritime, etc.) weather, mechanical, or other difficulties.

- The determination about the survivability of survivors, considering:

o Time elapsed since the incident;

o Environmental conditions; and

o Age and physical condition of potential survivors;

The IC may continue CISAR operations beyond the time when normally be suspended due to humanitarian considerations, large number of people involved, or forecast of greatly improved search conditions. However, CISAR responders should not be risked when the potential for saving life is minimal, or when their use may preclude their availability for other critical missions.

Suspension of CISAR Operations

Some CISAR operations may require extended searching. At some point, however, the Incident Commander, in consultation with the SMC, must make the difficult decision to suspend further CISAR operations pending the receipt of additional information.

As per the NSP: If the Incident Commander, in consultation with the SMC, determines the following:

1. Having thoroughly evaluated the effectiveness of the CISAR operation and all available information concerning the CISAR effort; and

2. Having made the determination that further CISAR efforts would be ineffective and most likely not result in additional lives saved; may

3. Temporarily suspend CISAR operations pending further developments.

The Incident Commander should promptly inform Federal, State, Tribal, Territorial, and local CISAR responders, authorities, or services which have been activated or notified.

Information subsequently received should be evaluated and CISAR operations resumed when justified on the basis of such information.

Reopening Suspended CISAR Operations

If significant new information is obtained, reopening CISAR operations should be considered. Reopening without good reason may lead to unwarranted use and risk of injury to CISAR responders, possible inability to respond to other emergencies, and false hope among relatives.

However, if there is any real possibility to save additional lives, CISAR efforts should be resumed.

Part 3: Supplemental Considerations

Section 3-1: Risk Assessment

Introduction

CISAR Risk Assessment: One Model, Two Parts

Part 1: Assess the Risk – GAR (Green, Amber, Red)

Part 2: CISAR Risk Mission Analysis

Introduction

SAR is inherently dangerous to both the survivor and the CISAR responder. The danger will most likely be greater in response to a catastrophic incident due to factors such as workload, environment, mission complexity, and inexperience. The CISAR environment and operational tempo make risk management even more critical for both operations management and force protection.

Risk assessment is a continuous process that should be conducted by the SMC and each CISAR responder. When a mission situation changes, risk should be reevaluated with any associated mitigation options discussed. These periodic updates are necessary anytime there is a significant change to either the environment or the SAR mission.

Every member of the CISAR response team should participate in the risk assessment process. Each person brings experiences to a risk assessment that others on the team may not have taken into account. This team approach to risk management also allows each member of the team to be made aware of the challenges and related risks of the mission.

The few moments CISAR responders take to assess the risks of a particular CISAR mission may be crucial in mitigating safety hazards.

CISAR Risk Assessment: One Model, Two Parts

CISAR responder risk assessments should be conducted in accordance with their respective agency procedures. However, if no such procedures are available, the following risk assessment process may be used.

There are two parts:

Part 1: Assess the Risk – Green, Amber, Red (GAR)

Part 1 guides CISAR responders and managers in determining risk for a particular CISAR mission. When using the GAR Risk Assessment tool, a numerical value is obtained and the risk is identified.

Part 2: Perform CISAR Mission Risk Analysis

In Part 2, the GAR identified risk is placed in Step 3 of the CISAR mission analysis flow chart.

Use this simple model or an agency specific process to assess risk. Assessing risk is an ongoing process and should be reassessed continually throughout the CISAR operation.

Part 1: Assess the Risk – GAR (Green, Amber, Red)

Factors identified in Table 3-1-1 on the next page should be used when assessing risk for a particular CISAR mission:

Table 3-1-1: Factors to Consider When Assessing CISAR Risk

Factor	Explanation	Assigned Risk Value (0 to 10)
Supervision	How closely do you need to supervise the SMC, CISAR responder team or crew? The higher the risk, the more a supervisor needs to focus on observing and checking.	
Planning/ Preparation	How much information is available? How clear is the information? How much time is available to plan and execute the mission?	
Crew Selection	Consider the experience of the CISAR responders performing the mission. If individuals are replaced during the mission, assess their experience level and ensure proper turnover.	
Crew Fitness	How tired are the CISAR responders? How many missions have they performed?	
Environment	Factors that affect personnel, unit readiness, and resource performance. These factors may include time of day, visibility, ceiling level, proximity to other external and geographic hazards and barriers and amount of infrastructure damage.	
Mission Complexity	Consider both the time and resources required to conduct the mission. The longer the exposure to hazards for the CISAR responder and the person(s) in need of assistance, the greater the risks involved. What is the precision level needed to successfully complete the mission?	
	Total:	

Calculating Risk: To identify the total degree of risk, for each factor in Table 3-1-1 above, assign a risk value of 0 (no risk) up to a value of 10 (maximum risk) for each factor.

(Note: These number values are subjective, based on the experience of the individual or group).

Add the values. Based on the risk number, use the criteria in Table 3-1-2 below to assign a GAR Risk Assessment.

Table 3-1-2: GAR Evaluation Scale

1-23	24-44	45-60
Green	Amber	Red
Low risk.	Medium risk. Consider procedures and actions to minimize	High risk. Implement measures to reduce risk prior to starting mission

Determine the CISAR mission risk based on
the criteria in Table 3-1-3 below.

Table 3-1-3: GAR Risk Assessment Tool			
	High Gain	Medium Gain	Low Gain
Low Risk (1-23)	Accept the Mission. Continue to monitor Risk Factors for any condition or mission changes.	Accept the Mission. Continue to monitor Risk Factors for any condition or mission changes	Accept the Mission. Reevaluate Risk vs. Gain should Risk Factors change.
Medium Risk (24-44)	Accept the Mission. Continue to monitor Risk Factors. Identify, and if possible, employ any potential options to mitigate identified hazards that exceed an acceptable degree of risk.	Accept the Mission. Continue to monitor Risk Factors. Identify, and if possible, employ any potential options to mitigate identified hazards that exceed an acceptable degree of risk.	Accept the Mission. Continue to monitor Risk Factors. Actively pursue any possible options to mitigate identified hazards to reduce Risk.
High Risk (45-60)	Accept the Mission only with parent agency endorsement. Communicate Risk vs. Gain to SMC and Incident Command. Actively pursue options to reduce risk.	Accept the Mission only with parent agency endorsement. Communicate Risk vs. Gain to SMC and Incident Command. Actively pursue options to reduce risk.	Do not accept the Mission. Notify parent agency, SMC and Incident Command. Wait until Risk Factors change; as possible, mitigate identified hazards to reduce risk.

Part 2: Perform CISAR Mission Risk Analysis

After the risk assessment is made (Part 1),
perform a CISAR Mission Risk Analysis by
using Figure 3-1-1 on the next page. Part 2
uses the risk value obtained in Part 1, which
is placed in Step 3 of Figure 3-1-1.

| Step 1:
Identify Mission Tasks | What does the task entail?
What do we have to do?
Are there other ways to do the task? |

| Step 2:
Identify Hazards | What can go wrong? (Equipment, Personnel, Environment)
How is the risk defined for us?
What safeguards exist? How effective are they? |

| Step 3:
Assess Risk
(Numerical value
from Part 1) |

Low Risk

Medium Risk

High Risk

| What are the effects? (Severity)
Can this happen to us? (Probability)
What is the event frequency or degree of involvement?
(Exposure) |

| Step 4:
Identify Options | Are risks acceptable or unacceptable?
What options can eliminate unacceptable risk (that which does not contribute to accomplishing the mission safely)?
What options reduce undesirable risk?
Can we modify mission to reduce risk?
Are any safeguards missing?
What new options should be consider? |

| Step 5:
Evaluate Risk vs. Gain | Did someone with authority validate the potential risks resulting from the options considered are worth the gain?
This risk decision must be made at the lowest appropriate level, considering experience and maturity. |

| Step 6:
Execute Decision | Implement the best options.
Have we allocated the necessary resources?
Have we initiated risk management procedures?
Does everyone know why we are doing this and the expected outcome? |

| Step 7:
Monitor Decision | Are the safeguards working?
Are participants accomplishing the mission or task objective?
Has the situation changed?
Risk management is a continuous process. |

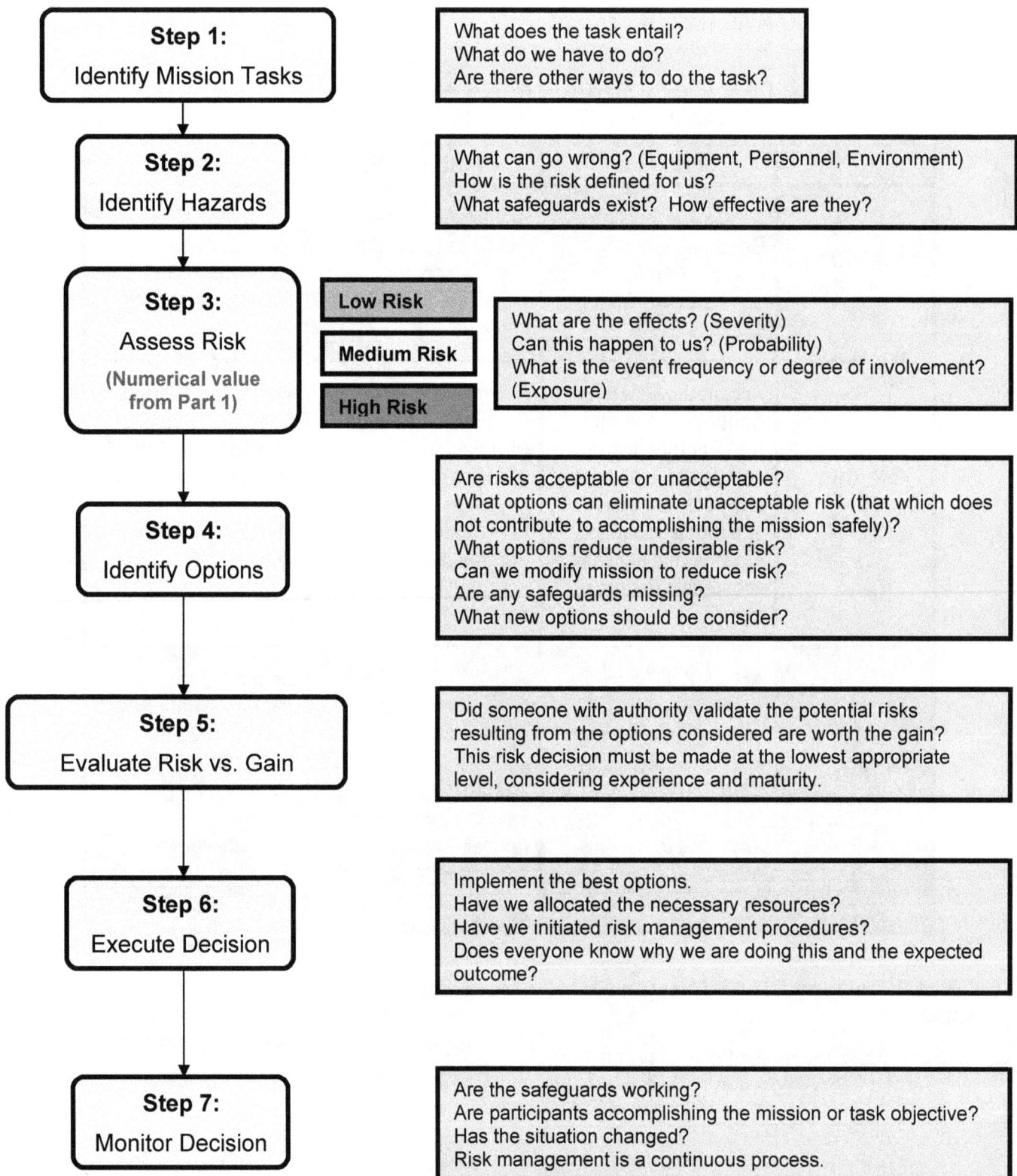

Figure 3-1-1: Part 2 - CISAR Mission Risk Analysis

Section 3-2: Health and Safety

Introduction

This Section details several health and safety issues related to CISAR operations. CISAR responders should understand and plan for the various health and safety risks associated with a CISAR environment.

Through proper risk assessment and the use of an effective safety plan, personal injuries and health risks can be more effectively mitigated.

CISAR responders who become sick or injured during operations should immediately contact the IC to obtain medical assistance. Additional assistance can be received from ESF #8 (Public Health and Medical Services) personnel.

Injury and Illness Prevention

Potential injuries and illnesses outlined in this Section and other potential hazards that a CISAR responder may encounter can be effectively mitigated by using appropriate personal protective equipment (PPE) for the hazard or environment encountered. The review of the Incident Safety Plan and ICS 215A Incident Action Plan Safety Analysis which outlines potential hazards and associated mitigations will assist the CISAR responders in determining what hazards will

be faced. Basic PPE equipment such as head (helmet), ear (ear plugs/muffs), hand (gloves), foot (boots), eye (safety glasses), flotation devices, and respiratory protection (dust mask or respirator) should be available and used by each CISAR responder. Additional PPE may be required as dictated by the hazards that may be encountered in a particular CISAR environment.

Safety is paramount. Personal safety should always be your first concern; your co-worker should be your second concern; and the victim's safety third.

Decontamination processes and procedures should be considered for all CISAR responders and included in any safety and health plan.

Injuries and illnesses should be reported to your immediate supervisor, incident safety officer, or commander.

Health and Safety Issues in Flooded Areas

The danger of flood does not end when the water stops rising. CISAR responders must work together and look out for one another to ensure safety in flooded areas.

Floodwaters may contain fecal material from overflowing sewage systems, as well as agricultural and industrial byproducts. There is always a risk of disease from eating or drinking anything contaminated with flood water.

Any open cuts, sores, or burns, no matter how minor, which can be exposed to contaminated flood water should be kept as clean as possible by washing with soap and clean water to control infection. If a wound develops redness, swelling, or drainage, seek immediate medical attention.

During CISAR operations, the risk of wounds may be increased. For this reason, CISAR responders should be sure they are up-to-date with tetanus vaccination, ideally before starting response operations. Being up-to-date for tetanus vaccine can greatly simplify the treatment for any wound that might occur.

Hepatitis

Hepatitis is an inflammation of the liver. Toxins, certain drugs, some diseases, heavy alcohol use, bacterial, and viral infections can cause hepatitis.

Hepatitis A is a liver disease caused by the Hepatitis A virus (HAV). HAV infection produces a self-limited disease that does not result in chronic infection or chronic liver disease. The fecal-oral route primarily transmits HAV infection, by either person-to-person contact or through consumption of contaminated food or water.

Contaminated water in a flooded area may contain high amounts of raw sewage that has leached from septic systems and municipal wastewater treatment infrastructure.

Hepatitis A vaccination is the most effective measure to prevent HAV infection. Good decontamination procedures, as well as washing the hands, body, and equipment

after flood water exposure will assist in preventing hepatitis A virus infection.

Excessive Noise

In CISAR operations, excessive noise from equipment such as chain saws, backhoes, tractors, pavement breakers, blowers, and dryers may cause ringing in the ears and possible subsequent hearing damage.

A good rule of thumb: If you must shout over noise to be heard, you should use agency approved hearing protection.

Proper hearing protection should be worn when operating equipment that causes excessive noise.

Mosquito Repellants

Prevent mosquito bites by wearing long pants, socks, and long-sleeved shirts. Use of an insect repellants is highly recommended.

Regardless of what insect repellant used, if you start to get mosquito bites reapply the repellent according to the label instructions or if possible, remove yourself from the area with biting insects.

The Environmental Protection Agency (EPA) recommends the following precautions when using insect repellents:

- Apply repellents only to exposed skin and/or clothing (as directed on the product label.);

- Do not use repellents under clothing;

- Never use repellents over cuts, wounds, or irritated skin;

- Do not apply repellants to eyes or mouth and apply sparingly around ears;

- When using sprays, do not spray directly on the face, but spray on hands first and then apply to face;

- Do not allow children to handle the repellant, apply to your own hands first and then put on the child (You may not want to apply to children's hands);

- Use just enough repellent to cover exposed skin and/or clothing because heavy application and saturation are generally unnecessary for effectiveness (If biting insects do not respond to a thin film of repellent, apply more);

- After returning indoors, wash treated skin with soap and water or bathe, especially when repellents are used repeatedly in a day or on consecutive days;

- Wash treated clothing before wearing again (This precaution may vary with different repellents. Check the product label.); and

- If you get a rash or other bad reaction from an insect repellent, stop using the repellent, wash the repellent off with mild soap and water, and seek medical guidance (If you seek medical attention because of the repellent, take the repellent with you to show the doctor).

Hazardous Materials

Beware of hazardous materials. Wear protective clothing and gear when handling hazardous materials. Wash skin that may have come in contact with hazardous chemicals. Contact the IC if you are not sure about hazardous materials you may come in contact with, or unsure of.

In flooded areas, be aware that the water may bury or move hazardous chemical containers of solvents or other industrial chemicals from their normal storage places.

If any compressed gas cylinders (e.g., 20-lb. tanks from a gas grill, or household propane tanks) are discovered, do not attempt to move them by yourself. Compressed gas cylinders represent a very real danger of fire or explosion.

Unstable Buildings and Structures

CISAR responders will be searching homes for persons in distress or have injuries. Be aware that buildings may be unstable. Leave immediately if you hear shifting or unusual noises or see visual signs that signal a potential structure collapse.

Wash Your Hands

Finding running water to wash your hands may become difficult. However, keeping your hands clean helps avoid getting sick. It is best to wash your hands with soap and water for 20 seconds. However, when water is not available, use alcohol-based hand products made for washing hands (sanitizers).

Water Sanitation

In a CISAR environment, water may not be safe to drink, clean, or bathe in. During and after a disaster, water can become contaminated with bacteria, sewage, agricultural or industrial waste, chemicals, and other substances that can cause illness or death. The following information will help to ensure the water is safe to use:

- Use only bottled, boiled, or treated water for drinking, cooking or preparing food, washing dishes, cleaning, brushing your teeth, washing your hands, making ice, and bathing until your water supply is tested and found safe;

- If your water supply is limited, use an alcohol-based hand sanitizer for hand washing;

- If you use bottled water, be sure it came from a safe source (if unsure whether the water came from a safe source, boil or treat the water before use);

- Boiling water is the preferred way to kill harmful bacteria and parasites:

  o Bringing water to a rolling boil for 1 minute will kill most organisms;

- o Boiling will not remove chemical contaminants; and

- o If you suspect or are informed that water is contaminated with chemicals, seek another source of water, such as bottled water.

- If you can't boil water, treat water with chlorine tablets, iodine tablets, or unscented household chlorine bleach (5.25% sodium hypochlorite):

  - o If chlorine or iodine tablets are used, follow the directions that come with the tablets;

  - o If household chlorine bleach is used, add 1/8 teaspoon (~0.75 mL of bleach per gallon of water if the water is clear;

  - o For cloudy water, add 1/4 teaspoon (~1.50 mL) of bleach per gallon;

  - o Mix the solution thoroughly and let it stand for about 30 minutes before use; and

  - o Treating water with chlorine tablets, iodine tablets, or liquid bleach will not kill many parasitic organisms; boiling is the best way to kill these organisms.

- Use water storage tanks and other types of containers with caution (fire truck storage tanks and previously used cans or bottles may be contaminated with microbes or chemicals);

- Water containers should be thoroughly cleaned, then rinsed with a bleach solution before use as per the following:

  - o Mix soap and clean water in the container; shake or stir to clean inside of container, then rinse;

  - o For gallon- or liter-sized containers, add approximately 1 teaspoon (4.9 mL) household bleach (5.25%) with 1

cup (240 mL) water to make a bleach solution;

- o Cover the container and shake the bleach solution thoroughly, allowing it to contact all inside surfaces; and

- o Cover and let stand for 30 minutes, then rinse with clean water.

- Practice basic hygiene:

  - o Wash hands with soap and bottled water or warm water that has been boiled or disinfected;

  - o Wash hands before preparing food or eating, after toilet use, after participating in clean-up activities, and after handling articles contaminated with floodwater or sewage; and

  - o Use an alcohol-based hand sanitizer to wash hands if you have a limited supply of clean water.

Rubble and Debris

Whether natural or manmade, CISAR responders will encounter significant rubble and debris that may require digging through while looking for survivors.

Caution must be considered in shifting, unstable, or overhead rubble which can cause strains, sprains, or other injuries.

Rubble piles are particularly hazardous because there are no tie-off points to secure safety ropes and harnesses. This becomes problematic when looking for survivors in the midst of rubble and debris.

CISAR responders normally have limited heavy construction equipment for rubble and debris removal. This type of equipment can present hazards to personnel, as well as to the equipment operators who may have to work in very tight quarters, uneven spaces, and poor visibility.

Use teams of two or more people to move bulky objects. One person should avoid lifting any material that weighs more than 50 pounds.

Hypothermia

When exposed to cold temperatures, the body begins to lose heat faster than can be produced. The result is hypothermia, or abnormally low body temperature. Body temperature that is too low affects the brain, making the victim unable to think clearly or move well. Hypothermia is particularly dangerous because a person may not know when it occurs.

Hypothermia occurs most commonly at very cold environmental temperatures, but can occur even at cool temperatures (above 40°F) if a person becomes chilled from rain, sweat, or submersion in cold water. Hypothermia warning signs include the following:

Adults:

- Shivering/exhaustion;
- Confusion/fumbling hands;
- Memory loss/slurred speech; and
- Drowsiness.

Infants:

- Bright red, cold skin; and
- Very low energy.

Hypothermia: What To Do

If signs of hypothermia are noticed, take the person's temperature. If below 95°, the situation is considered a medical emergency. If medical care is not available, begin warming the person, as follows:

- Place the victim in a warm room or shelter;
- Remove wet clothing;
- Warm the center of the body first - chest, neck, head, and groin (If necessary, use skin-to-skin contact under loose, dry layers of blankets, clothing, towels, or sheets);

- Warm beverages can help increase the body temperature (Do not give alcoholic beverages and do not try to give beverages to an unconscious person);

- After the victim's body temperature has increased, keep the person dry and wrapped in a warm blanket, including the head and neck; and

- Get medical attention as soon as possible.

A person with severe hypothermia may be unconscious, may not seem to have a pulse, or to be breathing. In this case, handle the victim gently, and get emergency assistance immediately. Even if the victim appears dead, CPR should be provided. CPR should continue while the victim is being warmed, until the victim responds, or medical aid becomes available. Hypothermia victims who appear to be dead could possibly be resuscitated successfully.

Frostbite

Frostbite is an injury to the body caused by freezing and leads to a loss of feeling and color in affected areas. Frostbite most often affects the nose, ears, cheeks, chin, fingers, or toes. Frostbite can permanently damage the body; severe cases can lead to amputation.

At the first signs of redness or pain in any skin area, protect exposed skin and if possible, get out of the cold. Any of the following signs may indicate frostbite:

- White or grayish-yellow skin area;

- Skin that feels unusually firm or waxy; and

- Numbness.

The signs of frostbite are often not recognized by the individual. If symptoms are detected, seek medical care.

Because frostbite and hypothermia both result from exposure, first determine whether the victim also shows signs of hypothermia. Hypothermia is a more serious medical condition and requires emergency medical assistance.

Frostbite: What To Do

If there is frostbite but no sign of hypothermia and immediate medical care is not available, proceed as follows:

- Place victim in a warm area as soon as possible;

- Unless absolutely necessary, do not walk on frostbitten feet or toes (This may increase frostbite damage);

- Immerse the affected area in warm (not hot) water (the temperature should be comfortable to the touch for unaffected parts of the body), or warm the affected area using body heat (e.g., the heat of an armpit can be used to warm frostbitten fingers);

- Do not rub with snow or massage the frostbitten area (This can cause more damage); and

- Do not use a heating pad, heat lamp, or the heat of a stove, fireplace, or radiator for warming (Affected areas are numb and can be easily burned).

Heat

CISAR responders must be aware of the potential for heat problems. Even short periods of high temperatures can cause serious health problems. Depending on the type of incident, a high-heat environment can tax the stamina of CISAR responders, who may be suited up in heavy protective garments and may work for long periods without breaks or equipment changes.

Heat Stroke

Heat stroke occurs when the body is unable to regulate its temperature. The body's temperature rises rapidly, the sweating mechanism fails, and the body is unable to cool down. Body temperature may rise to 106°F or higher within 10 to 15 minutes. Heat stroke can cause death or permanent disability if emergency treatment is not provided. Heat stroke becomes even more of a problem when medical facilities are not readily available during CISAR operations.

Recognizing Heat Stroke

Warning signs of heat stroke vary, but may include the following:

- An extremely high body temperature (above 103°F, orally);

- Red, hot, and dry skin (no sweating);

- Rapid, strong pulse;

- Throbbing headache;

- Dizziness;

- Nausea;

- Confusion; and

- Unconsciousness.

Heat Stroke: What To Do

If the signs of heat stroke are observed, a life-threatening emergency may be occurring. Have someone call for immediate medical assistance, begin cooling the victim, and do the following:

- Get victim to a shady area;

- Cool the victim using whatever methods you can;

- Monitor body temperature, and continue cooling efforts until the body temperature drops to 101-102°F;

- If emergency medical personnel or a medical evacuation is delayed, call the IC for further instructions;

- Do not give the victim fluids to drink; and

- Get medical assistance as soon as possible.

Sometimes a victim's muscles may begin to twitch uncontrollably as a result of heat stroke. If this happens, keep the victim from injuring himself, but do not place any object in the mouth and do not give fluids. If there is vomiting, make sure the airway remains open by turning the victim on his or her side.

Heat Exhaustion

Heat exhaustion is a milder form of heat-related illness that can develop after several days of exposure to high temperatures and inadequate or unbalanced replacement of fluids. It is the body's response to an excessive loss of water and salt contained in sweat.

Recognizing Heat Exhaustion

Warning signs of heat exhaustion include the following:

- Heavy sweating;
- Paleness;
- Muscle cramps;
- Tiredness;
- Weakness;
- Dizziness;
- Headache;
- Nausea or vomiting; and
- Fainting.

The skin may be cool and moist. The victim's pulse rate will be fast and weak, and breathing will be fast and shallow. If heat exhaustion is untreated, it may progress to heat stroke. Seek medical attention immediately if any of the following occurs:

- Symptoms are severe; and
- The victim has heart problems or high blood pressure.

Heat Exhaustion: What To Do

Cooling measures that may be effective include the following:

- Provide cool, nonalcoholic beverages;
- Rest;
- Cool shower, bath, or sponge bath;
- An air-conditioned environment; and
- Lightweight clothing.

Heat Cramps

Heat cramps usually affect people who sweat a lot during strenuous activity. This sweating depletes the body's salt and moisture. The low salt level in the muscles may be the cause of heat cramps. Heat cramps may also be a symptom of heat exhaustion.

Recognizing Heat Cramps

Heat cramps are muscle pains or spasms – usually in the abdomen, arms, or legs – that may occur in association with heavy activity. If you have heart problems or are on a low-sodium diet, get medical attention for heat cramps.

Heat Cramps: What to do

If medical attention is not necessary, take these steps:

- Stop all activity, and sit quietly in a cool place;
- Drink clear juice or sports beverage;
- Do not return to strenuous activity for a few hours after the cramps subside, because further exertion may lead to heat exhaustion or heat stroke; and
- Seek medical attention for heat cramps if they do not subside in one hour.

Hydration

CISAR responders must ensure they receive plenty of fluids during rescue operations. Generally, personnel recover more quickly by carefully choosing fluids and liquids after exertion to replace fluids and replenish glycogen.

(Note: Personnel should avoid alcoholic beverages prior to intense activity and appropriate amount of rest/sleep should be obtained.)

Recovery Fluids

- Hydration is optimized by drinking to quench thirst and then drinking additional fluids (as a guide, personnel should consume 150% of ounces of weight/fluid lost during activity);

- Sports drinks and carbohydrate-containing beverages replenish glycogen in the muscles and improve rate of absorption of water and sodium;

- Full fluid replacement may take as long as 24 to 48 hours following strenuous activity;

- Cool water is quickly absorbed and an excellent choice for fluid replacement;

- All participants should consume a minimum of 0.5 liters of water per hour of activity (water with sports drink additive should be rotated every other bottle);

- Careful attention should be made to avoid over hydration (Decrease consumption when activity decreases); and

- Solid foods containing carbohydrates should be consumed at least one hour before activity (allows time for absorption and prevents nausea);

References

The information in this Section was obtained from the following sources:

- Centers for Disease Control and Prevention, *Worker Safety After a Flood*;

- Centers for Disease Control and Prevention, *Floods: Sanitation and Hygiene*;

- Centers for Disease Control and Prevention, *Updated Information regarding Mosquito Repellants Fact Sheet*;

- Centers for Disease Control and Prevention, *Key Facts about Hurricanes and Flood Recovery: Protect Your Health and Safety After a Hurricane or Flood*;

- Centers for Disease Control and Prevention, *Clean Hands Save Lives: Emergency Situations*;

- Centers for Disease Control and Prevention, *After a Flood*;

- Centers for Disease Control and Prevention, *Keep Food and Water Safe after a Natural Disaster or Power Outage*;

- Centers for Disease Control and Prevention, *Extreme Heat: A Prevention Guide to Promote Your Personal Health and Safety*;

- International Fire Service Training Association, *Fire Department Safety Officer*; and

- National Fire Protection Association, *NFPA 1521 Standard for Fire Department Safety Officer*.

This page intentionally left blank.

Section 3-3: Traumatic Stress Reactions

Introduction

CISAR operations can be challenging, interesting, rewarding, and a source of personal and professional growth. There may be a dark side, however, if the stress associated with a CISAR operation is prolonged, extreme, or not resolved within a reasonable amount of time (usually about three to four weeks) after the operation.

Stress is a normal and natural state of elevated arousal in response to a stimulus. The greater the stimulus, the stronger will be the response. A heightened level of arousal alters the way we think or feel and it can have significant effects on our bodies and behaviors. At low to moderate levels of arousal, stress reactions are not harmful and can help us function safely and effectively. In extreme stress, deterioration in performance and physical and mental health is often the result.

Catastrophic incidents exert extraordinary demands on emergency response personnel. CISAR responders may be exposed to traumatic operations that can involve the treatment of severely injured children or adults, gory sights and disturbing sounds, the handling of dead bodies or body parts, or even the injury or death of colleagues. The long hours, intensity of operational demands, ambiguous roles, and exposure to human suffering can adversely affect even the most experienced professionals.

CISAR responders and managers identifying or experiencing symptoms associated with stress should consult Agency specific guidance. In addition, the Incident Command should be contacted to ensure ESF #8 (Public Health and Medical Services) personnel are notified, and the information contained in this Section is reviewed.

Normal Reactions to Disasters

Traumatic stress reactions to disaster begin as normal responses. Operations personnel often feel sadness, anger, and grief. Often, the need for rest is denied and they resist leaving the scene. Unresolved stress reactions are the source of future problems.

Too often, the stress experienced by CISAR responders is addressed as an afterthought. This approach is a formula for failure of supervisors and personnel to adequately manage traumatic stress. It virtually guarantees unnecessary stress-related suffering. It threatens unit cohesion and performance among emergency services personnel.

Stress Management Program Common Characteristics

The evidence thus far indicates that the very best stress management programs have all of the following characteristics:

- Comprehensive;
- Integrative;
- Systematic; and
- Multi-component.

A *comprehensive program* has elements in place before a traumatic event occurs, provides support during an incident, and offers a variety of services after the situation over.

An *integrative program* has elements that are strategically blended together and interrelated. This approach assures that the right support services are in place for the right target populations, at the best possible times, and that the people most likely to be successful in assisting emergency personnel provide these services.

A *systematic program* provides support services in a logical, sequential order. Simple, easy to apply services, such as, resting and feeding personnel, are usually quite effective in the early stages of a crisis. More complex interventions may be necessary if the traumatic stress reaction does not resolve in a reasonable time.

A *multi-component program* is one that has many interventions or "tools" available to trained crisis intervention personnel to support people engaged in emergency operations.

Fortunately, steps can be taken to minimize the effects of traumatic stress before and after it occurs. Stress prevention and management comprises two critical elements: 1) *organization* and 2) *individual*. Adopting a preventive perspective allows both workers and organizations to anticipate stressors and shape responses, rather than simply reacting to a crisis when it occurs.

Organizational Approaches for Stress Prevention and Management

What follows are several practical steps that can be taken organizationally to minimize the effects of traumatic stress:

- Provide effective stress management structure and leadership. Elements include:
  - o Policies and procedures to provide staff support services for the organization's personnel before, during, and after traumatic events;
  - o Clear chain of command and reporting relationships;
  - o Available and accessible supervisors;
  - o Disaster training for all responders;
  - o Access to a specially trained staff support team, including peer support personnel, during operations;
  - o Incorporate a staff support team into the incident command structure; and
  - o Place a staff support officer within the command post to coordinate psychological support functions.
- Shifts of no longer than 12 hours, followed by 12 hours off. Typically a two-hour work period is followed by a

half hour down time. Intense cold or heat or the intensity of the scene itself may cause alterations in the deployment cycles;

- o Periodically orient work crews to time;
- o Briefings at the beginning of shifts as CISAR responders enter the operation;
- o Disengage unnecessary personnel;
- o Provide logistical support; and
- o Communication tools (e.g, cell phones, radios, etc.).

• Provide education and information on critical incidents, critical incident stress and the traumatic stress response. People who are informed become more stress, resistant and are better able to manage critical incident stress when it strikes;

• Establish a well-trained staff support team and make sure it is trained to provide a variety of services to individuals and groups under different circumstances;

• Mobilize staff support services early in an operation;

• Define clear operational purpose and goals for each unit;

• Define clear strategies and tactics appropriate to assignment setting;

• Define operational roles by function;

• Nurture team support of one another;

• Create a buddy system to support and monitor stress reactions. Promote a positive atmosphere of support and tolerance with frequent praise; and

• Develop a plan for stress management:

- o Assess CISAR responders' functioning regularly;

- o Rotate workers between low-, mid-, and high-stress tasks;
- o Encourage breaks and time away from assignment;
- o Educate personnel about signs and symptoms of worker stress and coping strategies;
- o Provide individual support;
- o Provide immediate small group support (defusing) as required;
- o Provide Critical Incident Stress Debriefing at appropriate times after the incident; and
- o Develop an exit plan for CISAR responders leaving the operation, including an operations debriefing, stress support if necessary, re-entry information, opportunity to critique, and formal recognition for service.

International Critical Incident Stress Foundation (ICISF)

The ICISF, a non-profit organization recognized by the United Nations, coordinates a network of over 1,000 specially trained teams to provide support to emergency personnel. Each team has several mental health professionals who are familiar with the psychological needs of emergency personnel. The most important element of the team, however, is the use of peer support personnel who connect most with the operations personnel. Teams function as partnerships between mental health professionals and firefighters, police officers, emergency medical, and SAR peers. ICISF operates a 24-hour emergency hotline that can be a good source of emergency consultations and the deployment of crisis support teams to a traumatic event. Service fees are not charged to emergency services organizations, although assistance with expenses is usually expected. The 24-hour emergency contact number is: (410) 313-2473.

Free, down loadable disaster-related resource materials are located on the ICISF website: www.icisf.org.

Criteria for Disaster Crisis Teams

Support teams deployed to assist in a disaster should fulfill at least the following minimum criteria:

- Between 40 and 70 hours (minimum) training in crisis intervention for individuals and groups;

- A mental health professional as the team's clinical director;

- Trained peer support personnel;

- Assessment skills;

- Strategic planning skills;

- Staff support skills for individuals;

- Staff support skills for large groups;

- Staff support skills for small groups;

- Capacity to support emergency services significant others; and

- Follow-up and referral skills.

Services Offered by Staff Support Teams

Staff support teams offer a wide range of services to emergency personnel. They help to build stress resistance ("stress immunization") and enhance resiliency ("bounce back" capacity) by means of pre-incident education and stress management preparation.

Staff support teams assess both the severity of the incident and the intensity of the traumatic stress reactions in the personnel. They then develop an appropriate strategy to provide the best support services. A strategic approach to traumatic stress management should include the five 'Ts':

Target: Who requires support?

Type: What types of interventions?

Timing: When is the best time to provide assistance?

Theme: What key issues, concerns or circumstances need to be considered before a stress team takes action?

Team: Who would be the best people to provide the support services?

During a CISAR operation, the staff support team provides:

- Consultations to command personnel;

- Information updates;

- Individual crisis intervention; and

- Rest, Information, and Transition Sessions (RITS).

Individual support continues after a situation ends. Group support sessions such as, Critical Incident Stress Debriefings, may be held according to the needs of the various groups who worked the disaster. The primary aim of group sessions is to restore *group cohesion* and *unit performance*.

Significant other support and a variety of follow-up contacts are essential services offered by staff support teams. Phone calls, visits, and home contacts are some of the ways staff support teams provide follow-up. Occasionally, someone is seriously impacted by a traumatic experience and a referral to a mental health professional is required to assist in the recovery process.

Table 3-3-1 on the next page provides additional information on commonly used crisis intervention tactics.

Table 3-3-1: Summary of Commonly Used Crisis Intervention Tactics

INTERVENTION	TIMING	TARGET GROUP	POTENTIAL GOALS
Pre-event Planning/Preparation	Pre-event	Anticipated target/victim population	Anticipatory guidance, foster resistance, resilience
Surveillance and Assessment	Pre-intervention	Those directly and indirectly affected	Determine need for intervention(s)
Strategic planning	Both pre-event and during event; Also some after the event	Anticipated exposed and victim populations	Improve overall crisis response
Individual crisis intervention (including "psychological first aid") and SAFER-R	As needed	Individuals as needed	Assessment, screening, education normalization, reduction of acute distress, triage, and facilitation of continued support
Large group crisis intervention: Rest, Information, and Transition Session (RITS) {formerly known as "demobilization"}	Shift disengagement, end of deployment	Emergency personnel, large groups	Decompression, ease transition from intense to less intense work, screening, triage, education, and meet basis needs
*Respite center	Ongoing, large-scale events As needed	Usually Emergency Personnel	Respite, refreshment, screening, triage, and support
*Crisis Management Briefing (CMB); provide large group "psychological first aid"	On-going and post-event; may be repeated as needed	Heterogeneous large groups	Inform, control rumors, increase cohesion
Small group crisis intervention *Small Group Crisis Management Briefing (sCMB)	Ongoing events and post-events	Small groups seeking information and resources	Information, control rumors, reduce acute distress, increase cohesion, facilitate resilience, screening, and triage
*Immediate Small Group Support (ISGS) (also known as defusing), and a form of small group "psychological first aid"	(12 hours or less) Post-event:	Small homogeneous groups	Stabilization, ventilation, reduce acute distress, screening, information.
*Group debriefing (Powerful Event Group Support (PEGS) {also known as Critical Incident Stress Debriefing (CISD)}	*1-10 days for acute incidents; may be 3-4 weeks or even longer if group is in post-disaster recovery phases	Small homogeneous groups with equal trauma exposure (e.g., workgroups, emergency service and military)	Increase cohesion, and facilitate resilience. Increase cohesion, ventilation, information, normalization, reduce acute distress, facilitate resilience, screening and triage; Follow up is essential.
Family Crisis Intervention	Pre-event; as needed	Families	Wide range of interventions e.g., pre-event preparation, individual crisis intervention, CMB, PEGS (CISD), or other group process
Organization/Community Intervention, Consultation	Pre-event; as needed	Organizations affected by trauma or disaster	Improve organizational preparedness and response; leadership consultation
Pastoral Crisis Intervention	As needed	Individuals, small groups, large groups, congregations, and communities who desire faith-based presence/crisis intervention	Faith-based support
Follow-up and/or referral; facilitate access to continued care	As needed	Intervention recipients and exposed individuals	Assure continuity of care

G.S. Everly Jr. and J.T. Mitchell, *Integrative Crisis Intervention* (Ellicott City: Chevron Publishing Corporation, 2008). Adapted with permission.

Individual Approaches for Stress Prevention and Management

Managing personal stress will help each CISAR responder stay focused on hazards at the site and to maintain the constant vigilance required for personal safety. Often CISAR responders do not recognize the need to take time for themselves and to monitor their own emotional and physical health – especially when recovery efforts stretch into several weeks. The following simple guidelines are provided to support CISAR responders:

- Be calm and pace yourself. Think before reacting. Rescue and recovery efforts at the site may continue for days or weeks;

- Make work rotations from high stress to lower stress functions;

- Limit exposure to gory sights and sounds. Handling human remains is highly stressful for most people;

- Take frequent rest breaks. Rescue and recovery operations take place in extremely dangerous work environments. Mental fatigue over long shifts can place CISAR responders at greatly increased risk for injury. Breaks help to keep you alert;

- Four hours sleep in a 24-hour period is a minimum for health;

- Maintain a positive mental attitude;

- Maintain a positive sense of humor;

- Do not over control your emotions;

- Watch out for each other. Co-workers may be intently focused on a particular task and may not notice a hazard nearby or behind;

- Be aware of those around you. CISAR responders who are exhausted, feeling stressed or even temporarily distracted may place themselves and others at risk;

- Maintain as normal a schedule as possible. Regular eating and sleeping are crucial. Adhere to the team schedule and rotation;

- Eat nutritious foods. Avoid too much sugar, foods high in fat content, processed foods, and white bread;

- Try to eat a variety of foods and increase your intake of complex carbohydrates (e.g., breads and muffins made with whole grain, granola bars);

- Make sure plenty of fluids such as water and juices are consumed. Limit caffeine intake. Avoid alcohol use for several days after disaster work. It reduces rapid eye movement sleep patterns and which is not healthy;

- Whenever possible, take breaks away from the work area. Eat and drink in the cleanest area available;

- Recognize and accept what you cannot change (e.g., chain of command, organizational structure, waiting, equipment failures, etc.);

- Talk to people when you feel like it. You decide when you want to discuss your experience. Talking about an event may be uncomfortable. Choose you own comfort level. Listen to your colleagues;

- Use counseling assistance programs available through your agency;

- Recurring thoughts, dreams, or vivid, disturbing memories are normal. They will generally decrease over time. Get help if they become extreme, repetitive, or do not decrease over a month's time.

- Communicate with your loved ones at home as frequently as possible. Express your feelings to people you trust;

- If you are fit, physical exercise helps to reduce the chemicals of distress in a person's body. Even walking helps;

- Attend group support services when offered by a trained support team. They can "take the edge off" of a bad incident;

- Be careful with jokes about the incident. Others may be sensitive;

- Do not engage in unproductive criticism of others. Operational mistakes can be corrected later;

- Anger is a frequent emotion after disaster work. Don't take it personally. Anger should subside in a reasonable amount of time;

- Focus on the here and now. Telling old war stories of events that were worse than the current event is not always helpful;

- Listen to those who want to talk about their experience;

- Shedding tears after a painful event is perfectly normal. But, frequent uncontrolled crying spells accompanied by sleep disturbance and an inability to return to normal duties is an indication that a person needs additional assistance; and

- Extreme or prolonged (beyond a month) stress reactions may need professional assistance to reduce their impact. The good news is that several psychological therapies have a proven track record in facilitating the recovery of traumatized people. The earlier the therapy is begun, the better the results.

References

The information in this Section was obtained from the following sources: Centers for Disease Control, National Institute for Occupational Safety and Health, *Traumatic Incident Stress: Information for Emergency Response Workers*, DHHS (NIOSH) Publication Number 2002-107.

- Everly, G.S., Jr. and Mitchell, J.T. *Integrative Crisis Intervention and Disaster Mental Health.* Ellicott City, Chevron Publishing Corporation, 2008.

- Mitchell, J.T. *Psychological and Stress Problems in SAR Management.* In R.C. Stoffel. *The Textbook for Managing Land Search Operations.* Cashmere, Emergency Response International, Incorporated, 2006.

- Mitchell, J. T. *Group Crisis Support: Why it works, When and how to provide it.* Ellicott City: Chevron Publishing, 2007.

- Mitchell, J.T. Sustaining Staff. In D. Polk and J.T. Mitchell. *Prehospital Behavioral Emergencies and Crisis Response.* Boston: Jones and Bartlett, 2009.

- U.S. Department of Health and Human Services. *Emergency Mental Health and Traumatic Stress: Tips for Emergency and Disaster Response Workers.*

- U.S. Department of Health and Human Services. *Tips for Managing and Preventing Stress: A Guide for Emergency and Disaster Response Workers.*

Section 3-4: Persons with Special Needs

Vulnerability and Risks

Assistance

Tips for CISAR Responders

Special Response Teams (SRTs)

Vulnerability and Risks

Survivors sometimes have special needs or pose special risks.

A person at home alone is often a potential "victim of circumstance," especially if they are elderly or disabled. Distress situations can be worsened by an injurious fall, heart attack, withdrawal symptoms, robbery or bodily attack, or fear of death. Such people may be in a serious predicament and may be forced to just "lay there" until someone happens to show-up to check on or discover them. Health problems can be aggravated by stress or environmental conditions. An injured person who is alone may not be able to call for help and therefore may be among the last persons found during search operations.

Other special needs persons may need to be rescued from facilities such as hospitals, nursing homes or schools.

Prisoners may pose special risks and require continuity of custody.

Assistance

CISAR responders should be diligent to look for persons who may be unable to help themselves or even to call for help. When people with special needs are located, they should be carefully assessed to ensure their situation is not worsened.

CISAR responders should:

- Be sensitive and reassuring;

- Handle person(s) with special needs with extra care;

- Be alert to taking important medicines or service animals (see *Section 3-6: Animals*);

- Alert lily pad or place-of-safety personnel of special medical or other immediate needs of the person(s) rescued; and

- Help the person contact a loved one or care provider who may be able to provide appropriate supplemental information or assistance if necessary.

Special skills or equipment may be needed to assist some special needs persons, and large numbers of occupants of institutions such as hospitals or nursing homes may require special mass rescue assistance.

Tips for CISAR Responders

Tables 3-4-1 through 3-4-10 provide tips for CISAR responders concerning various special needs groups that may be encountered during CISAR operation. The information was obtained from Tips for First Responders (4th Edition), published by the Center for Development and Disability (CDD) and available on their website: http://cdd.unm.edu/dhpd/images/Fourth%20 Edition.pdf.

Table 3-4-1: Seniors

Always ask the person how you can best assist them.

Some elderly persons may respond more slowly to a crisis and may not fully understand the extent of the emergency. Repeat questions and answers if necessary. Be patient! Taking time to listen carefully or to explain again may take less time than dealing with a confused person who may be less willing to cooperate.

Reassure the person that they will receive medical assistance without fear of being placed in a nursing home.

Older people may fear being removed from their homes – be sympathetic and understanding and explain that this relocation is temporary.

Before moving an elderly person, assess their ability to see and hear; adapt rescue techniques for sensory impairments.

Seniors with a hearing loss may appear disoriented and confused when all that is really "wrong" is that they can't hear you. Determine if the person has a hearing aid. If they do, is it available and working? If it isn't, can you get a new battery to make it work?

(Refer to *3-4-6: People Who are Deaf or Hard of Hearing* for more information.)

If the person has vision loss, identify yourself and explain why you are there. Let the person hold your arm and then guide them to safety.

(Refer to *3-4-7: People Who Are Blind or Visually Impaired* for more information.)

If possible, gather all medications before evacuating. Ask the person what medications they are taking and where their medications are stored. Most people keep all their medications in one location in their homes.

If the person has dementia, turn off emergency lights and sirens if possible. Identify yourself and explain why you are there. Speak slowly, using short words in a calm voice. Ask "yes" or "no" questions: repeat them if necessary. Maintain eye contact.

Table 3-4-2: People with Service Animals

Traditionally, the term "service animal" referred to seeing-eye dogs. However, today there are many other types of service animals.

Remember: a service animal is not a pet.

Do not touch or give the animal food or treats without the permission of the owner.

When a dog is wearing its harness, it is on duty. In the event you are asked to take the dog while assisting the individual, hold the leash and not the harness.

Plan to evacuate the animal with the owner. Do not separate them!

Service animals are not registered and there is no proof that the animal is a service animal. If the person tells you it is a service animal, treat it as such. However, if the animal is out of control or presents a threat to the individual or others, remove it from the site.

A person is not required to give you proof of a disability that required a service animal. You should accept the claim and treat the animal as a service animal. If you have doubts, wait until you arrive at your destination and address the issue with the supervisor in charge.

The animal need not be specially trained as a service animal. People with psychiatric and emotional disabilities may have a companion animal, which is just as important to them as a service animal is to a person with a physical disability – please be understanding and treat the animal as a service animal.

A service animal must be in a harness or on a leash, but need not be muzzled.

Table 3-4-3: People with Autism

Communication

Speak calmly – use direct, concrete phrases with no more than one or two steps, or write brief instructions on a pad if the person can read.

Allow extra time for the person to respond.

The person may repeat what you have said, repeat the same phrase over and over, talk about topics unrelated to the situation, or have an unusual or monotone voice. This is their attempt to communicate, and is not meant to irritate you or be disrespectful.

Avoid using phrases that have more than one meaning such as "spread eagle" "knock it off" or "cut it out."

Visually check to see if there is a wrist or arm tattoo or bracelet that identifies the person as having an autism spectrum disorder.

Some people with autism don't show indications of pain – check for injuries.

Social

Approach the person in a calm manner. Try not to appear threatening.

The person may not understand typical social rules, so may be dressed oddly, invade your space, prefer to be farther away from you than typical, or not make eye contact. It's best not to try and point out or change these behaviors unless it's absolutely necessary.

The person may also look at you at an odd angle, laugh or giggle inappropriately, or not seem to take the situation seriously. Do not interpret these behaviors as deceit or disrespectful.

Because of the lack of social understanding, persons with autism spectrum disorders may display behaviors that are misinterpreted as evidence of drug abuse or psychosis, defiance or belligerence. Don't assume!

Sensory and behavior

If possible, turn off sirens, lights, and remove canine partners. Attempt to find a quiet location for the person, especially if you need to talk with them.

Avoid touching the person, and if necessary, gesture or slowly guide the person.

Table 3-4-3: (continued)

If the person is showing obsessive or repetitive behaviors, or is fixated on a topic or object, try to avoid stopping these behaviors or taking the object away from them, unless there is risk to self or others.

Make sure that the person is away from potential hazards or dangers (busy streets, etc.) since they may not have a fear of danger.

Be alert to the possibility of outbursts or impulsive, unexplained behavior. If the person is not harming themselves or others, wait until these behaviors subside.

Table 3-4-4: People who are Mentally Ill

You may not be able to tell if a person is mentally ill until you have begun the evacuation procedure.

If a person begins to exhibit behavior, ask if they have any mental health issues of which you need to be aware. However, be aware that they may or may not tell you. If you suspect someone has a mental health issue, use the following tips to help you through the situation.

In an emergency, the person may become confused. Speak slowly and in a normal, calm speaking tone.

If the person becomes agitated, help them find a quiet corner away from the confusion.

Keep your communication simple, clear, and brief.

If they are confused, don't give multiple commands – ask or state one thing at a time.

Be empathetic – show that you have heard them and care about what they told you. Be reassuring.

If the person is delusional, don't argue with them or try to "talk them out of it." Just let them know you are there to help them.

Ask if there is any medication they should take with them.

Try to avoid interrupting a person who might be disoriented or rambling – just let them know that you have to move quickly.

Don't talk down to them, yell, or shout.

Have a forward leaning body position – this shows interest and concern.

Table 3-4-5: People with Mobility Impairments

Always ask the person how you can help before beginning any assistance. Even though it may be important to evacuate quickly, respect their independence to the extent possible. Don't make assumptions about the person's abilities.

Ask if they have limitations or problems that may affect their safety.

Some people may need assistance getting out of bed or out of a chair, but CAN then proceed without assistance. Ask!

Here are some other questions you may find helpful:

- "Are you able to stand or walk without the help of a mobility device like a cane, walker, or a wheel chair?"
- "You might have to [stand] [walk] for quite awhile on your own. Will this be ok? Please be sure and tell someone if you think you need assistance."
- "Do you have full use of your arms?"

When carrying the person, avoid putting pressure on his or her arms, legs or chest. This may result in spasms, pain, and may even interfere with their ability to breathe.

Avoid the "fireman's carry." Use the one or two person carry techniques.

Crutches, Canes, or other Mobility Devices

A person using a mobility device may be able to negotiate stairs independently. One hand is used to grasp the handrail while the other hand is used for the crutch or cane. Do not interfere with the person's movement unless asked to do so, or the nature of the emergency is such that absolute speed is the primary concern. If this is the case, tell the person what you'll need to do and why.

Ask if you can help by offering to carry the extra crutch.

If the stairs are crowded, act as a buffer and run interference for the person.

Evacuating Wheelchair Users

If the conversation will take more than a few minutes, sit or kneel to speak to the person at eye level.

Wheelchair users are trained in special techniques to transfer from one chair to another. Depending on their upper body strength, they may be able to do much of the work themselves.

Before you assume you need to help, or what that help should be, ask the person what help they need.

Table 3-4-5: (continued)

Carry Techniques for Non-Motorized Wheelchairs

The in-chair carry is the most desirable technique to use, if possible.

One-person assist

- Grasp the pushing grips, if available.
- Stand one step above and behind the wheelchair.
- Tilt the wheelchair backward until a balance (fulcrum is achieved).
- Keep your center of gravity low.
- Descend frontward.
- Let the back wheels gradually lower to the next step.

Two-person assist

- Position the second rescuer: stand in front of the wheelchair and face the wheelchair.
- Stand one, two, or three steps down (depending on the height of the other rescuer).
- Grasp the frame of the wheelchair.
- Push into the wheelchair.
- Descend the stairs backwards.

Motorized Wheelchairs

Motorized wheelchairs may weigh over 100 lbs unoccupied, and may be longer than manual wheelchairs. Lifting a motorized wheelchair and user up or down stairs requires two to four people.

People in motorized wheelchairs probably know their equipment much better than you do! Before lifting, ask about heavy chair parts that can be temporarily detached, how you should position yourselves, where you should grab hold, and what, if any, angle to tip the chair backward.

Turn the wheelchair's power off before lifting it.

Most people who use motorized wheelchairs have limited arm and hand motion. Ask if they have any special requirements for being transported down the stairs.

Table 3-4-6: People who are Deaf or Hard of Hearing
There is a difference between hard of hearing and deaf. People who are hearing impaired vary in the extent of hearing loss they experience. Some are completely deaf, while others can hear normally with hearing aids.
Hearing aids do not guarantee that the person can hear and understand speech. They increase volume, not necessarily clarity.
If possible, flick the lights when entering an area or room to get their attention.
Establish eye contact with the individual, not with the interpreter, if one is present.
Use facial expressions and hand gestures as visual cues.
Check to see if you have been understood and repeat if necessary.
Offer pencil and paper. Write slowly and let the individual read as you write.
Written communication may be especially important if you are unable to understand the person's speech.
Speak slowly and clearly, but do not over-enunciate.
Do not block your mouth with your hands or an object when speaking.
Do not allow others to interrupt you while conveying the emergency information.
Be patient – the person may have difficulty understanding the urgency of your message.
Provide the person with a flashlight to signal their location in the event they are separated from the rescue team. This will facilitate lip-reading or signing in the dark.
While written communication should work for many people who are deaf. Keep instructions simple, in the present tense and use basic vocabulary. American Sign Language (ASL) is its own syntax and grammar. Native ASL users may read and write English as a second language.

Table 3-4-7: People who are Visually Impaired
There is a difference between visual impairment and blindness. Some people who are "legally blind" have some sight, while others are totally blind.
Announce your presence, speak out, and then enter the area.
Speak naturally and directly to the individual.
Do not shout.
Don't be afraid to use words like "see," "look," or "blind."
State the nature of the emergency and offer them your arm. As you walk, advise them of any obstacles.
Offer assistance but let the person explain what help is needed.
Do not grab or attempt to guide them without first asking them.
Let the person grasp your arm or shoulder lightly for guidance.
They may choose to walk slightly behind you to gauge your body's reactions to obstacles.
Be sure to mention stairs, doorways, narrow passages, ramps, etc., before you come to them.
When guiding someone to a seat, place the person's hand on the back of the chair.
If leading several individuals with visual impairments, ask them to guide the person behind them.
Remember that you'll need to communicate any written information orally.
When you have reached safety, orient the person to the location and ask if any further assistance is needed.
If the person has a service animal, don't pet it unless the person says it is ok to do so. Service animals must be evacuated with the person. (Refer to *Table 3-4-2: People with Service Animals* for more information.)

Table 3-4-8: People with Cognitive Disabilities

Say:

- My name is... I'm here to help you, not hurt you.
- I am a... *(name your job)*.
- I am here because... *(explain the situation)*.
- I look different than my picture on my badge because... *(for example, if you are wearing protective equipment)*.

Show:

- Your picture identification badge *(as you say the above)*.
- That you are calm and competent.

Give:

- Extra time for the person to process what you are saying and to respond.
- Respect for the dignity of the person as an equal and as an adult *(example: speak directly to the person)*.
- An arm to the person to hold as they walk. If needed, offer your elbow for balance.
- If possible, quiet time to rest *(as possible, to lower stress and fatigue)*.

Use:

- Short sentences.
- Simple, concrete words.
- Accurate, honest information.
- Pictures and objects to illustrate your words. Point to your ID picture as you say who you are, point to any protective equipment as you speak about it.

Predict:

- What will happen *(simply and concretely)*?
- When events will happen *(Tie to common events in addition to numbers and time. For example, "By lunch time..." "By the time the sun goes down...")*.
- How long this will last – when things will return to normal *(if you know)*.
- When the person can contact or rejoin loved ones *(For example: calls to family, re-uniting pets.)*.

Table 3-4-8: (continued)

Ask for/Look for:

- An identification bracelet with special health information.
- Essential equipment and supplies *(For example: wheelchair, walker, orxygen, batteries, communication devices [head pointers, alphabet boards, speech synthesizers, etc.])*.
- Medication.
- Mobility aids *(For example, assistance or service animal)*.
- Special health instructions *(for example: allergies)*.
- Special communication information *(For example, is the person using sign language?)*.
- Contact information.
- Signs of stress and/or confusion *(For example, the person might say he or she is stressed, looked confused, withdraw or start rubbing their hands together)*.
- Conditions that people might misinterpret *(For example, someone might mistake Cerebral Palsy for drunkenness)*.

Repeat:

- Reassurances *(For example, "You may feel afraid. That's ok. We're safe now.")*.
- Encouragement *(For example: "Thanks for moving fast. You are doing great. Other people can look at you and know what to do.")*.
- Frequent updates on what's happening and what will happen next. Refer to what you predicted will happen *(For example, "Just like I said before, we're getting into my car now. We'll go to... now.")*.

Reduce: Distractions (For example: lower volume of radio, use flashing lights on vehicle only when necessary.).

Explain:

- Any written material (including signs) ini everyday language.
- Public address system announcements in simple language.

Share: The information you've learned about the person with other workers who'll be assisting the person.

Table 3-4-9: Childbearing Women and Newborns	Table 3-4-9: (continued)
Tips for childbearing women	If you must transport a pregnant woman, regardless of whether she is in labor or not: • Transport her lying on her side, not flat on her back. • Ask her if she has a copy of her pregnancy/prenatal records; if she does, make sure they are brought with her.
Usually, pregnancy is not an emergency. In fact, if the pregnant women is otherwise healthy, it's likely that she can be included in any plans for evacuation or sheltering for the general population.	**Tips for just after a baby has been born**
However, if the women has had a cesarean section ("C-Section") at any time in the past, or if she has any of the following problems now or in the previous three hours, she is at a higher risk. • Steady bleeding "like a period" from the vaginia. • Convulsion or a really bad (unusual) headache that will no go away with Tylenol. • Constant strong belly or back pain with hardness in her pregnant belly. • Strong pains and hardening belly that comes and goes every couple of minutes and a "due date" three weeks away or more.	Dry and rub the baby gently to keep baby warm and to stimulate breathing.
	Place the naked baby on mother's skin between her breasts and cover both mom and baby.
	Cutting the cord is not an emergency. The cord should only be cut when you have sterile tools (scissor, knife blade, etc.). It's better to wait rather than cut the cord with a non-sterile blade.
	Usually, the placenta (afterbirth) will follow the baby on its own in about a half an hour or less. After it comes, it can be put in a plastic bag, wrapped with the baby or left behind, depending on the circumstances.
If she has had any of these problems, she should be taken to a hospital (if hospital access is available) or other health care facility for an assessment. If taking her to a facility is not possible, she should be helped to find a comfortable position and not be left alone.	Monitor bleeding from the vagina. Some bleeding is normal – like a heavy period. If should slow down to a trickle in 5 to 10 minutes. If it doesn't, the woman needs medical care.
If she has not had these problems, the hospital is often not the best place to take pregnant women, women in labor, or new mothers with newborns due to danger from infections or other exposures. Remember: a normal birth is not an illness.	Encourage mom to put baby to breast. The baby's hands should be free to help find the breast. Point baby's nose toward mom's nipple and the baby's tummy toward mom's.
A woman who has one or more of the symptoms below may be in labor and about to give birth. Do not move her – it is better to have a birth where you are than on the way to somewhere else. • Making grunting sounds every one to three minutes. • She says, "yes," if you ask, "Is the baby pushing down?" or she says, "The baby's coming." • You see bulging out around the vagina when she grunts or bears down.	If you need to transport a mother and her newly born baby: • Keep the mom and newborn together: baby in mom's arms or on her belly. • Take diapers, baby clothes and formula and bottles (if mom is bottle feeding the baby) if they are available.
Give pregnant women and new moms lots of fluids to drink (water or juice is best).	
Be as calming as possible; expectant mothers may be especially anxious in emergency situations. Reassure them you will do everything you can for them.	
Try not to separate expectant or newly delivered moms and their family, even if transporting.	

Table 3-4-10: People with Multiple Chemical Sensitivities

Reassure the person that you understand he or she is chemically sensitive and will work with him or her in providing care. Be sure to ask what the person is sensitive to, including his or her history of reactions to various drugs you may have to administer.

Flag the person's chart or other written information that he or she is chemically sensitive.

Whenever possible, take the person's own medical supplies and equipment with them, including oxygen mask and tubing, medications, food and water, bedding, clothing and soap – he or she may be sensitive to these items if issued at a shelter or hospital.

If you do administer drugs:

- Administer low doses with caution.
- Use IV fluid bottle in glass without dextrose if possible – many people react to corn-based dextrose.
- Capsules are generally better than tablets – they have fewer binders, fillers and dyes.
- If administering anesthesia, use short-acting regional rather than general anesthesia whenever possible and try to avoid the use of halogenated gas anesthetics.

If the person is taken to an emergency shelter or a hospital, help protect him or her from air pollution. Some suggestions:

- Avoid placing the person in rooms with recent pesticide sprays, strong scented disinfectants or cleaners, new paint or carpet, or other remodeling.
- Place a sign on the door stating that the person inside has chemical sensitivities.
- Assign caregivers who are not wearing perfume or fabric softener on clothes and who are not smokers.
- Allow the person to wear a mask or respirator, use an air filter, or open a window as needed.
- Keep the door to the person's room closed, if possible.
- Reduce time the person spends in other parts of the hospital, if possible, by performing as many procedures and evaluations as possible in his or her room.

Table 3-4-11: People with Seizure Disorders

Some types of seizures have warning symptoms while others do not. Warning symptoms may include visual or auditory hallucinations, or the person notices a burning smell. If the person senses an upcoming seizure, suggest they lie down and provide help if asked.

Stay calm – talk with the person softly, and rub the person's arm or back gently.

If possible, look at a watch or a clock to time the duration of the seizure. After the seizure is over, give this information to the person. If the seizure lasts more than five minutes or the person does not resume consciousness, call 911.

Attempt to turn the individual on her/his side; preferably the left side to allow saliva or other substances to drain from the mouth and keep the airway open.

Move any nearby objects away from the person that could lead to injury if the person hits the object, or see if the person can be moved if they are near hard objects too heavy to move. You may place a pillow, towel, coat, or other soft object underneath the person's head to protect it.

Special Response Teams (SRTs)

SRTs may be used to support evacuations of large or small numbers of persons with special needs (usually early in the CISAR operation), or persons who require special handling, such as prisoners. In a notice event (e.g., hurricane landfall), State, Tribal, Territorial, or local emergency managers should have already identified and prioritized institutions and locations for which SRTs will be required.

Rescue of persons from institutions such as hospitals or nursing homes should be coordinated as closely as possible with responsible local medical and care providers and planners. Many of these persons may need to be rescued in wheelchairs or even in prone positions, and may have life support equipment that must also be evacuated.

SRTs may also be needed to respond to enclaves of non-English speaking survivors.

For impending distress situations (e.g. hurricanes) where there is advance notice and reasonable certainty that a facility will need to be evacuated, partial or whole pre-event evacuations may be prudent.

Assigned resources should report conditions and resource needs to the IC as soon as that information is available.

Persons rescued from facilities such as nursing homes, hospitals, or prisons may need to be transported to other comparable facilities either directly, or after they arrive at a lily pad, or place of safety.

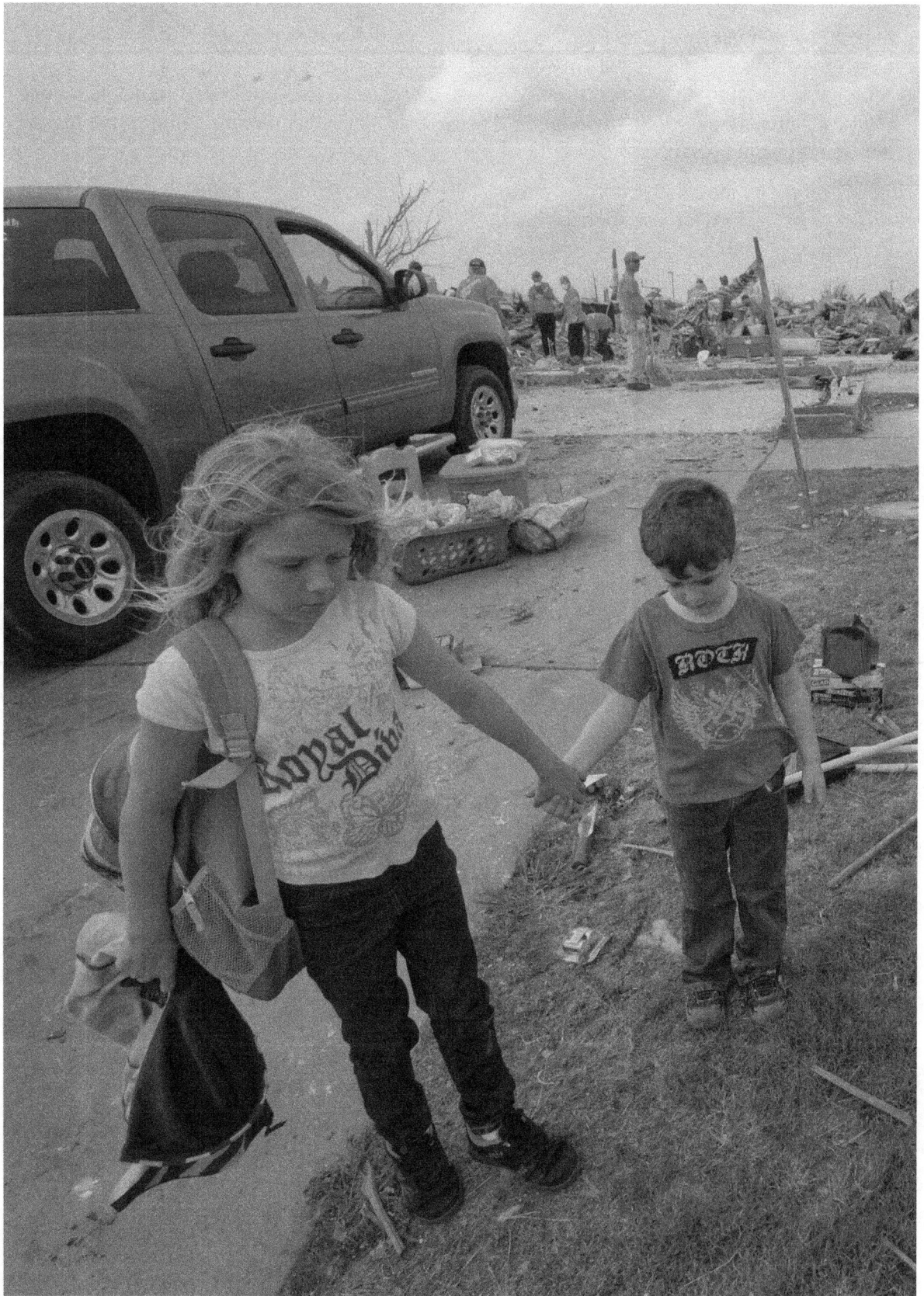

Section 3-5: CISAR and Children

Vulnerability

Initiate a Continuum of Care

General Guidance

Child Tracking

Considerations for Unaccompanied Minors

Promoting Resilience

National Emergency Child Locator Center (NECLC)

Additional Resources

Vulnerability

Multiple issues affecting children emerge during a disaster, including needs for evacuation and shelter, pediatric medical treatment, mental health services, family reunification, emergency child care, legal services, restoration of child care, school, and recreational services/infrastructures.

During Hurricane Katrina the nation witnessed how little attention our disaster response infrastructure had paid to understanding the needs of children, much less to developing resources and capabilities to meet those needs.

Among the most critical problems arising in the immediate aftermath of Hurricane

Katrina was the separation of over 5,000 children from their parents or guardians in the chaos of the evacuations and rescue procedures. For both legal and humanitarian reasons, these unaccompanied minors posed a special problem for local, State, and Federal Governments and response agencies, and it was six months before all these children were reunited with their families.

In order to promote positive outcomes, CISAR planning and response should include attention to three fundamental obligations concerning children (Table 3-5-1 below):

Table 3-5-1: Children in a Catastrophic Incident	
Recognize and Plan for Children's Distinct Needs	Children have unique medical, psychosocial, physical, and judicial needs in disaster situations, many of which differ from the needs of adults and require special protocols, procedures, and tools.
Foster Resiliency	Emergency rescue and disaster relief responders should intentionally develop resources and response capabilities based on current understandings of how to foster resiliency under various crisis conditions encountered in disasters.
Provide a Continuum of Care	The ultimate goal of disaster relief agencies, whether private or governmental, should be to "hold safe" the children in their care until families, community agencies, child care centers, schools, churches, and other local primary caregivers for children can resume their responsibilities of care after a disaster. The "continuum of care" reflects the distinct needs of children from the immediate first-response relief through the rebuilding of normal lives and living conditions.
(Reference: Lenore T. Ealy and Paige T. Ellison-Smith, "To Hold Safe: Framing a New Era of Disaster Child Care," Project K.I.D. White Paper (Carmel, IN: Project K.I.D., 2007). Available online at www.project-kid.org.)	

Initiate a Continuum of Care

CISAR responders are the first link in the "Continuum of Care" for children who are victims of disasters.

CISAR typically takes place in the immediate hours and days after a disaster and requires attention to provision of Emergency Child Care.

CISAR responders should be trained in the essentials of addressing child trauma, and emergency medical response should have access to pediatric personnel, dosages, and equipment.

Of particular concern at this stage are procedures for preventing family separation and for addressing the needs of unaccompanied minors. These procedures are important during all SAR activities, but can be particularly important during a catastrophic incident when many children are affected.

Table 3-5-2 explains the Continuum of Care that could be implemented during a catastrophic incident.

Table 3-5-2: Disaster Child Care Continuum

Phase I "Emergency Care"	Phase II "Respite Care"	Phase III "Temporary Care"	Phase IV "Long-Term Recovery"
		Infrastructure Recovery	
Emergency/first response. Immediately post-disaster until situation stabilizes	As situation stabilizes until need no longer exists	Temporary care begins as families stabilize and return to work routines and continues until full recovery is complete Damage surveys and rebuilding begin.	Ongoing infrastructure recovery will also attend to supporting ongoing medical, mental health and psychosocial needs of children as they return to school and child care environments.
Key Need: Engagement with calm, competent adults and support of protective factors.	**Key Need:** Normalization through engaging activities and support for children's primary caregivers.	**Key Need:** Restoration of normal interpersonal bonds and routines in both family and community.	**Key Need:** Reframing of traumatic experience and subsequent events and construction of new reality.
Key Goal: Address the emergency medical, evacuation, and legal protection needs of children. May include sheltering unaccompanied minors.	**Key Goal:** Provide a short respite from childcare for parents and child-centered stress relief for children.	**Key Goal:** Normalize and restore children to stable childcare environments as quickly as possible, enabling parents to return to work and focus on restoring normalcy to family and home.	**Key Goal:** Restore full permanent childcare and schooling capacity and support ongoing psychosocial and mental health recovery.
Key Actions: Disaster child care can support first responders, local law enforcement, and families by providing emergency "safe havens" where children may be kept safe and cared for until they can be returned to parents or provided other custodial care. Medical care, adjudication issues, and respite activities can be provided.	**Key Actions:** May include activating child care services for first responders, as well as establishing drop-in play areas in critical areas to help children address emerging psychosocial, nutritional, and possible medical and legal needs. May include ongoing shelter of unaccompanied minors.	**Key Actions:** Disaster child care can support local churches, schools, shelters, businesses, and other sites that might provide space for temporary child care and educational facilities (mobile or modular units may be required). Help pre-existing providers with minimal damage return to business. Opportunities for programmed psycho-social interventions may emerge at this point.	**Key Actions:** Support local community in working toward full facility, personnel, and service recovery of Head Start facilities, child care facilities, schools and other child/youth-oriented organizations and facilities.

©Project K.I.D., Inc., 2005

General Guidance

The following general guidance is provided for CISAR responders concerning children in catastrophic incidents.

- If possible, keep children with parents. CISAR responders should make every effort to keep children with parents/guardians during extraction/rescue;

- If it becomes necessary to separate children from their parents, record the parent and child's personal information and make every effort to ensure the parent is delivered to the same location as the child; and

- If safety necessitates removal of the children from parents, rescuers should ensure that parent's information is collected and that children are immediately tagged and tracked with an appropriate system.

Child tracking

Low-tech child tracking solutions may include writing parent's information on the child's arm or back with a Sharpie marker. Two-piece tyvek wristbands could also be used to attach parent's information to the child and vice-versa.

High-tech systems could include RFID or barcoded wristbands linked to tracking software. Parents can also be provided a card with information on the National Emergency Child Locator Center, which will likely be activated in a catastrophic incident. CISAR responders should share information on children rescued separately from parents in order to ensure that parents can be delivered to the same location as their child.

Considerations for Unaccompanied Minors

Current systems for addressing the needs of children in catastrophic incidents are largely based on societal assumptions that often no longer hold true.

Provision for evacuation and mass care still largely assumes that children will be in the company of responsible guardians during an emergency. While this is true for most children, many will be dislocated from families both by circumstances of the disaster (on school/work days most children will need to be reunited with their families) or by chaos and/or inadequate controls at schools, child care facilities, shelters and evacuation points.

Further challenges to the safety of children in emergencies arise from the ill intentions

of a few people who take advantage of the chaos. For example, during the Hurricane Katrina response, there were many credible reports of predatory behavior toward children, whether by relatives or strangers.

Local, State, Tribal, and Territorial, and CISAR plans should provide for the security, care, and initial casework for rescued children, especially those who have been separated from their families.

During the first 72 hours of a catastrophic incident, CISAR responders must create "safe havens" for children where they will be cared for by personnel familiar with the potential medical, emotional, and security needs of children. IC staff should include a "go-to" person to coordinate appropriate resources for children. Disaster child care responders should be available to receive children from emergency personnel and to begin addressing their needs for medical care, food, shelter, and reunion with families.

Holding children safe in disasters requires establishing identification and tracking policies and systems for children, especially those in mass care environments as well as those who have been or will be separated from their parents/guardians even for a short time during disasters.

The following considerations are provided for CISAR responders when rescuing unaccompanied children:

- Ensure unaccompanied minors are in close proximity to personnel who will keep them safe and initiate necessary recovery procedures;

- If child is rescued unaccompanied, observe and collect the child's personal information. Ensure the child's information is provided to the next caregiver who receives the child. Such data may include:

- o Child's name;

- o Brief description of the child (age, race, sex, clothing, etc.);

- o Rescue date;

- o Rescue time;

- o Rescue location, including geo-positioning data if available;

- o Brief description of circumstances in which the child was found (e.g., description of building, other people present, mode of rescue, medical condition, emotional state, etc.); and

- o Names and contact information for rescuing unit and rescue personnel.

- • Deliver children to place of safety and ensure personal handoff to another responsible adult. Effort should be made to deliver children to the most secure sites possible, with recognition that conditions at temporary survivor drop-off locations ("Lily Pads") are generally very bad for children. CISAR responders should seek clear guidance on designated places of safety for rescued children.

Table 3-5-3 (Unaccompanied Minor Intake Questionnaire) is provided at the end of this Section to assist CISAR responders when locating unaccompanied minors.

(Note: Organizations coming into contact with children during disasters should have adequate child protection policies as well as child identification and tracking procedures in place and should train for emergency situations. Child protection guidelines should include protocols for volunteer access to and interaction with children. In most cases, provision should be made requiring appropriate credentialing for child care volunteers that includes a criminal background check. In addition, legally defensible protocols should be established for validating custodial rights and the identity of parents and guardians.)

Promoting Resilience

When children are exposed to traumatic experiences such as the devastation following a natural disaster or similar disruption to their environment, they are often immersed in situations marked by physical destruction and emotional fear. Even when children are not separated from their parents, their caregivers are often themselves in panic, strained by the immediate need to ensure survival and protection for their family members and overwhelmed in facing the challenge of rebuilding their lives.

Even though CISAR responders may not spend much time with the displaced individuals, their interactions can greatly benefit children and their parents by helping lay early stepping stones for resilience. Consider the following:

(Note: The following guidelines (Figure 3-5-1) were adapted from New York University Child Study Center, "Caring for Kids After Trauma, Disaster," and, "Death: A Guide for Parents and Professionals," Second Edition (NYU, 2006), from Project K.I.D.'s PlayCare Training materials, and with input from child resilience expert Heather Wood Ion. FEMA and the American Red Cross have compiled additional information on traumatic stress responses in children. Their document, Helping Children Cope With Disaster, may be accessed online at http://www.fema.gov/pdf/library/children.pdf.)

Try to make a personal connection.	→	Introduce yourself in a friendly manner and let the child know what they can call you. Ask the child their name, how old they are, and other conversational questions to distract them from the trauma they may have experienced during the disaster or the rescue. Questions about where they go to school or the names of their best friends will be helpful as well as calming. Riding in a basket up to a helicopter may be either a fun or a frightening experience, for instance. Try to gauge the child's fear or excitement and respond to it personally.
Ask children to help.	→	Resilience is improved when people can regain a sense of efficacy and control during stressful times. Depending on the child's age, maturity, and emotional state at the time, consider whether there is a little thing you can ask a child to do, such as hold something or give you a hand. Do not pressure the child to help, but when possible let them help in some small and meaningful way to the extent of their ability. Especially in circumstances where the child and the caregivers do not share a common language, tasks are calming (and can be communicated through gestures).
Provide clear and factual information.	→	Uncertainty adds to fear and stress. Provide children and adults clear and factual information about what you are doing, where they are going, and what is going to happen to them. If you do not know the answers to their questions, respond with "I don't know yet, but let's find out." Talk to them in simple and age-appropriate language. Give a teenager more information than a younger child. If they ask questions, answer with patience and respect and encourage their curiosity.
Remain patient and calm.	→	If the child is visibly upset, try to stay patient and calm. Younger children may be distracted and calmed if you offer a small toy to play with. If in the immediate situation there are no toys, ask one child to help another. If they are too distraught to focus on your instructions or pose a danger to themselves or others, try speaking in a soft voice so they have to listen closely to hear you. Singing lullabies is deeply reassuring to the child. If you are a rock of stability and can provide reassurance, your calmness may be catching.
Protect children from invasive media coverage.	→	While there may be photographers and journalists present at rescue points or delivery sites, protect children from invasive and aggressive contacts that may be upsetting. Direct media away from children and distressed families.
Ensure secure delivery of survivors (continuum of care).	→	When delivering or transferring children to a place of safety introduce them by name to the next responsible adult and ensure all information has been handed off and that they are in secure custody of the next responsible party in the continuum of care. If you succeeded in making an emotional connection with the child, be sure to give them a respectful goodbye. Get down on their eye level; say their name and assure them that their new custodian will take good care of them; shake hands or give them a hug. Don't prolong the goodbye, but convey that you have been glad to meet them and wish them well.
Do not impose false expectations of resolution on any child or adult.	→	In a catastrophic incident, you as a first responder or volunteer may have expectations of response which cannot be met under the immediate and actual circumstances. It is very important that you do not convey to the child or any adult in your care an expectation that may prove false. Timelines are especially disappointing. Instead of saying the situation will be resolved in a specific time frame, offer to help the time go quickly, or ask the child for suggestions of things to do while waiting. Do not make promises that you personally cannot fulfill.

Figure 3-5-1: Guidelines for Helping Children Cope with a Disaster

Figure 3-5-1 is continued on the next page.

Constantly reassure in words and deeds.	→	Everyone in a catastrophic incident needs to know that they themselves can contribute to the rebuilding of normalcy, and that they are valued for that contribution. Reassuring others helps to begin the process of recovery even prior to rescue. Physical tasks are particularly helpful in providing reassurance "Let's build a shelter", "Shall we find out who needs dry socks" can all keep things practical and build confidence.

Figure 3-5-1: Guidelines for Helping Children Cope with a Disaster (continued)

National Emergency Child Locator Center (NECLC)

In October 2006, Congress and President George W. Bush established the National Emergency Child Locator Center (NECLC). NECLC will assist in the location of children and the reunification of families resulting from the disaster or subsequent evacuations.

In the event of a natural disaster, the National Emergency Child Locator Center will:

- Establish a toll-free hotline to receive reports of displaced children;

- Create a website to provide information about displaced children;

- Deploy staff to the location of a declared disaster area to gather information about displaced children;

- Provide information to the public about additional resources;

- Partner with federal, state, and local law enforcement agencies; and

- Refer reports of displaced adults to the Attorney General's designated authority and the National Emergency Family Registry and Locator System.

Additional Resources

Since Hurricane Katrina, new capabilities are emerging at multiple levels to address the needs of children in disaster.

At the Federal Government level the National Commission on Children and Disasters, FEMA's Working Group on Children in Disasters, and the Administration on Children and Families, Department of Health and Human Services, may all be good resources of information on addressing children's needs in disasters.

Private sector organizations such as Project K.I.D., the National Institute for Urban Search and Rescue (NIUSR), Disaster Children's Services (Church of the Brethren), the National Center for Missing and Exploited Children, the American Red Cross, the National Center for Disaster Preparedness (Columbia University) and Save the Children may also be of assistance.

Project K.I.D. in particular is working to develop state and local based PlayCare Response Teams to provide guidance and assistance to first responders in meeting the needs of children during the earliest hours and days of a disaster. Project K.I.D. has a variety of documents and after-action reports available. Project K.I.D. can be reached at holdsafe@gmail.com or toll free at 877-846-7529.

Table 3-5-3: Unaccompanied Minor Intake Questionnaire

ARRIVAL INFORMATION

ARRIVAL DATE:_____ARRIVAL TIME _____

FROM _____

RESPONDER INTERVIEW:

DESCRIBE CIRCUMSTANCES IN WHICH CHILD WAS FOUND (location, other people, mode of rescue, etc.):

_____

_____

_____

_____

_____

_____
Signature of Person Delivering Child   Date

_____
Agency

CHILD INTERVIEW:

1. CHILD'S NAME _____

2. SEX _____ DOB _____   BEST AGE GUESS IF NO ID _____

3. TYPE OF IDENTIFICATION _____

4. WITH WHOM DOES THE CHLD LIVE (MOTHER, FATHER, GRANDMOTHER...?)

_____

_____

5. WHERE DO PARENTS/GUARDIAN WORK?

_____

_____

6. PHONE NUMBERS OFPARENTS/GUARDIAN (or other numbers the child knows)

_____   # (____) _____
Name   Phone

_____   # (____) _____
Name   Phone

_____   # (____) _____
Name   Phone

7. HOME ADDRESS

_____

_____

8. RELATIVES (WHO-WHERE DO THEY LIVE)

_____

_____

9. WHERE DOES THE CHILD GO TO SCHOOL? TO CHURCH? TO SHOP?

_____

_____

10. PHYSICAL APPEARANCE

     EYE COLOR _____ HAIR COLOR _____ HAIR STYLE _____

     TYPE OF CLOTHES (COLOR- STYLE-CONDITION)
     TOP_____

     PANTS _____

     SHOES _____

11. PERSONAL POSSESSIONS – IF NO ID (DESCRIBE TYPE, COLOR, QUANTITY)

_____

12. SPOKEN LANGUAGE _____

13. IDENTIFYING EXPRESSIONS OR PHYSICAL CHARECTORISTICS, FAVORITE TOY OR HOBBY

_____

_____

14. ANY OTHER IDENTIFYING INFORMATION THE CHILD CAN PROVIDE

_____

_____

15. DOES THE CHILD PRESENT ANY EMERGENCY MEDICAL NEEDS

_____

_____

16. DOES THE CHILD HAVE ANY PRE-EXISTING HEALTH CONDITIONS? REGULARLY TAKE MEDICATION?

_____

_____

NOTIFICATION OF AUTHORITIES/CASE NOTES

Please maintain a detailed record of authorities contacted and response

Contact Made With   Date/Time

Discussion

_____

_____

_____

Result

_____

_____

_____

Contact Made With   Date/Time

Discussion

_____

_____

_____

Result

_____

_____

Contact Made With   Date/Time

Discussion

_____

_____

_____

Result

_____

_____

_____

Contact Made With   Date/Time

Discussion

_____

_____

_____

Result

_____

_____

_____

_____

Contact Made With   Date/Time

Discussion

_____
_____
_____
_____

Result

_____
_____
_____

Contact Made With   Date/Time

Discussion

_____
_____
_____
_____

Result

_____
_____
_____

Section 3-6: Animals

General Guidance

Service and Companion Animals

    Identifying Service Animals

Livestock, Wildlife and Captive Animals

CISAR Responder Practical Considerations

Animal Response Teams

General Guidance

Animal Rescue

No animal rescue activities should be attempted by CISAR responders on scene when, in their judgment, such activities would risk the lives or safety of themselves or others, or possibly create critical delays in rescuing persons in distress. Otherwise, animals should be rescued or assisted as practicable.

Any attempt to capture animals can be potentially dangerous. Always use caution when approaching animals. Some animals may bite or show aggression even if not provoked. It is best if trained disaster personnel experienced in animal behavior attempt to rescue animals. Therefore, it is prudent to develop a current list of qualified animal rescuers who can assist CISAR responders.

State, Tribal, Territorial, and local plans should provide for handling and care of rescued animals. For example, local animal control officials may be able to provide trained and equipped personnel to assist with animal control and rescue, including aggressive and difficult–to-access animals and unusual or exotic pets such as snakes.

Ultimately, the person in charge of the CISAR aircraft, boat, or response team should make the final decision concerning the rescue of animals.

Animal Transportation Risks

Great care must be exercised in transporting animals in a CISAR operation.

Transport of animals by CISAR responder aircraft, boats, and response teams can pose a problem, depending upon the size, health, and temperament of the animal(s).

CISAR responders should not be placed at increased risk to rescue animals.

Animals in Table 3-6-1 below may be encountered during CISAR operations:

Table 3-6-1: Animal Categories

Animal Type	Definition
Companion Animals	Household Pet: A domesticated animal, such as a dog, cat, bird, rabbit, rodent, or turtle that is traditionally kept in the home for pleasure rather than for commercial purposes, can travel in commercial carriers, and be housed in temporary facilities. Household pets do not include reptiles (except turtles), amphibians, fish, insects/arachnids, farm animals (including horses), and animals kept for racing purposes.
Service/Assistance Animals	Any guide dog, signal dog, or other animal individually trained to provide assistance to an individual with a disability including, but not limited to, guiding individuals with impaired vision, alerting individuals with impaired hearing to intruders or sounds, providing minimal protection or rescue work, pulling a wheelchair, or fetching dropped items.
Livestock and Farm Animals	Mainly provide food for human or animal consumption.
Wildlife	Wildlife primarily lives independent of human control and rely on individual ability to obtain food or water.
Captive Animals	Captive animals live in zoos or aquariums and that might otherwise be endangered wild animals, and in research facilities, and which are totally dependent on humans for survival.

References:
[1] Disaster Assistance Policy DAP9523.19: Eligible Costs Related to Pet Evacuations and Sheltering (24 Oct 2007)
[2] Department of Justice, Americans with Disabilities Act (ADA), 42 USC 1201 et seq, implementing regulations at 28 CFR 36.104

Service and Companion Animals

Ideally, service or companion animals should be rescued along with their owners, with priority given to service animals that are essentially an extension of a disabled person. Rescue of service animals can mean the difference between a person who requires assistance from shelter staff and a person who can function independently.

Safety of human life is always the first consideration even when there is a service animal involved. If the only choice is between transporting a service animal with an individual who has a disability or rescuing the disabled individual and another person, the human always has first priority.

Ideally, the service animal can be retrieved later and reunited with its owner.

If service or companion animals cannot be rescued along with their owners, CISAR responders should provide information on the animals and their location for subsequent recovery by animal rescue services and rejoined with their owners.

Identifying Service Animals

Some, but not all service animals wear special collars and harnesses. Some of these animals are licensed or certified and have identification papers. However, availability of relevant documentation should not be a condition of providing SAR services. The person with the animal should be asked whether the animal is required because of a disability.

Regardless of the animal's category, any animal should be left behind that poses a direct threat to the health or safety of people or the transport vehicle, such as when the animal is actually exhibiting anxiety or potentially vicious behavior.

Livestock, Wildlife, and Captive Animals

There are many situations where livestock, wildlife, or captive animals might be encountered during CISAR operations. For example, animals that are an apparent threat to human safety should be reported and avoided, or be confined, sedated, or killed if necessary. Sound judgment must be exercised in dealing with such animals.

Livestock or confined animals might be able to avoid threats such as fire or flooding if they were free. The simplest solution may be to release these types of animals for capture later. CISAR responders should seek advice from the Incident Command as required.

CISAR Responder Practical Considerations

The following are some practical considerations CISAR responders can take when animals are encountered:

- Possible warning signs that an animal is about to attack may include tail high and stiff, ears up, hair on back standing up, barking, and showing teeth (Even the friendliest dog can bite or attack when in fear and/or in pain);

- For non-aggressive animals, use an approved muzzle, slip collar, leash and/or food lures;

- For aggressive, unpredictable animals, use snare poles, restraints, and humane traps, or sedation when appropriate;

- To help avoid animal bites, remain as still as possible, avoid direct eye contact, and put something between you and the aggressive animal such as a trash can lid;

- If you fall and an animal attacks, protect your head, curl into a ball with your hands over your ears, and remain motionless;

- If attacked or bitten by a dog, use repellent, wash wounds immediately with soap, seek immediate medical help, and secure and observe offending dog, if practicable;

- Capture animals using humane live traps, catch poles, leashes, cages and appropriate strength ropes; and

- Use appropriate bite-resistant gloves, and maintain a properly equipped first-aid kit.

Animal Response Teams

The Humane Society of the United States (HSUS) is a good resource of information about organizations that can assist with animal rescue and care needs.

The HSUS maintains National Disaster Animal Response Teams (NDART) that provide assistance during catastrophic incidents. These teams:

- Serve as a resource for individuals, animal-related organizations, government agencies, and others concerned about the urgent needs of animals before, during, and after disasters; and

- Assist with animal rescue, handling, and transport.

This page intentionally left blank.

Section 3-7: Handling of Human Remains

Introduction

Handling of Human Remains

Physical Health Risks from Human Remains

Direct Contact with Human Remains

Precautions When Handling Human Remains

References

Introduction

CISAR responder discovery of human remains is likely in large-scale incidents. A key principle in CISAR operations is to give priority to the living over the dead.

Human Remains and the CISAR Responder

In CISAR operations, if people are in distress and require assistance, the CISAR responder must make saving lives the priority. Notify the Incident Command of the location any human remains and continue CISAR operations.

Handling human remains may be subject to Agency specific guidance.

Handling of Human Remains

CISAR responders may be required to handle human remains. Searching for and recovering bodies should be conducted with respect and dignity, according to applicable customs, laws, and regulations. The location of any remains should be marked and reported to the Incident Command to ensure that the proper authorities are notified for recovery, identification, and disposition.

SAR authorities may also need to have plans and agreements in place with other entities to handle human remains. Such arrangements should be incorporated into State, Tribal, Territorial, and local emergency plans.

Physical Health Risks from Human Remains

Past disasters have demonstrated that the risk of epidemic disease transmission from human remains is negligible. Unless the affected population was already experiencing a disease suitable for epidemic development, a CISAR environment should not create such a situation. Most disaster victims die from traumatic events and not from pre-existing disease.

Typical pathogens in the human body normally die off when the host dies, although not immediately. Risk of transmission is no greater than that for routine handling of human remains. However, water supplies contaminated with decaying human remains can serve as a method of transmission of illnesses.

Direct Contact with Human Remains

Human remains may contain blood-borne viruses and bacteria. These viruses and bacteria do not pose a risk to someone walking nearby, nor do they cause significant environmental contamination.

In flood water, bacteria and viruses from human remains are a minor part of the overall contamination that can include uncontrolled sewage, a variety of soil and water organisms, and household and industrial chemicals. There are no additional practices or precautions for floodwater related to human remains, beyond what is normally required for safe food and drinking water, standard hygiene, and first aid.

However, for people who must directly handle remains, such as CISAR responders in the course of conducting CISAR operations and recovery personnel, there can be a risk of exposure to such viruses or bacteria.

Precautions When Handling Human Remains

CISAR responders who must handle human remains should use the following precautions:

- Protect your face from splashes of body fluids and fecal material. Use a plastic face-shield or a combination of eye protection (indirectly vented safety goggles are a good choice if available; safety glasses provide only limited protection) and a surgical mask. In extreme situations, a cloth tied over the nose and mouth can be used to block splashes (be advised that a cloth can absorb splashes);

- Protect your hands from cuts, puncture wounds, or other injuries that might break the skin and allow for direct contact with body fluids. A combination of cut proof inner layer glove and a latex or similar outer layer is preferable. Wash your hands with soap and water, or with an alcohol-based hand cleaner immediately after you remove your gloves. Appropriate footwear should similarly protect against sharp debris;

- Appropriate PPE should be worn during procedures that are likely to generate splashes of blood or other body fluids;

- Give prompt care, including immediate cleansing with soap and clean water. A tetanus booster should be provided to personnel when wounds are sustained during work with human remains; and

- Notify the Incident Command if exposed to hazardous conditions.

Table 3-7-1 on the next page summarizes the primary issues concerning handling of human remains.

References

The information in this Section was obtained from the following sources:

- U.S. Department of Health and Human Services, *Interim Health Recommendations for Workers who Handle Human Remains*; and

- World Health Organization, *Disposal of dead bodies in emergency situations.*

- Pan American Health Organization, *Management of Dead Bodies After Disasters: A Field Manual for First Responders* (Washington D.C.: PAHO).

Table 3-7-1: Human Remains - Infectious Disease Risks Summary

Overview	After most natural disasters there is fear that dead bodies will cause epidemics. However, human remains do not normally cause epidemics after natural disasters.
Infections and dead bodies	At the time of death, victims are not likely to be sick with epidemic-causing infections (i.e., plague, cholera, typhoid, and anthrax). A few victims may have chronic blood infections (hepatitis or HIV), tuberculosis, or diarrheal disease. Most infectious organisms do not survive beyond 48 hours in a dead body. An exception is HIV which has been found six days postmortem.
Risk	Individuals handling human remains may have a small risk through contact with blood and feces (bodies often leak feces after death) from the following: <ul><li>Hepatitis B and C;</li><li>HIV;</li><li>Tuberculosis; and</li><li>Diarrheal disease.</li></ul> Individuals working in hazardous environments (e.g., collapsed buildings and debris) and may also be at risk of injury and tetanus (transmitted via soil).
Safety precautions	Basic hygiene protects workers from exposure to diseases spread by blood and certain body fluids. Workers should use the following precautions: <ul><li>Use gloves and boots, if available;</li><li>Wash hands with soap and water after handling bodies and before eating;</li><li>Avoid wiping face or mouth with hands;</li><li>Wash and disinfect all equipment, clothes, and vehicles used for transportation of bodies;</li><li>Face masks are unnecessary, but should be provided if requested to avoid anxiety; and</li><li>The recovery of bodies from confined, unventilated spaces should be approached with caution. After several days of decomposition, potentially hazardous toxic gases can build-up. Time should be allowed for fresh air to ventilate confined spaces.</li></ul>

This page intentionally left blank.

Section 3-8: Handling of Animal Remains

Introduction

Health Risks from Animal Remains

Decomposing Animal Remains

References

Introduction

CISAR responders may come in contact with dying or dead animals. The cleanup of animal carcasses is not normally a primary responsibility of CISAR responders, but it is necessary to understand the risks associated with the handling of animal remains.

Many of the issues about human corpses directly correspond to those relating to animal corpses. Animals however, do spread a number of diseases among humans.

During lifesaving operations, the location of animal remains should be reported to the Incident Command to ensure appropriate authorities are notified for proper disposition.

Health Risks from Animal Remains

In most cases, the bodies of dead animals pose little risk to humans. Animal corpses constitute a public health hazard only in specific conditions. However, animals that have had a specific communicable disease, may pose a risk to humans.

Zoonoses[1] are becoming an increasing threat to human populations. However, most zoonotic infections do not survive in the dead body of an animal. Like diseases that survive in the human remains, zoonotic diseases from animal carcasses must occur in an endemic area for that disease if they are to present any risk. If the area is not endemic for the disease, the probability of animal carcass-to-human transmission is very low.

Two situations exist in which an animal carcass can be a risk for humans:

- The presence of specific infectious agents; and

- The contamination of water by feces and discharge from lesions.

In the two above scenarios, there is still little risk to CISAR responders. A series of coexisting factors must be present for the animal bodies to constitute a risk for humans:

- First, the animal must be infected with a disease that can be transmitted to humans;

- Second, the germ must be able to survive the death of the host; and

- Third, the environment must facilitate the spread of the infectious agent (e.g., contaminated water).

Any interruption in this chain of events results in there being a minimal public health hazard. Moreover, the presence of animal carcasses alone should not be associated with the spread of infectious diseases.

---

[1] Zoonose is any infectious disease that is able to be transmitted from animals, both wild and domestic, to humans or from humans to animals.

Decomposing Animal Remains

There is only a short window of time for proper disposal of animal carcasses following their death. Within 7-10 days of death, animal carcasses become too decomposed to handle easily with disposal equipment such as front loaders.

Depending on the type of disaster and location, CISAR responders may encounter problems with decomposing animal carcasses as areas are searched for human survivors. Although animal corpses pose a minimal health risk, the proper disposal of animal remains is important after the initial response to the disaster.

References

The information in this Section was obtained from the following sources:

- Pan American Health Organization, *Management of Dead Bodies in Disaster Situations* (Washington D.C.: PAHO,);

- Dee B. Ellis, *Carcass Disposal Issues in Recent Disasters, Accepted Methods, and Suggested Plan to Mitigate Future Events* (Southwest Texas State University,).

Section 3-9: Media and Public Relations

Introduction

During CISAR operations, the public and media should be informed, within the limits of confidentiality, of ongoing CISAR operations. Some potential benefits of early release of information include:

- The possibility to obtain additional, useful information from the public to enable more effective use of CISAR resources;

- Fewer time-consuming requests from the news media; and

- Reduction in inaccurate public speculation about ongoing operations.

A CISAR operation often creates great interest with the general public and with radio, television, and newspapers.

IAMSAR Manual (Volume 2) and FEMA's NIMS training have considerable information relevant to media and public relations for CISAR responders. This information can help enhance media and public relations effectiveness and avoid troublesome mistakes.

Relationships

SAR-related contacts with media, which can take many forms, are normally the responsibility of managers, public affairs specialist, or assigned to a Joint Information Center (JIC). It is important to establish a good relationship between the media and response authority to ensure that information reaching the public is factual and complete. This relationship should be established prior to any major incident. The responsible CISAR authority should partner with the media to communicate the overarching message, services provided, and impact on the community.

Release of names to the media can be a sensitive issue. Names of casualties should not be released until every effort has been made to contact family relatives. Until the relatives have been notified, normally only the number of deceased, survivors, and injured survivors should be released. Names of military casualties should be released only by the military service to which the casualties belong. Names of survivors should not be released until positive identification has been accomplished.

The SMC should be aware of the concerns of the relatives of missing persons. Waiting during searches and lack of information can be stressful for family members of those in distress. During the search, the SMC or staff should maintain regular contact with relatives to provide information and outline future plans. If possible, contact telephone numbers should be issued for relatives. These steps assist the relatives in accepting the SMC's decision to conclude CISAR operations even if the missing persons are not located.

Ensure that the media knows who is in charge of coordinating CISAR operations.

NIMS Public Information Guidance

In NIMS, the Public Information Officer (PIO) and the Joint Information Center (JIC) help the Incident Command and SMC manage the flow of public information.

Public Information Officer (PIO)

The PIO is a key member of the command staff. The PIO advises the Incident Command on public information matters related to the management of the incident, including media and public inquiries, emergency public information and warnings, rumor monitoring and control, media monitoring, and other functions required to coordinate, clear with proper authorities, and disseminate accurate and timely information related to the incident. The PIO ensures that decision makers and the public are fully informed throughout a domestic incident response.

Joint Information Center (JIC)

Establishment of a JIC is an effective means of disseminating public information. Using the JIC as a central location, CISAR information can be coordinated and integrated between the CISAR response organizations. Prompt establishment of a JIC away from the SMC will help to achieve this goal. The JIC can establish proper procedures for releasing information to the public and how the information will be released. Since the information may be sensitive, it is critical that everyone communicates the same information.

Media On Scene

The media is a 24-hour global market, with news broadcast worldwide. The media will find a way to get to the scene for first hand information, pictures, and video. By providing media transportation to the scene, safety can be improved and the media can be more effectively informed and supported.

Section 3-10: CISAR Exercises

Purpose

Objectives

Planning

Purpose

Since opportunities to handle actual incidents involving CISAR operations are rare and challenging, exercising CISAR plans is particularly important. Mass evacuation and CISAR operations may be difficult and costly, and the number of authorities involved in the response leads to complexity.

Objectives

CISAR exercises should ideally achieve the following objectives:

- Test implementation of planned command authorities and functions;

- Account for all survivors until they are delivered to a place of safety and can return to their homes;

- Identify and task available SAR resources and local resources such as hospitals, fire departments, and other community and transportation resources;

- Evaluate notification processes, resource availability, timeliness of initial response, real-time elements, conference capabilities and overall co-ordination;

- Ensure all agency roles are sorted out, understood and properly implemented;

- Test capabilities of potential OSCs and aircraft coordinators and ability to transfer OSC duties;

- Evaluate span of control;

- Evacuate an area or facility;

- Co-ordinate activities and achieve information exchanges;

- Communication by all available;

  o Information for all concerned (identify, merge, purge, retrieve and transfer to the right place in the right form at the right time);

  o New communication and information management technologies;

  o Media and next-of-kin; and

  o Test all communication links that may be needed for notification, co-ordination and support.

- Safely transfer and care for passengers;

- Conduct medical triage and provide first aid;

- Exercise co-ordination with local response agencies;

- Provide food, water, and protective clothing to survivors;

- Test plans for mass rescue operations;

- Assess how effectively earlier lessons learned have been accounted for in updated plans and how well these lessons were disseminated; and

- Exercise external affairs, such as international and public relations:

o Necessary participants involved;

o JIC established quickly and properly staffed; and

o Press briefings handled effectively (e.g., consistent information from different sources).

- Rescued persons tracked, kept informed and needs monitored, and reunited with belongings.

Planning

The following steps are normally carried out during exercise planning:

- Agree on the exercise scenario, goals and extent;

- Assembly a multi-disciplinary planning team and agree on objectives for each aspect of the exercise;

- Develop the main events and associated timetables;

- Confirm availability of agencies to be involved, including any media representatives or volunteers;

- Confirm availability of transportation, buildings, equipment, aircraft, ships or other needed resources;

- Test all communications that will be used, including tests of radio and mobile phones at or near the locations where they will be used;

- Identify and brief all participants and people who will facilitate the exercise, and ensure that facilitators have good independent communications with person who will be controlling the exercise;

- Ensure that everyone involved knows what to do if an actual emergency should arise during the exercise;

- If observers are invited, arrange for their safety, and to keep them informed about the exercise progress;

- For longer exercises, arrange for food and toilet facilities;

- Use "exercise in progress" signs, advance notifications and other means to help ensure that person not involved in the exercise do not become alarmed;

- Schedule times and places for debriefs;

- Agree and prepare conclusions and recommendations with the entity responsible for handling each recommendation along with the due date for any actions;

- Prepare a clear and concise report and distribute it as appropriate to the participating organizations; and

- Consider the outcome of this exercise in planning future exercises.

Part 4: Natural Disasters

Overview

National efforts to ensure resilience in the U.S. are focusing on improving existing catastrophic event preparedness capabilities, but with a renewed conviction to plan for the most extreme disasters. For ESF #9 and the response to catastrophic incidents, activation of ESF #9 for the conduct of CISAR operations are routinely due to natural disasters.

As a result, the CISAR Addendum has included this Part to detail the most significant natural disasters that Federal, State, Tribal, Territorial, and local SAR authorities may encounter.

Five natural disasters that continue to require Federal ESF #9 CISAR support are discussed:

- Earthquakes.

- Flooding.

- Hurricanes.

- Tornadoes.

- Tsunamis.

The response to each type of natural disaster is unique. Having a basic understanding of the type of disaster and the potential impact the disaster can have on a community, can substantially assist the CISAR responder in the conduct of lifesaving operations.

The information provided for each disaster is not all-inclusive, but only a limited overview of each disaster type and some pertinent guidance for CISAR responders who will be conducting lifesaving operations in these potentially hazardous environments.

In no way should this information be used instead of agency specific policy and doctrine, thorough preparedness, best practices, training, and exercises.

CISAR responders knowing that time is of the essence when responding to these disasters, must still take the time to understand the environment, work together, and remain safe.

Section 4-2: Earthquakes

Overview

An earthquake is the vibration, sometimes violent, of the Earth's surface that follows a release of energy in the Earth's crust. This energy can be generated by a sudden dislocation of segments of the crust, volcanic eruption, or by manmade explosions.

Most destructive quakes, however, are caused by dislocations of the crust. The crust may first bend and then, when the stress exceeds the strength of the rocks, break and "snap" to a new position. In the process of breaking, vibrations called "seismic waves" are generated.

These waves travel outward from the source of the earthquake along the surface and through the Earth at varying speeds depending on the material through which they move. Some of the vibrations are of high enough frequency to be audible, while others are of very low frequency. These vibrations cause the entire planet to quiver or ring like a bell or tuning fork.

A *fault* is a fracture in the Earth's crust along which two blocks of the crust have slipped with respect to each other. Faults are divided into three main groups, depending on how they move (Figure 4-2-1).

- *Normal faults* occur in response to pulling or tension; the overlying block moves down the dip of the fault plane.

- *Thrust (reverse) faults* occur in response to squeezing or compression; the overlying block moves up the dip of the fault plane.

- *Strike-slip (lateral) faults* occur in response to either type of stress; the blocks move horizontally past one another.

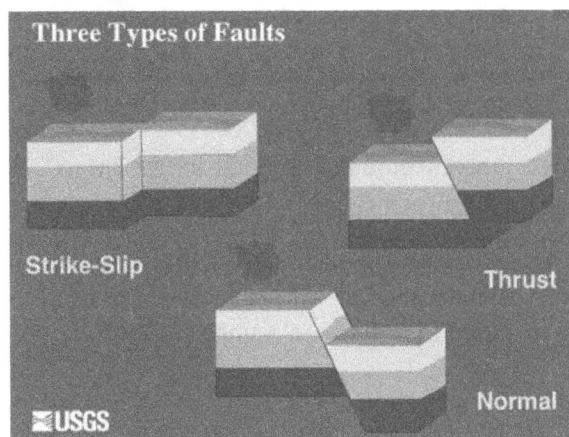

Figure 4-2-1: Normal, Thrust and Strike-Slip Faults (Photo: USGS)

Most faulting along spreading zones is normal, along subduction zones is thrust, and along transform faults is strike-slip.

Geologists have found that earthquakes tend to reoccur along faults, which reflect zones of weakness in the Earth's crust. Even if a fault zone has recently experienced an earthquake, however, there is no guarantee that all the stress has been relieved. Another earthquake could still occur. Furthermore, relieving stress along one part of the fault may increase stress in another part of the fault.

The majority of earthquakes originate at depths not exceeding tens of kilometers. Earthquakes occurring at a depth of less than 70 km are classified as 'shallow-focus' earthquakes, while those with a focal-depth between 70 and 300 km are commonly termed 'mid-focus' or 'intermediate-depth' earthquakes. Deep-focus earthquakes may occur at much greater depths (ranging from 300 up to 700 kilometers).

Earthquake Risk

Earthquakes are unpredictable and pose considerable risks to society. The U.S. averages:

- Several thousand earthquakes every year that range up to 5.9 in magnitude;

- Six earthquakes per year in the 6.0 to 6.9 magnitude range;

- One earthquake per year in the 7.0 to 7.9 magnitude range; and

- One earthquake every 14 years above magnitude 8.0 – the last of which took place in American Samoa in 2009.

Richter Scale

The Richter Scale, named after Dr. Charles F. Richter of the California Institute of Technology, is the best known scale for measuring the magnitude of earthquakes. The scale is logarithmic so that a recording of 7, for example, indicates a disturbance with ground motion 10 times as large as a recording of 6. A quake of magnitude 2 is the smallest quake normally felt by people. Earthquakes with a Richter value of 6 or more are commonly considered major; great earthquakes have magnitude of 8 or more on the Richter scale.

According to the U.S. Geological Survey (USGS):

- Earthquakes are one of the most costly natural hazards faced by the U.S., posing a significant risk to 75 million Americans in 39 States;

- Although earthquakes occur less frequently in the Eastern U.S., studies show that urban areas in the East could face devastating losses because severe shaking would affect a larger area than a similar earthquake in the Western U.S. (Additionally, most structures in the Eastern U.S. are not designed to resist earthquakes. In addition, population density is also high in the Eastern U.S., and residents are not as well prepared for earthquakes as communities in the west);

- In Alaska and the Pacific Northwest, the effects of a destructive earthquake can extend well beyond local impact by potentially creating far-reaching tsunamis and resulting in economic losses that could exceed any that have occurred from previous earthquakes or tsunamis (See, *Section 4-6: Tsunamis,* for additional information);

- Landslides triggered by earthquakes often cause more destruction than the earthquakes themselves;

- People can be affected in three major ways by earthquake hazards:
  - o Injured or killed by falling or collapsing objects;
  - o By objects thrown into the air; and/or
  - o By earthquake-induced fires or flooding.

- The damage caused by an earthquake is dependent on the intensity, depth, distance from a population center, soil type, building construction, and many other factors.

Generally, earthquakes are projected to cause damage based on magnitude, as shown in Table 4-2-1 below.

Table 4-2-1: Earthquake Occurrence and Damage to Structure by Magnitude			
Magnitude	**Description**	**Earthquake Effects**	**Frequency of Occurrence (Globally)**
Less than 2.0	Micro	Micro earthquakes, not felt.	Est. 8,000 per day
2.0 - 3.9	Minor	May be felt, rarely cause damage.	Est. 1,000 per day
4.0 - 4.9	Light	Noticeable shaking of indoor items, rattling noises; significant damage unlikely.	Est. 6,200 per year
5.0 - 5.9	Moderate	Can cause major damage to poorly constructed buildings near epicenter; no or slight damage to well-designed buildings.	800 per year
6.0 - 6.9	Strong	Can be destructive up to about 100 miles from epicenter	120 per year
7.0 – 7.0	Major	Can cause serious damage over large areas	18 per year
8.0 – 8.9	Great	Can cause serious damage across several hundred miles	1 per year
9.0 – 9.9	Great	Devastating in areas over a thousand miles across	1 per 20 years
Ref: FEMA, *Federal Interagency Response Plan – Earthquake 2011*, Version 3.5 (March 11, 2011): B-6			

Earthquake Response Phases

The Federal Government executes its roles and responsibilities in all phases of the response effort, which are delineated as per Figure 4-2-2 below.

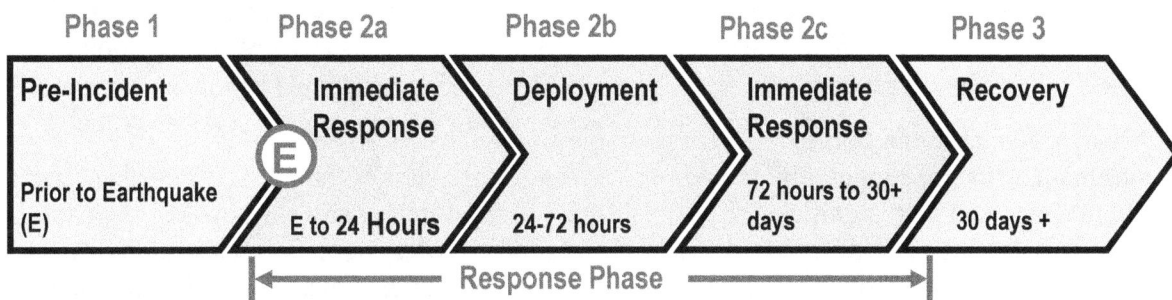

Figure 4-2-2: Earthquake Response Time Phases
(Reference: FEMA, *Federal Interagency Response Plan – Earthquake 2011*, Version 3.5 (March 11, 2011): 6.)

The Response Phase focuses on an immediate, coordinated, and effective Federal response to save lives and reduce casualties following a catastrophic earthquake, in support of survivors, communities, and affected governments.

Key points to consider for CISAR Responders in an earthquake disaster response include the following:

- Normally, FEMA US&R will assume ESF #9 overall Primary Agency for a catastrophic earthquake response due to the conduct of structural collapse CISAR operations;

- CISAR responders have the operational objective of completing all primary searches within 72 hours after the earthquake;

- In many earthquake incidents, no clear transition exists from one phase to the next, and phases may run concurrently;

- During incidents that affect multiple States and/or FEMA Regions, different jurisdictions may transition through the phases at various paces depending on the impact to the respective geographical area;

- Due to the nature of earthquakes and the likelihood of aftershocks, the progression of phases may relapse;

- A catastrophic earthquake may require mobilizing and deploying Federal SAR assets before they are requested;

- Multiple incidents may occur simultaneously or sequentially, including aftershocks, that may significantly impede/delay the Federal response and/or the availability of Federal resources;

- Response activities will probably begin without a detailed or complete situation and critical needs assessment; and

- Evacuation plans and response operations will be impacted by people who refuse to comply with evacuation orders and/or attempt to return to the area prior to notification that the area is safe to return to.

(Note: Further information on the time phases of an earthquake response can be found in FEMA's Federal Interagency Response Plan – Earthquake 2011)

Impact

- Earthquakes can cause buildings and bridges to collapse, disrupt gas, electric, and phone service, and sometimes trigger landslides, avalanches, flash floods, fires, and tsunamis;

- Buildings with foundations resting on unconsolidated landfill, old waterways, or other unstable soil are most at risk;

- Buildings or trailers and manufactured homes not tied to a reinforced foundation anchored to the ground are at risk since they can be shaken off their mountings;

- When internal load bearing elements fail, a building will collapse into itself and exterior walls are pulled into the falling structure, which may result in a dense debris field with a small footprint;

- Most earthquake-related injuries result from collapsing walls, flying glass, and falling objects as a result of ground shaking, or people trying to move more than a few feet during the shaking;

- Aftershocks (smaller earthquakes that follow the main earthquake) can cause further damage to weakened buildings (Be aware that some earthquakes are actually foreshocks, and a larger earthquake might occur.).

Considerations for CISAR Responders

Structural collapse creates many hazardous conditions for CISAR responders. The best methods to reduce the risk of injuries or illness during a CISAR operation are prevention and avoidance. CISAR responders can reduce the risks inherent in CISAR operations through knowledge and awareness of potential hazards. Consider the following:

- Be aware of aftershocks that can cause further damage to weakened buildings and pose considerable risk to CISAR responders;

- Collapsed structures may present the risk of falling, which could result in injury or death;

- Working in elevated situations may require the use of barrier lines to prevent access to dangerous edges, the use of safety lines to belay persons at risk of falling, wearing safety equipment, and the designation of a Safety Officer;

- Any surface that must be negotiated where there is the potential for a fall or loss of control is unstable and may not have the strength to support a CISAR responder's weight (Any questionable or untested surface must be considered hazardous. These areas must be identified and either made safe or avoided.);

- Unstable surfaces may be caused by slippery materials such as water or oil on a concrete or metal surface (These problems may be mitigated by identification, using safety equipment, avoidance of the area, reducing exposure in the area, or removal of the hazard.); and

- CISAR responders need to be aware of the following hazards and proceed with caution when entering a collapsed structure:

  o Water system breaks may flood basement areas;

  o Exposure to pathogens from sanitary sewer system breaks;

  o Exposed and energized electrical wiring;

  o Exposure to airborne smoke and dust (e.g., asbestos, silica, etc.);

  o Exposure to blood borne pathogens;

  o Exposure to hazardous materials (e.g., ammonia, battery acid, leaking fuel, etc.);

  o Natural gas leaks creating a flammable and toxic environment;

  o Structural instability;

  o Insufficient oxygen;

  o Confined spaces;

  o Slip, trip, or fall hazards from holes, protruding rebar, etc;

  o Being struck by a falling object;

  o Fire;

  o Proximity to heavy machinery such as cranes;

  o Sharp objects such as glass and debris;

  o Secondary collapse from aftershock, vibration, and explosions; and

  o Unfamiliar surroundings.

Earthquake safety

Tables 4-2-2 and 4-2-3 on pages 4-10 and 4-11, respectively, provide safety information for anyone that may be caught in an earthquake, as well as after an earthquake occurs.

Table 4-2-2: What to Do *During* an Earthquake

Stay as safe as possible during an earthquake. Be aware that some earthquakes are actually foreshocks and a larger earthquake might occur. Minimize your movements to a few steps to a nearby safe place and if you are indoors, stay there until the shaking has stopped and you are sure exiting is safe.

If indoors	• **DROP** to the ground; take **COVER** by getting under a sturdy table or other piece of furniture; and **HOLD ON** until the shaking stops. If there isn't a table or desk near you, cover your face and head with your arms and crouch in an inside corner of the building. • Stay away from glass, windows, outside doors and walls, and anything that could fall, such as lighting fixtures or furniture. • Stay in bed if you are there when the earthquake strikes. Hold on and protect your head with a pillow, unless you are under a heavy light fixture that could fall. In that case, move to the nearest safe place. • Use a doorway for shelter only if it is in close proximity to you and if you know it is a strongly supported, load bearing doorway. • Stay inside until the shaking stops and it is safe to go outside. Most injuries occur when people inside buildings attempt to move to a different location inside the building or try to leave. • Be aware that the electricity may go out or the sprinkler systems or fire alarms may turn on. • DO NOT use the elevators.
If outdoors	• Stay there. • Move away from buildings, streetlights, and utility wires. • Once in the open, stay there until the shaking stops. The greatest danger exists directly outside buildings, at exits and alongside exterior walls. Many of the 120 fatalities from the 1933 Long Beach earthquake occurred when people ran outside of buildings only to be killed by falling debris from collapsing walls. Ground movement during an earthquake is seldom the direct cause of death or injury. Most earthquake-related casualties result from collapsing walls, flying glass, and falling objects.
In a moving vehicle	• Stop as quickly as safety permits and stay in the vehicle. Avoid stopping near or under buildings, trees, overpasses, and utility wires. • Proceed cautiously once the earthquake has stopped. Avoid roads, bridges, or ramps that might have been damaged by the earthquake.
If trapped under debris	• Do not light a match. • Do not move about or kick up dust. • Cover your mouth with a handkerchief or clothing. • Tap on a pipe or wall so rescuers can locate you. Use a whistle if one is available. Shout only as a last resort. Shouting can cause you to inhale dangerous amounts of dust.

Table 4-2-3 on the next page, although applying to earthquake survivors, are tips that will benefit CISAR responders as they conduct SAR operations in damaged buildings and structures looking for survivors.

Table 4-2-3: What to Do *After* an Earthquake	
Expect aftershocks.	These secondary shockwaves are usually less violent than the main quake but can be strong enough to do additional damage to weakened structures and can occur in the first hours, days, weeks, or even months after the quake.
Listen to a battery-operated radio or television.	Listen for the latest emergency information.
Open cabinets cautiously.	Beware of objects that can fall off shelves.
Stay away from damaged areas.	Stay away unless your assistance has been specifically requested by police, fire, or relief organizations. Return home only when authorities say it is safe.
Be aware of possible tsunamis if you live in coastal areas.	These are also known as seismic sea waves (mistakenly called "tidal waves"). When local authorities issue a tsunami warning, assume that a series of dangerous waves is on the way. Stay away from the beach.
Help injured or trapped persons.	Remember to help your neighbors who may require special assistance such as infants, the elderly, and people with disabilities. Give first aid where appropriate. Do not move seriously injured persons unless they are in immediate danger of further injury. Call for help.
Clean up spilled medicines, bleaches, gasoline or other flammable liquids immediately.	Leave the area if you smell gas or fumes from other chemicals.
Inspect the entire length of chimneys for damage.	Unnoticed damage could lead to a fire.
Inspect utilities.	• **Check for gas leaks.** If you smell gas or hear blowing or hissing noise, open a window and quickly leave the building. Turn off the gas at the outside main valve if you can and call the gas company from a neighbor's home. If you turn off the gas for any reason, it must be turned back on by a professional. • **Look for electrical system damage.** If you see sparks or broken or frayed wires, or if you smell hot insulation, turn off the electricity at the main fuse box or circuit breaker. If you have to step in water to get to the fuse box or circuit breaker, call an electrician first for advice. • **Check for sewage and water lines damage.** If you suspect sewage lines are damaged, avoid using the toilets and call a plumber. If water pipes are damaged, contact the water company and avoid using water from the tap. You can obtain safe water by melting ice cubes.

Additional Information

Table 4-2-4 on the next page provides websites for additional information on earthquakes and earthquake preparedness.

Although not exhaustive, these Internet links will provide additional links to other Earthquake information websites.

Table 4-2-4: Earthquake Information Websites	
FEMA Earthquake Preparedness and Information	http://www.fema.gov/hazard/earthquake/index.shtm
National Earthquakes Hazard Reduction Program	http://www.nehrp.gov/
U.S. Geologic Survey Earthquake Information	http://earthquake.usgs.gov/
Occupational Safety and Health Administration	http://osha.gov/SLTC/emergencypreparedness/guides/structural.html http://osha.gov/SLTC/emergencypreparedness/guides/earthquakes.html

Section 4-3: Flooding

Overview

Floods are one of the most common hazards in the U.S. Flood effects can be local, impacting a neighborhood or community, or very large, affecting entire river basins and multiple States.

Flooding: Nashville, Tennessee, May 4th, 2010 (David Fine/FEMA)

However, all floods are not alike. Some floods develop slowly, sometimes over a period of days. Flash floods, however, can develop quickly, sometimes in just a few minutes and without any visible signs of rain. Flash floods often have a dangerous wall of roaring water that carries rocks, mud, and other debris and can sweep away most things in its path. Overland flooding occurs outside a defined river or stream, such as when a levee is breached, but still can be destructive. Flooding can also occur when a dam breaks, producing effects similar to flash floods. Other natural processes, such as hurricanes, weather systems, and snowmelt can also cause floods.

On average, floods kill about 140 people each year and cause $6 billion in property damage.

Although loss of life due to floods during the past half-century has declined, mostly because of improved warning systems, economic losses have continued to rise due to increased urbanization and coastal development.

Flood Facts

- More than half of all fatalities during floods are auto related, usually the result of drivers misjudging the depth of water on a road and the force of moving water (A car can float in just a few inches of water.);

- Six inches of water will reach the bottom of most passenger cars causing loss of control and possible stalling;

- One foot of water will float many vehicles;

- Two feet of rushing water can carry away most vehicles including sport utility vehicles (SUVs) and pick-up trucks;

- The principal causes of floods in the Western U.S. are snowmelt and rainstorms;

- The principal causes of floods in the Eastern U.S. and Gulf Coast are hurricanes and storms; and

(Note: See Section 4-4: Hurricanes (Typhoons), for additional information on these dangerous storms.)

- Flash floods (a rapid flooding of low lying areas in less than six hours, normally caused by intense rainfall from a thunderstorm or several thunderstorms, collapse of a man-made structure, ice dam, etc.) are the number one weather-related killer in the U.S. since they can roll boulders, tear out trees, and destroy buildings and bridges.

U.S. Flood Risks

General flood risks in the U.S. include the following:

- Many areas in the Western States (Arizona, California, Idaho, Nevada, Oregon, and Washington) are at an increased flood risk due to wildfires. The charred ground where vegetation has burned away cannot easily absorb rainwater, increasing the risk of flooding and mudflows over a number of years;

- Several areas of the U.S. are at heightened risk for flooding due to heavy rains (excessive rainfall and flooding can occur throughout the year.);

- Long cold spells can cause the surface of rivers to freeze, leading to ice jams (When a rise in the water level or a thaw breaks the ice into large chunks, these chunks can become jammed at man-made and natural obstructions, resulting in severe flooding.);

- Levees are designed to protect against a certain level of flooding, however, levees can and do decay over time, making maintenance a serious challenge (Levees can also be overtopped, or even fail during large floods.);

- Mudflows are rivers of liquid, flowing mud on normally dry land, often caused by a combination of brush loss and subsequent heavy rains. Mudflows can develop when water saturates the ground, such as from rapid snowmelt or heavy or long periods of rainfall, causing a thick liquid downhill mudflow;

- A midwinter or early spring thaw can produce large amounts of runoff in a short period of time. Because the ground is hard and frozen, water cannot penetrate and be reabsorbed. The water then runs off the surface and flows into lakes, streams, and rivers, causing excess water to spill over their banks; and

- During the spring, frozen land prevents melting snow or rainfall from seeping into the ground. Each cubic foot of compacted snow contains gallons of water and once the snow melts, it can result in the overflow of streams, rivers, and lakes.

CISAR Responder Considerations

The following information is provided for CISAR responders in the conduct of flooding disaster operations. While not all inclusive, this list provides key safety and health information.

(Note: See Section 2-14: Boat Operations Management, for additional information on conducting boat operations during flooding.)

- Flood water may be electrically charged from underground or down power lines;

- Stay clear of downed power lines (report downed power lines to the SMC/IC);

- Unless a CISAR responder is properly equipped and trained, avoid moving flood water;

- Be aware of areas where floodwaters have receded (Roads may have weakened and could collapse under the weight of a car.);

- Use caution when entering buildings (there may be hidden damage, particularly in foundations);

- Flood waters can rearrange and damage natural walkways, as well as sidewalks, parking lots, roads, buildings, and open fields. Never assume that water-damaged structures or ground are stable. Buildings that have been submerged or have withstood rushing flood waters may have suffered structural damage and could be dangerous;

- Fire can pose a major threat to an already badly damaged flood area for several reasons: inoperative fire protection systems, hampered fire department response, inoperable firefighting water supplies, and flood-damaged fire protection systems;

- Assume flood water is contaminated by oil, gasoline, garbage, fecal material from raw sewage, agricultural and industrial waste, etc., and should be considered hazardous material (ICs should consider instituting decontamination protocols for the conduct of flood CISAR operations.);

- Chemical contamination of floodwaters can also occur and contamination levels may be higher nearer to sources such as industrial locations;

- Although skin contact with floodwater does not, by itself, pose a serious health risk, CISAR responders should avoid direct contact with standing water when possible to minimize the chance for infection;

- There is a risk of disease from eating or drinking anything contaminated with floodwater (The most likely symptoms from an infection are stomach-ache, fever, vomiting, and diarrhea.);

- Properly clean and disinfect clothes and equipment that got wet, which can contain sewage and chemicals;

- Wearing wet gloves or PPE can cause skin irritation. Long exposures to wet conditions can compromise the function of the skin barrier. Repeated use of impermeable gloves, especially in hot and humid conditions, can aggravate skin rashes. Cotton liners are sometimes used under protective gloves to improve comfort and to prevent dermatitis. Latex gloves should be avoided because of the risk of developing skin sensitivity or allergy;

- If skin contact with flood water does occur, use soap and water to clean exposed areas. Waterless alcohol-based hand rubs can be used when soap or clean water is not available. Hands should be washed after removal of gloves;

- CISAR responders in flood events should use personal flotation devices when conducting operations within 10 feet of flood water;

- Consider that lifejackets and wetsuits used in polluted flood water as throw away items at the end of the operation;

- CISAR responders in flood operations should have Hepatitis inoculations and current tetanus;

- Don't drink the tap water. Floods are notorious for fouling municipal water systems. Ensure there is plenty of drinking water available. The combination of potentially cold water and exertion may not only exhaust CISAR responders, but cause dehydration as well. Drinking water during flood operations can quickly become a concern;

- Flood water poses drowning risks for CISAR responders and persons in distress, regardless of their swimming ability. Swiftly moving shallow water can be deadly, and even shallow standing water can be dangerous for small children. Rules of thumb:

  o Moving Water: Six inches of water can sweep most people off their feet and into the current; and

  o Standing Water: Operations in standing water greater than "knee deep," will require amphibious vehicle, small boat, and/or helicopter operations.

- Flood waters can displace animals, insects, and reptiles. Be alert and avoid contact;

- Open wounds and rashes exposed to flood waters can become infected. Cover open wounds with a waterproof bandage. Keep open wounds as clean as possible by washing well with soap and clean water. If a wound develops redness, swelling, or drainage, seek immediate medical attention; and

- Protracted CISAR operations in cold water, cold weather, or rain can quickly

sap the efficiency of CISAR responders. Injuries are often the result of exposure and exhaustion. Standing or working in water that is cooler than 75 degrees F (24 degrees C) will remove body heat more rapidly than it can be replaced, resulting in hypothermia. To reduce the risk of hypothermia, wear high rubber boots, ensure that clothing and boots have adequate insulation, avoid working alone, take frequent breaks *out of the water*, and change into dry clothing when possible.

Person in Distress vs. Isolated Person

Person in distress: There is reasonable certainty that a person is threatened by grave and imminent danger and requires immediate assistance.

Isolated person: In an ESF #9 incident only, any non-distressed person or persons stranded within a specific area or residence by incident conditions where *immediate* assistance is determined not to be required.

In flooding disasters, SAR planners must weigh the operational risks to SAR responders concerning immediate assistance to persons in distress vs. isolated persons who are stranded due to the disaster. Isolated persons still require support, especially to ensure that over time they do not become persons in distress. However, SAR planners should give higher priority to survivors in distress situations.

Flooding Safety Information

Table 4-3-1 on the next page provides a brief description of flooding safety information.

Table 4-3-1: Flood Watches and Warnings	
Flood Watch	• Conditions are favorable for flash flooding in and close to the watch area. • May cover States, counties, rivers, portions of States, portions of counties, or portions of rivers (e.g., one or more forecast points). • Issued on a county by county basis by the local Weather Forecast Office and are generally in effect for up to 6 hours.
Flood Warning	• General or areal flooding of streets, low-lying areas, urban storm drains, creeks and small streams is occurring, imminent, or highly likely. • Issued for flooding that occurs more than 6 hours after the excessive rainfall. • Issued on a county by county basis by the local Weather Forecast Office and are generally in effect for 6 to 12 hours.
Flash Flood Warning	• Flash flooding is occurring, imminent or highly likely. • A flash flood is a flood that occurs within 6 hours of excessive rainfall and that poses a threat to life and/or property. Ice jams and dam failures can also cause flash floods. • These warnings are issued on a county by county basis by the local Weather Forecast Office and are generally in effect for up to 6 hours.

Additional Information

Table 4-3-2 below provides websites for additional information on flooding.

Table 4-3-2: Flooding Information Websites	
U.S. Geological Survey (USGS)	http://www.usgs.gov/
USGS Surface Water Information	http://water.usgs.gov/osw/
FEMA	http://www.fema.gov/areyouready/flood.shtm
Centers for Disease Control and Prevention	http://www.bt.cdc.gov/disasters/floods/

Section 4-4: Hurricanes (Typhoons)

Overview

A hurricane is the most severe category of the meteorological phenomenon known as the "tropical cyclone" ("hurricane" and "typhoon" are regionally specific tropical cyclone names).

Tropical cyclones are warm core, non-frontal low pressure systems that develop over tropical or subtropical waters and have a definite organized surface circulation.

The ingredients for a tropical cyclone include a pre-existing weather disturbance, warm tropical oceans, moisture, and relatively light winds aloft. If the right conditions persist long enough, they can combine to produce violent winds, large waves, storm surge, torrential rains, and floods that are associated with hurricanes.

Tropical depressions, tropical storms, and hurricanes are all forms of tropical cyclones, differentiated only by the intensity of the associated winds (Table 4-4-1 below).

Table 4-4-1: Hurricane Development	
Tropical Depression	An organized system of clouds and thunderstorms with a defined surface circulation and maximum sustained winds[1] of 38 mph or less.
Tropical Storm	An organized system of strong thunderstorms with a defined surface circulation and maximum sustained winds of 39-73 mph.
Hurricane	An intense tropical weather system of strong thunderstorms and a well-defined surface circulation and maximum sustained winds of 74 mph or higher.
[1] Sustained winds are defined as a 1-minute average wind measured at about 33 ft (10 meters) above the surface. [2] 1 knot = 1 nautical mile per hour or 1.15 statute miles per hour. Abbreviated as "kt."	

Hurricane Katrina: Hurricane Katrina makes landfall in the bayou near Buras, Louisiana, August 29th, 2005. (NOAA)

Hurricane Origin and Life Cycle

(Note: Information in this Section obtained from: Hurricane Basics (NOAA, May 1999); available on the Internet at: http://hurricanes.noaa.gov/pdf/ hurricanebook.pdf)

In the Atlantic, tropical cyclones form over warm waters from pre-existing disturbances. These disturbances typically emerge every three or four days from the coast of Africa as "tropical waves" that consist of areas of unsettled weather. Tropical cyclones can also form from the trailing ends of cold fronts and occasionally from upper-level lows.

The process by which a tropical cyclone forms and subsequently strengthens into a hurricane depends on at least three conditions:

1. A pre-existing disturbance with thunderstorms;

2. Warm (at least 80°F) ocean temperatures to a depth of about 150 feet; and

3. Light upper level winds that do not change much in direction and speed throughout the depth of the atmosphere (low wind shear).

Heat and energy for the storm are gathered by the disturbance through contact with the warm ocean waters. The winds near the ocean surface spiral into the disturbance's low pressure area. The warm ocean waters add moisture and heat to the air which rises. As the moisture condenses into drops, more heat is released, contributing additional energy to power the storm. Bands of thunderstorms form, and the storm's cloud tops rise higher into the atmosphere. If the winds at these high levels remain relatively light (little or no wind shear), the storm can remain intact and continue to strengthen.

In these early stages, the system is an unorganized cluster of thunderstorms. If weather and ocean conditions continue to be favorable, the system can strengthen and become a tropical depression. At this point, the storm begins to take on the familiar spiral appearance due to the flow of the winds and the rotation of the earth.

If the storm continues to strengthen to tropical storm status, the bands of thunderstorms contribute additional heat and moisture to the storm.

The storm becomes a hurricane when winds reach a minimum of 74 mph. At this time, the cloud-free hurricane eye typically forms because rapidly sinking air at the center dries and warms the area.

Hurricanes begin to decay due to a number of reasons:

- Wind shear can tear a hurricane apart;

- Moving over cooler water or drier areas can lead to weakening; and

- Landfall typically shuts off a hurricane's main moisture source, and the surface circulation can be reduced by friction when it passes over land.

Generally, a weakening hurricane or tropical cyclone can reintensify if it moves into a more favorable region or interacts with mid-latitude frontal systems.

Hurricane Description

- Typical hurricanes are approximately 300 miles wide, although they can vary considerably in size;

- The eye at a hurricane's center is relatively calm, clear area approximately 20-40 miles across;

Eye of Hurricane Isabel (2003). (NASA)

- The eyewall surrounding the eye is composed of dense clouds that contain the highest winds in the storm;

- A storm's outer rainbands (often with hurricane/tropical storm winds) are comprised of dense bands of thunderstorms ranging from a few miles to tens of miles wide and 50-300 miles long;

- Hurricane-force winds can extend outward to about 25 miles in a small hurricane and to more than 150 miles for a large one;

- Tropical storm force winds can stretch out as far as 300 miles from the center of a large hurricane;

- In the Northern Hemisphere, the right side of a hurricane (facing in the direction toward which the storm is moving) is the most dangerous in terms of storm surge, winds, and tornadoes (known as the "dangerous semicircle"):

  o The actual wind speed in the dangerous semicircle is greater than on the left side of the hurricane (known as the "less dangerous semicircle" or "navigable semicircle") due to the pressure gradient alone, since it is augmented by the forward motion of the storm. Because wind speed is greater in the dangerous semicircle, the seas are higher than in the less dangerous semicircle; and

  o The direction of the wind and sea is such as to carry a vessel into the path of the storm (in the forward part of the semicircle).

(Note: In the Southern Hemisphere, the dangerous semicircle is to the left of the storm track, and the less dangerous semicircle is to the right of the storm track.)

- A hurricane's speed and path depend on complex ocean and atmospheric interactions, including the presence or absence of other weather patterns, making it difficult to predict it's speed and direction.

Saffir-Simpson Scale

The U.S. utilizes the Saffir-Simpson hurricane intensity scale for the Atlantic and Northeast Pacific basins to give an estimate of the potential flooding and damage to property given a hurricane's estimated intensity. Table 4-4-2 on the next page is the Saffir-Simpson Scale with likely effects.

Table 4-4-2: Saffir-Simpson Scale[1]	
Category	Wind Speed and Likely Effects
One	**Winds 74-95 mph:** No real damage to building structures, Damage primarily to unanchored mobile homes, shrubbery, and trees. Also, some coastal road flooding and minor pier damage.
Two	**Winds 96-110 mph:** Some roofing material, door, and window damage to buildings. Considerable damage to vegetation, mobile homes, and piers. Small craft in unprotected anchorages break moorings.
Three	**Winds 111-130 mph:** Some structural damage to small residences and utility buildings with a minor amount of curtainwall failures, Mobile homes are destroyed. Flooding near the coast destroys smaller structures with larger structures damaged by floating debris. Terrain may be flooded well inland.
Four	**Winds 131-155 mph:** More extensive curtainwall failures with some complete roof structure failure on small residences. Major erosion of beach areas. Major damage to lower floors of structures near the shore Terrain may be flooded well inland.
Five	**Winds greater than 155 mph:** Complete roof failure on many residences and industrial buildings. Some complete building failures with small utility buildings blown over or away. Major damage to lower floors of all structures located near the shoreline. Massive evacuation of residential areas may be required.

[1] In operational use, the scale corresponds to the 1-minute average sustained wind speed as opposed to gusts which could be 20% higher or more.

Hurricane Hazards

The main hazards associated with tropical cyclones and especially hurricanes are storm surge, high winds, heavy rain, flooding, and tornadoes. Table 4-4-3 below is a summary of hurricane hazards that CISAR responders can anticipate occurring when a hurricane approaches the coast and after landfall.

Table 4-4-3: Hurricane Hazards Summary	
Storm Surge	• Storm surge is the greatest potential threat to life and property associated with hurricanes. • A storm surge is a large dome of water, 50 to 100 miles wide, that sweeps across the coastline near where a hurricane makes landfall. It can be more than 15 feet deep at its peak. • The level of surge in a particular area is primarily related to the intensity of the hurricane and slope of the continental shelf. • The Sea, Lake, and Overland Surges from Hurricanes (SLOSH) model is used by communities to evaluate storm surge threat from different categories of hurricanes striking from various directions. • Because storm surge has the greatest potential to kill more people than any of the other hurricane hazards, it is wise to err on the conservative side by planning for a storm that is one category more intense than is forecast.
High Winds	• Typically, the more intense the storm (in terms of the Saffir-Simpson Hurricane Scale), the more wind damage a community will sustain, particularly if it does not have an effective mitigation program and has not prepared in advance for the storm. • Tropical storm-force winds (39-73 mph) can also be dangerous, and it is wise to have evacuations completed before they reach your area.

Table 4-3-3 is continued on the next page.

Table 4-4-3: Hurricane Hazards Summary (continued)	
Heavy Rains	• Hurricanes (and some tropical storms) typically produce widespread rainfall of 6 to 12 inches or more, often resulting in severe flooding. • Rains are generally heaviest with slower moving storms (less than 10 mph). • The heaviest rain usually occurs to the right of the cyclone track in the period 6 hours before and 6 hours after landfall. • However, storms can last for days, depending on what inland weather features they interact with.
Tornadoes	• Tornadoes are most likely to occur in the right-front quadrant of the hurricane. However, they are also often found elsewhere in the rainbands. • Typically, the more intense a hurricane is, the greater the tornado threat. • Tornado production can occur for days after landfall. • Most tornadoes occur within 150 miles of the coast. • The National Weather Service's Doppler radar systems can provide indications of tornados from a few minutes to about 30 minutes in advance. Consequently, preparedness is critical.
Flooding	• Inland flooding has been the primary cause of tropical cyclone-related fatalities over the past 30 years. • Inland flooding can be a major threat to communities hundreds of miles from the coast due to intense rainfall. • Large amounts of rain can occur more than 100 miles inland where flash floods and mudslides are typically the major threats.

Hurricane Timeline

(Note: Information in this section obtained from: NOAA, Office of Oceanic and Atmospheric Research, Atlantic Oceanographic and Meteorological Laboratory, FAQ H5: What's it like to go through a hurricane on the ground? What are the early warning signs of an approaching tropical cyclone? (13 Aug 2004); available on the Internet at: http://www.aoml.noaa.gov/hrd/tcfaq/ H5.html.)

Figure 4-4-1 on the next page is a generic timeline that provides a general description of a hurricane before, during, and after landfall.

Each hurricane is different. This timeline is provided to help CISAR planners understand the potential impacts and conditions that can be anticipated for a hurricane CISAR response.

Figure 4-4-1: Generic Hurricane Timeline

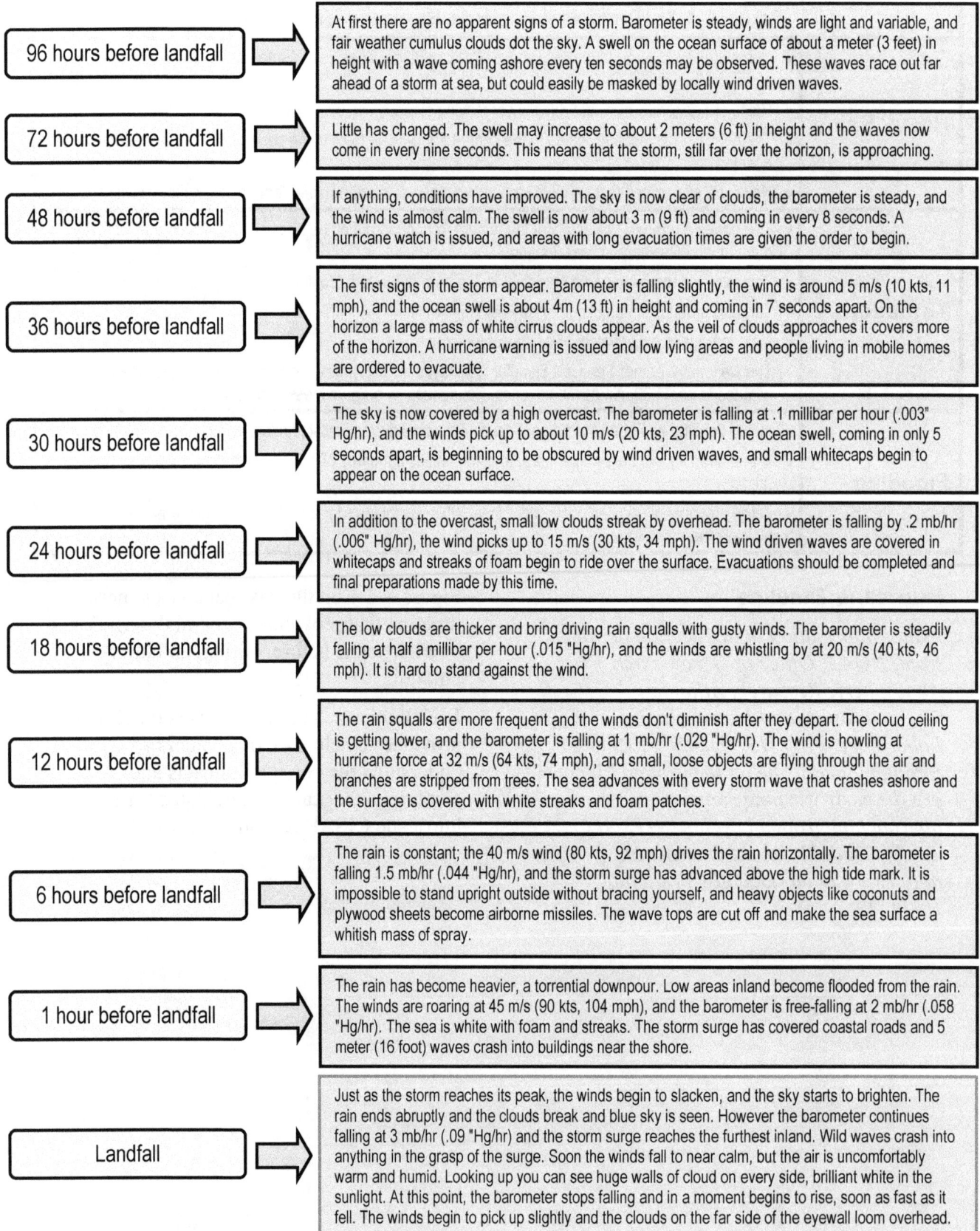

96 hours before landfall →
At first there are no apparent signs of a storm. Barometer is steady, winds are light and variable, and fair weather cumulus clouds dot the sky. A swell on the ocean surface of about a meter (3 feet) in height with a wave coming ashore every ten seconds may be observed. These waves race out far ahead of a storm at sea, but could easily be masked by locally wind driven waves.

72 hours before landfall →
Little has changed. The swell may increase to about 2 meters (6 ft) in height and the waves now come in every nine seconds. This means that the storm, still far over the horizon, is approaching.

48 hours before landfall →
If anything, conditions have improved. The sky is now clear of clouds, the barometer is steady, and the wind is almost calm. The swell is now about 3 m (9 ft) and coming in every 8 seconds. A hurricane watch is issued, and areas with long evacuation times are given the order to begin.

36 hours before landfall →
The first signs of the storm appear. Barometer is falling slightly, the wind is around 5 m/s (10 kts, 11 mph), and the ocean swell is about 4m (13 ft) in height and coming in 7 seconds apart. On the horizon a large mass of white cirrus clouds appear. As the veil of clouds approaches it covers more of the horizon. A hurricane warning is issued and low lying areas and people living in mobile homes are ordered to evacuate.

30 hours before landfall →
The sky is now covered by a high overcast. The barometer is falling at .1 millibar per hour (.003" Hg/hr), and the winds pick up to about 10 m/s (20 kts, 23 mph). The ocean swell, coming in only 5 seconds apart, is beginning to be obscured by wind driven waves, and small whitecaps begin to appear on the ocean surface.

24 hours before landfall →
In addition to the overcast, small low clouds streak by overhead. The barometer is falling by .2 mb/hr (.006" Hg/hr), the wind picks up to 15 m/s (30 kts, 34 mph). The wind driven waves are covered in whitecaps and streaks of foam begin to ride over the surface. Evacuations should be completed and final preparations made by this time.

18 hours before landfall →
The low clouds are thicker and bring driving rain squalls with gusty winds. The barometer is steadily falling at half a millibar per hour (.015 "Hg/hr), and the winds are whistling by at 20 m/s (40 kts, 46 mph). It is hard to stand against the wind.

12 hours before landfall →
The rain squalls are more frequent and the winds don't diminish after they depart. The cloud ceiling is getting lower, and the barometer is falling at 1 mb/hr (.029 "Hg/hr). The wind is howling at hurricane force at 32 m/s (64 kts, 74 mph), and small, loose objects are flying through the air and branches are stripped from trees. The sea advances with every storm wave that crashes ashore and the surface is covered with white streaks and foam patches.

6 hours before landfall →
The rain is constant; the 40 m/s wind (80 kts, 92 mph) drives the rain horizontally. The barometer is falling 1.5 mb/hr (.044 "Hg/hr), and the storm surge has advanced above the high tide mark. It is impossible to stand upright outside without bracing yourself, and heavy objects like coconuts and plywood sheets become airborne missiles. The wave tops are cut off and make the sea surface a whitish mass of spray.

1 hour before landfall →
The rain has become heavier, a torrential downpour. Low areas inland become flooded from the rain. The winds are roaring at 45 m/s (90 kts, 104 mph), and the barometer is free-falling at 2 mb/hr (.058 "Hg/hr). The sea is white with foam and streaks. The storm surge has covered coastal roads and 5 meter (16 foot) waves crash into buildings near the shore.

Landfall →
Just as the storm reaches its peak, the winds begin to slacken, and the sky starts to brighten. The rain ends abruptly and the clouds break and blue sky is seen. However the barometer continues falling at 3 mb/hr (.09 "Hg/hr) and the storm surge reaches the furthest inland. Wild waves crash into anything in the grasp of the surge. Soon the winds fall to near calm, but the air is uncomfortably warm and humid. Looking up you can see huge walls of cloud on every side, brilliant white in the sunlight. At this point, the barometer stops falling and in a moment begins to rise, soon as fast as it fell. The winds begin to pick up slightly and the clouds on the far side of the eyewall loom overhead.

Figure 4-4-1: Generic Hurricane Timeline (continued)

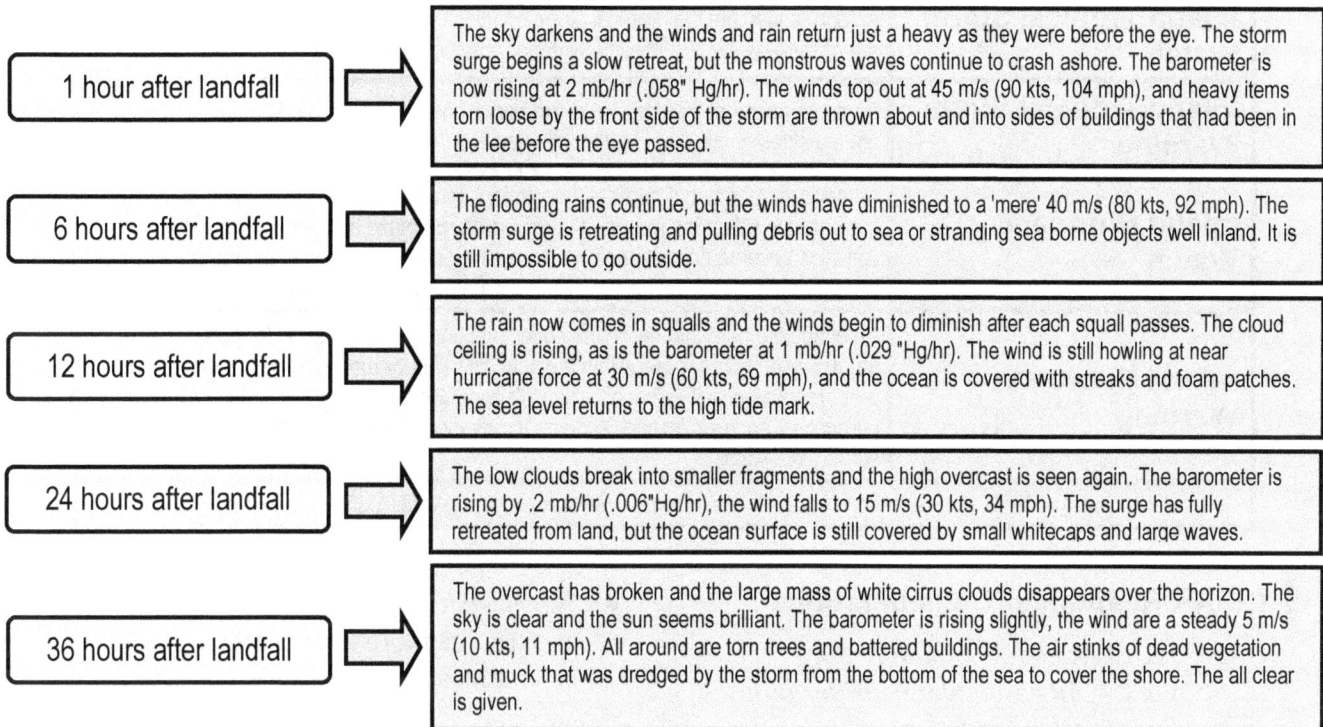

1 hour after landfall ⇒	The sky darkens and the winds and rain return just a heavy as they were before the eye. The storm surge begins a slow retreat, but the monstrous waves continue to crash ashore. The barometer is now rising at 2 mb/hr (.058" Hg/hr). The winds top out at 45 m/s (90 kts, 104 mph), and heavy items torn loose by the front side of the storm are thrown about and into sides of buildings that had been in the lee before the eye passed.
6 hours after landfall ⇒	The flooding rains continue, but the winds have diminished to a 'mere' 40 m/s (80 kts, 92 mph). The storm surge is retreating and pulling debris out to sea or stranding sea borne objects well inland. It is still impossible to go outside.
12 hours after landfall ⇒	The rain now comes in squalls and the winds begin to diminish after each squall passes. The cloud ceiling is rising, as is the barometer at 1 mb/hr (.029 "Hg/hr). The wind is still howling at near hurricane force at 30 m/s (60 kts, 69 mph), and the ocean is covered with streaks and foam patches. The sea level returns to the high tide mark.
24 hours after landfall ⇒	The low clouds break into smaller fragments and the high overcast is seen again. The barometer is rising by .2 mb/hr (.006"Hg/hr), the wind falls to 15 m/s (30 kts, 34 mph). The surge has fully retreated from land, but the ocean surface is still covered by small whitecaps and large waves.
36 hours after landfall ⇒	The overcast has broken and the large mass of white cirrus clouds disappears over the horizon. The sky is clear and the sun seems brilliant. The barometer is rising slightly, the wind are a steady 5 m/s (10 kts, 11 mph). All around are torn trees and battered buildings. The air stinks of dead vegetation and muck that was dredged by the storm from the bottom of the sea to cover the shore. The all clear is given.

Hurricane Safety Information

Table 4-4-4 provides a brief description of hurricane safety information:

Table 4-4-4: Tropical Storm and Hurricane Watches and Warnings	
Tropical Storm Watch	An announcement that tropical storm conditions (sustained winds of 39 to 73 mph) are *possible* within the specified coastal area within 48 hours.
Tropical Storm Warning	An announcement that tropical storm conditions (sustained winds of 39 to 73 mph) are *expected* somewhere within the specified coastal area within 36 hours.
Hurricane Watch	An announcement that hurricane conditions (sustained winds of 74 mph or higher) are *possible* within the specified coastal area. Because hurricane preparedness activities become difficult once winds reach tropical storm force, the hurricane watch is issued 48 hours in advance of the anticipated onset of tropical-storm-force winds.
Hurricane Warning	An announcement that hurricane conditions (sustained winds of 74 mph or higher) are *expected* somewhere within the specified coastal area. Because hurricane preparedness activities become difficult once winds reach tropical storm force, the hurricane warning is issued 36 hours in advance of the anticipated onset of tropical-storm-force winds.

Table 4-4-4 is continued on the next page.

Table 4-4-4: Tropical Storm and Hurricane Watches and Warnings (continued)	
Inland Tropical Storm Watch	An announcement that tropical storm conditions (sustained winds of 39 to 73 mph) are *possible* within the specified interior area within 48 hours.
Inland Tropical Storm Warning	An announcement that tropical storm conditions (sustained winds of 39 to 73 mph) are *expected* somewhere within the specified interior area within 36 hours.
Inland Hurricane Watch	An announcement that hurricane conditions (sustained winds of 74 mph or higher) are *possible* within the specified interior area. Because hurricane preparedness activities become difficult once winds reach tropical storm force, the hurricane watch is issued 48 hours in advance of the anticipated onset of tropical-storm-force winds.
Inland Hurricane Warning	An announcement that hurricane conditions (sustained winds of 74 mph or higher) are *expected* somewhere within the specified interior area. Because hurricane preparedness activities become difficult once winds reach tropical storm force, the hurricane warning is issued 36 hours in advance of the anticipated onset of tropical-storm-force winds.
Ref: National Hurricane Center Hurricane Preparedness Website (http://www.nhc.noaa.gov/HAW2/english/forecast/warnings.shtml)	

CISAR Responder Considerations

(Note: See Section 4-3: Flooding, for additional CISAR responder considerations in flooding disaster operations.)

June 1 to November 30 is the official hurricane season.

During an average season, ten tropical storms will develop in the Atlantic Basin, of which six will become hurricanes. Many of these will remain over the ocean and not affect any landmass. However, about five hurricanes strike the U.S. coastline every three years. Of these five, two will be major hurricanes (Category 3 or greater on the Saffir Simpson Hurricane Scale). All five storms will move inland to begin a decaying process, producing torrential rains, flooding, and flash flooding many miles from their impact points on the coast.

In addition to wind, tornadoes, and storm surge flooding, CISAR responders need to be aware of inland freshwater flooding. Tropical cyclones can have life-threatening effects hundreds of miles inland.

Factors affecting inland flooding caused by hurricanes include:

- *Storm speed* – the slower the system moves, the more time for the rains to fall over a location.

- *Orography* – Lifting of the warm, moist tropical air over geographical barriers such as hills and mountains. Also the gradual increase in elevation as the system moves inland amplifies and intensifies the rain.

- *Interaction with other weather features* – For example, Hurricane Agnes (1972) fused with another storm system and produced floods in the northeast that caused 122 deaths.

- *Antecedent conditions* – The wetness or dryness of the soil, the existing capacity of streams, rivers, ponds, lakes, and reservoirs.

Additional Information

- Table 4-4-5 on the next page provides websites for additional information on hurricanes and hurricane preparedness.

Table 4-4-5: Hurricane Information Websites	
NOAA	http://www.noaa.gov/
National Hurricane Center	http://www.nhc.noaa.gov/index.shtml
National Weather Service	http://www.nws.noaa.gov/
USGS Natural Hazards Website	http://www.usgs.gov/natural_hazards/
Centers for Disease Control and Prevention	http://www.bt.cdc.gov/disasters/hurricanes/index.asp

Section 4-5: Tornadoes

Overview

(Note: Disasters caused by tornadoes do not normally include ESF #9 Federal SAR assistance. The information in this Section is provided to assist in tornado disaster response planning.)

Although tornadoes occur in many parts of the world, they are found most frequently in the U.S. east of the Rocky Mountains during the spring and summer months. In an average year, 800 tornadoes are reported nationwide, resulting in 80 deaths and over 1,500 injuries. The most violent tornadoes are capable of tremendous destruction with wind speeds of 250 mph or more. Damage paths can be in excess of one mile wide and 50 miles long.

Thunderstorms develop in warm, moist air in advance of eastward-moving cold fronts. These thunderstorms often produce large hail, strong winds, and tornadoes. Tornadoes in the winter and early spring are often associated with strong, frontal systems that form in the Central States and move east. Occasionally, large outbreaks of tornadoes occur with this type of weather pattern.

Several States may be affected by numerous severe thunderstorms and tornadoes.

Tornado: Union City, Oklahoma, May 24th, 1973 (NOAA Photo Library, NOAA Central Library; OAR/ERL/National Severe Storms Laboratory (NSSL))

During the spring in the Central Plains, thunderstorms frequently develop along a "dryline," which separates very warm, moist air to the east from hot, dry air to the west. Tornado-producing thunderstorms may form as the dryline moves east during the afternoon hours.

Along the front range of the Rocky Mountains, in the Texas panhandle, and in the southern High Plains, thunderstorms frequently form as air near the ground flows "upslope" toward higher terrain. If other favorable conditions exist, these thunderstorms can produce tornadoes.

Tornadoes occasionally accompany tropical storms and hurricanes that move over land. Tornadoes are most common to the right and ahead of the path of the storm center as it comes onshore.

Enhanced Fujita Scale (EF-Scale)

The EF-Scale, which became operational on February 1, 2007, is used to assign a tornado a "rating" based on estimated wind speeds and related damage. When tornado-related damage is surveyed, it is compared to a list of Damage Indicators (DIs) and Degrees of Damage (DOD) which help estimate better the range of wind speeds the tornado likely produced. From that, a rating (from EF0 to EF5) is assigned.

The EF-Scale was revised from the original Fujita Scale to reflect better examinations of tornado damage surveys so as to align wind speeds more closely with associated storm damage. The new scale is based on how most structures are designed (Table 4-5-1 below).

Table 4-5-1: EF-Scale	
EF Rating	3 Second Gust (mph)
0	68-85
1	86-110
2	111-135
3	136-165
4	166-200
5	Over 200

(Note: The EF-Scale is still a set of wind estimates (not measurements) based on damage. Its uses three-second gusts estimated at the point of damage based on a judgment of 8 levels of damage to 28 damage indicators. The 3 second gust is not the same wind as in standard surface wind observations, which are taken by weather stations in open exposures, using a directly measured "one minute mile" speed.)

Description

- Tornadoes are the most violent of all atmospheric storms;

- A tornado is defined as a violently rotating column of air extending from a thunderstorm to the ground;

- Tornadoes may strike quickly, with little or no warning;

- Tornadoes may form during the early stages of rapidly developing thunderstorms (most common along the front range of the Rocky Mountains, the Plains, and the Western States);

- Tornadoes may appear nearly transparent until dust and debris are picked up;

- Tornadoes can accompany tropical storms and hurricanes as they move onto land;

- Occasionally, two or more tornadoes may occur at the same time;

- Most tornadoes (but not all) in the Northern Hemisphere spin counterclockwise;

- In the Southern States, peak tornado occurrence is in March through May, while peak months in the Northern States are during the summer. In some States, a secondary tornado maximum occurs in the fall;

- Tornadoes are most likely to occur between 3 and 9 p.m., but have been known to occur at all hours of the day or night;

- The average tornado moves from southwest to northeast, but tornadoes have been known to move in any direction;

- The average forward speed is 30 mph but may vary from nearly stationary to 70 mph; and

- Waterspouts:

  o Waterspouts are weak tornadoes that form over warm water.

  o Waterspouts are most common along the Gulf Coast and southeastern States;

  o In the western U.S., waterspouts occur with cold late fall or late winter storms, during a time when tornado development is least expected;

  o Waterspouts occasionally move inland becoming tornadoes causing damage and injuries.

Table 4-5-2 below categorizes the various tornado shapes and sizes.

Table 4-5-2: Tornado Shapes/Sizes	
Weak Tornadoes	• 69% of all tornadoes. • Less than 5% of tornado deaths. • Lifetime: 1-10+ minutes. • Winds less than 110 mph.
Strong Tornadoes	• 29% of all tornadoes. • Nearly 30% of all tornado deaths. • Lifetime: May last 20 minutes or longer. • Winds 110-205 mph.
Violent Tornadoes	• Only 2% of all tornadoes. • 70% of all tornado deaths. • Lifetime: Can exceed 1 hour. • Winds greater than 205 mph.

Surviving a Tornado

Some tornadoes strike rapidly without time for a tornado warning, and sometimes without a thunderstorm in the vicinity. When watching for rapidly emerging tornadoes, it is important to know that you cannot depend on seeing a funnel: clouds or rain may block your view. The following weather signs may mean that a tornado is approaching:

- A dark or green-colored sky;

- A large, dark, low-lying cloud;

- Large hail; and/or

- A loud roar that sounds like a freight train.

If any of these weather conditions are observed, take cover immediately, and keep tuned to local radio and TV stations or to a NOAA weather radio.

If a funnel cloud is observed in the vicinity, take shelter immediately. If a tornado is observed far away, notify the authorities or local TV station before taking shelter.

Tornado Safety Information

Table 4-5-3 below provides a brief description of tornado watches and warnings:

Table 4-5-3: Tornado Watches and Warnings	
Tornado Watch	• Issued by the National Weather Service when conditions are favorable for the development of tornadoes in and close to the watch area. Their size can vary depending on the weather situation. • Tornado Watches are usually issued for a duration of 4 to 8 hours. • Issued well in advance of the actual occurrence of severe weather. During the watch, people should review tornado safety rules and be prepared to move a place of safety if threatening weather approaches. • A Tornado Watch is issued by the Storm Prediction Center (SPC) in Norman, Oklahoma. Prior to the issuance of a Tornado Watch, SPC will usually contact the affected local National Weather Forecast Office (NWFO) and they will discuss what their current thinking is on the weather situation. Afterwards, SPC will issue a preliminary Tornado Watch and then the affected NWFO will then adjust the watch (adding or eliminating counties/parishes) and then issue it to the public. After adjusting the watch, the NWFO will let the public know which counties are included by way of a Watch Redefining Statement. During the watch, the NWFO will keep the public informed on what is happening in the watch area and also let the public know when the watch has expired or been cancelled.
Tornado Warning	• Issued when a tornado is indicated by the WSR-88D radar or sighted by spotters; therefore, people in the affected area should seek safe shelter immediately. • Tornado Warnings can be issued without a Tornado Watch being already in effect. • Tornado Warnings are usually issued for a duration of around 30 minutes. • Issued by the local National Weather Service office (NWFO). It will include where the tornado was located and what towns will be in its path. If the tornado will affect the nearshore or coastal waters, it will be issued as the combined product--Tornado Warning and Special Marine Warning. If the thunderstorm which is causing the tornado is also producing torrential rains, this warning may also be combined with a Flash Flood Warning. If there is an ampersand (&) symbol at the bottom of the warning, it indicates that the warning was issued as a result of a severe weather report. • After the Tornado Warning has been issued, the affected NWFO will followed it up periodically with Severe Weather Statements. These statements will contain updated information on the tornado and they will also let the public know when warning is no longer in effect.

CISAR Responder Considerations

(Note: See Section 4-2: Earthquakes, for CISAR considerations for collapsed structures.)

Additional Information

Table 4-5-4 below provides websites for additional information on tsunamis and tsunami preparedness.

Table 4-5-4: Tornado Information Websites	
NOAA's National Weather Service	http://www.nws.noaa.gov/om/brochures/tornado.shtml
NOAA Storm Prediction Center	http://www.spc.noaa.gov/faq/tornado/
NOAA National Severe Storms Laboratory	http://www.nssl.noaa.gov/primer/tornado/tor_basics.html
CDC Prevention	http://emergency.cdc.gov/disasters/tornadoes/

Section 4-6: Tsunamis

Overview

A "tsunami" (soo-NAH-mee) is a series of traveling ocean waves of extremely long length generated primarily by earthquakes occurring below or near the ocean floor (Figure 4-6-1 below).

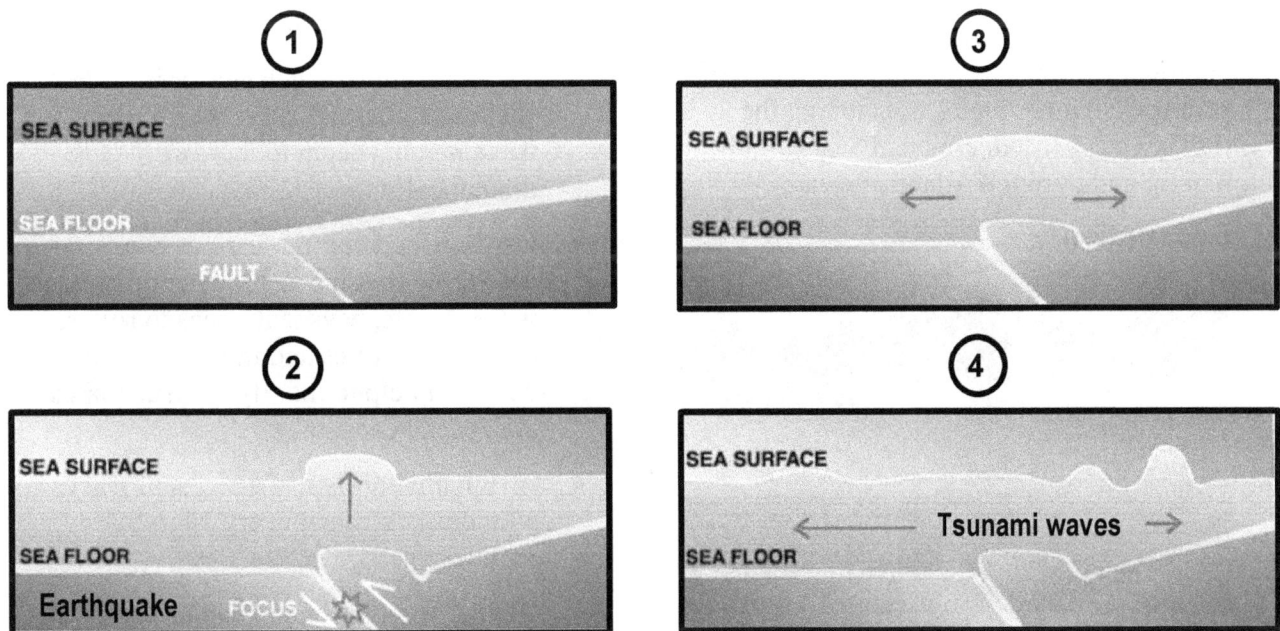

The earthquake focus is the point in the earth where the rupture first occurs and where the first seismic waves originate. The epicenter is the point on the Earth's surface directly above the focus.

Sudden displacement moves overlying column of water generating tsunami wave(s). As a tsunami wave enters shallow water, the wave speed slows and its height increases, creating destructive, live-threatening waves.

Figure 4-6-1: Earthquake Generated Tsunami

Underwater volcanic eruptions and landslides can also generate tsunamis. In the deep ocean, the tsunami waves propagate across the deep ocean with a speed exceeding 500 miles per hour, and a wave height of only 1 foot or less.

Tsunami waves are distinguished from ordinary ocean waves by their great length between wave crests, often exceeding 60 miles or more in the deep ocean, and by the time between these crests, ranging from 10 minutes to an hour. As they reach the shallow waters of the coast, the waves slow down and the water can pile up into a wall of destruction 30 feet or more in height. The effect can be amplified where a bay, harbor or lagoon funnels the wave as it moves inland. Large tsunamis have been known to rise over 100 feet. Even a tsunami 3-6 meters high can be very destructive and cause many deaths and injuries.

Although 60% of all tsunamis occur in the Pacific, they can also threaten coastlines of countries in other regions, including the Indian Ocean, Mediterranean Sea, Caribbean region, and even the Atlantic Ocean.

Description

Tsunami: Hilo Hawaii; April 1st, 1946 (NOAA)

- Tsunamis (also called "seismic sea waves," or incorrectly "tidal waves") are caused generally by earthquakes, less commonly by submarine landslides, infrequently by underwater volcanic eruptions and very rarely by large meteorite impacts in the ocean;

- All ocean regions of the world can experience tsunamis, but in the Pacific Ocean and its marginal seas, there is a much more frequent occurrence of large, destructive tsunamis because of the many large earthquakes along the margins of the Pacific Ocean;

- In the deep ocean, destructive tsunamis can be small – often only a few tens of centimeters or less in height – and cannot be seen or felt on ships at sea (the tsunami reaches shallower coastal waters, wave height can increase rapidly);

- Where the ocean is over 6,000 meters deep, unnoticed tsunami waves can travel at the speed of a commercial jet aircraft (over 500 mph), traveling across the Pacific Ocean in less than a day;

- Tsunamis travel much slower in shallower coastal waters where their wave heights begin to increase dramatically;

- Offshore and coastal features can determine the size and impact of tsunami waves (reefs, bays, entrances to rivers, undersea features and the slope of the beach all help to modify the tsunami as it attacks the coastline);

- A tsunami consists of a series of waves with crests arriving every 10 to 60 minutes;

- Danger from a tsunami can last for several hours after the arrival of the first wave;

- Tsunami waves typically do not curl or break;

- Tsunamis that strike coastal locations are almost always caused by earthquakes.

- When a tsunami hits the coast, often as a wall of water, sea levels can rise many meters. In extreme cases, the water level has risen to more than 50 feet for tsunamis of distant origin and over 100 feet for tsunami waves generated near the epicenter;

- The first tsunami wave may not be the largest in a series of waves;

- One coastal community may see no damaging wave activity while in another nearby community destructive waves can be large and violent;

- Sometimes coastal waters are drawn out into the ocean just before the tsunami strikes. When this occurs, more shoreline may be exposed than even at the lowest tide and is a warning;

- Flooding from tsunamis can extend inland by 0.5 miles or more, covering large expanses of land with water and debris; and

- Tsunamis can occur at any time, day, or night.

Impact

(Photo: NOAA)

- Deaths occur by drowning and physical impact or other trauma when people are caught in the turbulent, debris-laden tsunami waves;

- Strong tsunami-induced currents have led to the erosion of foundations and the collapse of bridges and seawalls;

- Flotation and drag forces can demolish frame buildings and other structures;

- Considerable damage may be caused by floating debris, including boats, cars, and trees that become dangerous projectiles that may crash into building, piers, and other vehicles;

- Ships and port facilities may be damaged by surge action caused by even weak tsunamis;

- Fires resulting from oil spills or combustion from damaged ships in port, or from ruptured coastal oil storage and refinery facilities, can cause damage greater than that inflicted by the tsunami wave; and

- Other secondary damage can result from sewage and chemical pollution following the destruction.

Surviving a Tsunami

(Note: Information in this Section obtained from: Brian F. Atwater, Marco Cisternas V., Joanne Bourgeois, Walter C. Dudley, James W. Hendley II, and Peter H. Stauffer, Surviving a Tsunami – Lessons Learned from Chile, Hawaii, and Japan, Circular 1187, Version 1.1 (U.S. Geological Survey, 2005); available on the Internet at: http://pubs.usgs.gov/circ/c1187/.)

CISAR Responders in a Potential Earthquake/Tsunami Impact Region

The most common cause of tsunamis is earthquakes.

CISAR responders in potential earthquake/tsunami prone regions must heed tsunami advisories in case an earthquake or other disaster occurs that generates life-threatening tsunami waves.

The worst case scenario is for an earthquake to occur, CISAR responders begin lifesaving operations, and a tsunami wave impacts the affected area.

In addition, CISAR responders need to understand *there will be more than one tsunami wave that impacts the affected area minutes to hours later.* The initial wave may not be the largest wave.

CISAR responders MUST heed disaster/earthquake/ tsunami warning information.

The following information is provided for surviving a tsunami.

- Heed natural warnings. Strong ground shaking serves as a warning that a tsunami is coming, as well as an unusual or rapid fall or rise of coastal waters;

- Heed official warnings. Even if warnings seem ambiguous or you think the danger has passed;

- Expect:
  o Many tsunami waves;
  o The next tsunami wave to be bigger than the previous; and
  o The tsunami waves to continue for many hours.

- Head for higher ground and stay there. Move uphill or at least inland, away from the coast;

- Abandon belongings. Save your life, not your possessions;

- Don't count on the roads. You may find roads broken or blocked;

- If unable to reach higher ground, go to an upper story or roof of a sturdy building;

- If trapped in the impacted area, as a last resort climb a strong tree;

- If swept up by a tsunami wave, look for something as big as possible to use as a raft;

- Expect tsunami waves to carry debris (e.g., houses, cars, etc.), which can cause injury;

- Expect earthquakes to lower nearby coastal regions, allowing for permanent tidal water flooding; and

- Expect company. Shelter your neighbors.

Tsunami Rule of Thumb:

"If you can see a tsunami, you're too close."

Tsunami Safety Information

The International Tsunami Warning System monitors the world's oceans through a network of buoys and scientific instruments. When a major earthquake is detected with the potential for a tsunami to occur, warnings are issued to local authorities who can order the evacuation of low-lying areas, as required.

NOAA's National Weather Service operates two tsunami warning centers:

- West Coast/Alaska Tsunami Warning Center (WC/ATWC), Palmer, Alaska (Serves Alaska, Washington, Oregon, California, the U.S. Atlantic and Gulf of Mexico coasts, Puerto Rico, the U.S. Virgin Islands and Canada); and

- Pacific Tsunami Warning Center (PTWC), Ewa Beach, Hawaii (Serves Hawaii and the U.S. Pacific Territories, and as an international warning center for the Pacific and Indian Oceans and the Caribbean Sea).

Table 4-6-1 provides a brief description of tsunami safety information:

Table 4-6-1: Tsunami Advisories, Watches, and Warnings	
Tsunami Advisory	<ul><li>An earthquake occurred, which might generate a tsunami and produce strong currents or waves dangerous to those in or near the water.</li><li>Coastal regions historically prone to damage due to strong currents induced by tsunamis are at the greatest risk.</li><li>The threat is likely to continue for many hours after the arrival of the initial wave, but significant widespread inundation is not expected for areas under an advisory.</li><li>Local officials may close beaches, evacuate harbors and marinas, and reposition ships to deep waters when there is time to safely do so.</li><li>Advisories are normally updated to continue the advisory, expand/contract affected areas, upgrade to a warning, or cancel the advisory.</li></ul>
Tsunami Information Statement	<ul><li>An earthquake occurred or a tsunami watch, advisory, or warning was issued for another section of the ocean.</li><li>In most cases, information statements are issued to indicate there is no threat of a destructive tsunami and to prevent unnecessary evacuations as the earthquake may have been felt in coastal areas.</li><li>An information statement may, in appropriate situations, caution about the possibility of destructive local tsunamis.</li><li>The National Weather Service will update information statements as required. However, a watch, advisory, or warning may be issued for the impacted area after analysis and/or updated information becomes available.</li></ul>
Tsunami Watch	<ul><li>A tsunami was or may have been generated, but is at least two hours travel time to the area in watch status.</li><li>The watch area may be upgraded to an advisory or warning or canceled based on updated information and analysis. Therefore, emergency management officials and the public should prepare to take action.</li><li>Watches are normally issued based on seismic information without confirmation that a destructive tsunami is underway.</li></ul>
Tsunami Warning	<ul><li>A potential tsunami with significant widespread inundation is imminent or expected.</li><li>Warnings alert the public that widespread, dangerous coastal flooding accompanied by powerful currents is possible and likely to continue for many hours after arrival of the initial wave.</li><li>Warnings also alert emergency management officials to take action for the entire tsunami warning zone.</li><li>Local officials may evacuate low-lying coastal areas and reposition ships to deep water when there is time to safely do so.</li><li>Warnings are updated, adjusted geographically, downgraded, or canceled.</li><li>To provide the earliest possible alert, initial warnings are normally based only on seismic information.</li></ul>

CISAR Responder Considerations

(Note: See Section 4-3: Flooding for CISAR responder considerations in flooding disaster operations.)

In addition to the information found in *Section 4-4: Flooding*, the following information is provided for CISAR responders in the conduct of tsunami disaster operations.

- The majority of deaths associated with tsunamis are related to drowning, but traumatic injuries are also a primary concern. Injuries such as broken limbs and head injuries are caused by the physical impact of people being washed into debris such as houses, trees, and other stationary items;

- As the water recedes, the strong suction of debris being pulled into large populated areas can further cause injuries and undermine buildings and services;

- Tsunami flood waters can pose health risks such as contaminated water and food supplies;

- Water quality:

  - As the ocean water comes ashore, drinking water wells can become submerged and potentially contaminated with microorganisms (bacteria, viruses, parasites) and chemicals that can adversely affect human health (sea salts associated with saltwater flooding of coastal drinking water supplies are not an immediate health threat);

  - Because of the unpleasant taste of saltwater, most people will not ingest (swallow) a large enough amount to cause immediate health problems; and

  - Disease-causing microorganisms spread by the flood do not normally produce a strong taste. If water containing disease-causing microorganisms is ingested, even in small amounts, it may cause immediate, life-threatening health problems such as chronic diarrhea, cholera, and serious infections.

- Expect communication problems. Cell phone, radio, and repeater towers may be disabled. Although the actual tsunami waves may have little effect on tower legs and telephone poles (water may flow around and through towers), the debris carried by the tsunamis (e.g., cars, buildings, trees, refrigerators, etc.) smashing into the towers may cause failures;

- Search high as well as low. People may be thrown into trees, on top of buildings, as well as buried under piles of debris;

- Roads may be blocked by debris or washed away. Rescue vehicles may not be able to get far into the affected area. Expect to do a lot of walking. You may have to carry what you need; and

- CISAR responders may find large numbers of bodies, or may not find any bodies at all. The tsunami waves could very well sweep bodies out to sea.

Additional Information

Table 4-6-2 on the next page provides websites for additional information on tsunamis and tsunami preparedness.

Table 4-6-2: Tsunami Information Websites	
FEMA Tsunami Preparedness and Information	http://www.fema.gov/hazard/tsunami/index.shtm
NOAA Tsunami Information	http://www.tsunami.noaa.gov/
National Weather Service Pacific Tsunami Warning Center	http://www.weather.gov/ptwc/
National Weather Service West Coast and Alaska Tsunami Warning Center	http://wcatwc.arh.noaa.gov/
National Weather Service Tsunami Ready Community Website	http://www.tsunamiready.noaa.gov/
International Tsunami Information Center	http://www.tsunamiwave.info/
U.S. Geological Survey Tsunamis and Earthquakes Homepage	http://walrus.wr.usgs.gov/tsunami/
Centers for Disease Control and Prevention	http://emergency.cdc.gov/disasters/tsunamis/

Part 5: CBRNE Incident

This page intentionally left blank.

Section 5-1: Chemical, Biological, Radiological, Nuclear and High-Yield Explosive (CBRNE) Incident – Introduction

Overview

Specialized CBRNE Response Teams

Weapons of Mass Destruction-Civil Support Team (WMD-CST)

CBRNE Enhanced Response Force Package (CERFP)

Homeland Response Force (HRF)

DoD CBRN Response Force (DCRF)

Command and Control CBRN Response Element (C2CRE) A/B

Overview

CISAR operations may be conducted within the context and environment associated with a CBRNE incident, which may have been caused either by accident or act of terrorism. In such situations, CISAR responders should:

- Be concerned with personnel safety;

- Understand and be aware of impacts on SAR capabilities;

- Be aware of the impact these types of situations have on government and public behavior; and

- Understand and adhere to instructions from experts and authorities in charge.

CBRNE incidents represent particular challenges for the traditional ICS structure. Events that are not site specific, are geographically dispersed, or evolve over longer periods of time will require extraordinary coordination between Federal, State, local, tribal, private-sector, and nongovernmental organizations. An area command may be established to oversee the management of such incidents.

Specialized CBRNE Response Teams

Relevant emergency plans should identify available specialized teams that can be used to assist with CISAR operations within a CBRNE environment.

Many Federal, State, Tribal, Territorial, and local agencies and organizations have such teams; however, their immediate availability varies.

A primary resource available to assist in a CBRNE incident is the National Guard, which can mobilize two types of special response teams:

- Weapons of Mass Destruction (WMD) – Civil Support Teams (CSTs);

- CBRN Enhanced Response Force Packages (CERFP);

- Homeland Response Forces (HRFs);

- DoD CBRN Response Force (DCRF); and

- Command and Control CBRN Response Element (C2CRE (A/B)).

Weapons of Mass Destruction – Civil Support Team (WMD-CST)

There are 57 WMD-CSTs assigned to the NG. This is a high priority, rapid response unit made up of 22 fulltime Title 32 Active Guard Reserve NG personnel. The unit is jointly manned by both Army and Air National Guard personnel. These units were established by Congress, certified by SECDEF to Congress, and are assigned to every State, Territory and the District of Columbia with two WMD-CSTs in California, Florida and New York.

By statute, each WMD-CST operates under the control of the Governor, and can be employed as a State asset without DoD authorization. Under presidential mobilization the WMD-CST could be employed as part of a Title 10 response force package.

Each WMD-CST is prepared to conduct an immediate response deployment within 1.5 hours of notification for the Advanced Echelon (ADVON) and 3 hours for the main body. WMD-CSTs will:

- Usually be the first DoD responders to support civil authorities at any domestic CBRN incident site or event;

- May provide a response capability for an intentional or unintentional release of nuclear, biological, radiological, toxic or poisonous chemical materials; or,

- Provide a response capability to man-made or natural disasters in the U.S. that result in, or could result in catastrophic loss of life or property.

The WMD-CST mission is to identify CBRN agents/substances, assess current or projected consequences, advise on response measures and assist with appropriate requests for state support to facilitate additional resources.

On site, the WMD-CST is prepared to conduct and sustain continuous operations for 72 hours.

CBRN Enhanced Response Force Package (CERFP)

There are 17, approximately 186-person CERFPs currently authorized by Congress. CERFP units are sourced by the NG and are a DoD capability under the control of the State Governors unless federalized. CERFPs are a key element of the DoD's overall program to provide military support to civil authorities in the event of an intentional or accidental CBRNE incident in the U.S.

Within the national response continuum, CERFP teams are required to be ready for deployment in 6 hours. CERFPs are provided additional special training and equipment to plan and conduct personnel casualty search and extraction; emergency medical triage, treatment, and patient stabilization; and mass casualty decontamination.

Homeland Response Force (HRF)

Ten HRFs will be established in 2012 and will incorporate a CERFP as the core CBRN response element. There will be one 566-person HRF within each FEMA Region. HRFs are a DoD capability under the control of the Governors unless federalized.

The HRF includes a security capability which will be tasked by the Command and Control (C2) HRF element and will respond directly to State tasking in support of local incident command and possibly other State security missions.

The HRF construct will also include a Command and Control (C2) element operating as directed by the State Joint Force Headquarters (JFHQ) and/or Joint Task Force – State (JTF-State).

DoD CBRN Response Force (DCRF)

The 5,200 person DCRF is the Title 10 Active Component, Title 10 Federal, and Reserve Component Title 10 Federal, element designated as the lead DoD Federal response to a CBRN event within the USNORTHCOM or USPACOM AOR and responding to SECDEF direction through the CDR, USNORTHCOM.

The headquarters element of the DCRF is Joint Task Force Civil Support (JTF-CS), which is assigned to USNORTHCOM under the operational control of the USNORTHCOM Joint Forces Land Component Commander (JFLCC).

There is only one DCRF. The Title 10 DCRF includes additional capability besides multiple Immediate Response Forces (IRFs) to provide robust command and control, aviation, level II and III medical care, transportation and engineering units in a responsive capability in order to enhance life-saving operations early in any CBRNE catastrophic event. In a similar manner, the Title 10 Task Force operations of the DCRF will operate under the C2 of the Title 10

Joint Task Force commander managing work that has been requested in support of the Federal response. The C2 element must also be prepared to direct and support additional CERFPs, WMD-CSTs, or other military units.

Command and Control CBRN Response Element (C2CRE) A/B

C2CRE A/B are designated to respond to subsequent/additional CBRN events, augment the DCRF, or respond to other simultaneous events. C2CREs respond to SECDEF direction through the CDR, USNORTHCOM.

The headquarters element of a C2CRE is a Contingency Command Post (CCP) headquarters assigned to U.S. Army North (USARNORTH). This CCP may be CCP 1, CCP2, or another JTF, if so directed.

A C2CRE can provide minimal immediate life-saving capability by the assigned Technical Support Force (TSF) IRF and as the core capability for additional contingency Request for Forces (RFF) units, also known as Follow-on Forces.

This page intentionally left blank.

Section 5-2: CBRNE – Chemical Incident

Nature of Chemical Incidents

Chemical agents are poisonous vapors, aerosols, liquids, and solids that have toxic effects on people, animals, or plants. Chemical agents can be:

- Released by industrial or transportation accident;

- Released by bombs or sprayed from aircraft, boats, and vehicles;

- Used as a liquid to create a hazard to people and the environment; and

- Odorless, tasteless, and have an immediate (within a few seconds or minutes) or a delayed (2 to 48 hours) effect.

While potentially lethal, chemical agents are difficult to deliver in lethal concentrations. Outdoors, the rate that some chemical agents dissipate varies depending upon type of agent, whether it is a liquid or vapor and what types of surfaces it deposits upon. For example, grass, sand and soil surfaces may absorb a chemical agent in times ranging from a few minutes to approximately four hours. Painted surfaces also sorb liquid chemical agents, but over a longer timeframe. This results in a liquid hazard which varies from up to one hour on most painted surfaces to as long as six hours for some paints. Impervious surfaces, such as glass or unpainted metal, retain a liquid hazard for longer periods unless the agent is removed. Chemical agents also are difficult to produce, transport, and deliver.

Indicators of Chemical Agents

Table 5-2-1 on the next page identifies possible indicators of chemical agents.

Table 5-2-1: Possible Indicators of Chemical Agents:
Unexplained dead or dying animals or lack of insects.
Unexplained casualties: multiple victims, serious illness, nausea, disorientation, breathing difficulty, convulsions, or other chemical-indicative casualty patterns.
Unusual liquid, spray or vapor, droplets, oily film; unexplained odor, low flying clouds, or fog unrelated to weather.
Suspicious devices or packages, unusual metal debris; abandoned spray devices or unexplained munitions.

Chemical Incident Symptoms

A chemical attack could come without warning. Onset of symptoms could become present in seconds, hours, or even days. Signs of a chemical release include people having difficulty breathing, experiencing eye irritation, losing coordination, becoming nauseated, or having a burning sensation in the nose, throat, and lungs. This may also include burning, itching, red skin, prominent tearing/burning/redness of eyes, eyelid edema, shortness of breath, nausea and vomiting, cough, chest tightness, or sore throat.

Chemical Incident Decontamination

Decontamination is required within minutes of exposure to minimize health consequences, or as soon as possible. Do not leave the safety of a shelter to go outdoors to help others until authorities announce it is safe to do so.

Chemical exposure requires immediate professional medical attention. If medical help is not immediately available, decontaminate yourself and assist in decontaminating others by using the following procedures in Figure 5-2-1.

Chemical Incident Decontamination Procedures

Using extreme caution, remove all clothing and other items in contact with the body.

(Note: Contaminated clothing normally removed over the head should be cut off to avoid contact with eyes, nose, and mouth.)

Put contaminated clothing and items into a plastic bag and seal it. Decontaminate hands using soap and water. Remove eyeglasses or contact lenses. Put glasses in a pan of household bleach (dilution 0.5%) to decontaminate, then rinse and dry.

↓

Flush eyes with water.

↓

Gently wash face and hair with soap and water before thoroughly rinsing with water.

↓

Decontaminate other body areas likely to have been contaminated with a cloth soaked in soapy water and rinse with clear water.

↓

Change into uncontaminated clothes (clothing stored in drawers or closets is likely to be uncontaminated).

↓

Proceed to a medical facility for screening and professional treatment.

Figure 5-2-1: Chemical Incident Decontamination Procedures

Clothing Disposal

After washing yourself, place your clothing inside a plastic bag. Avoid touching contaminated areas of the clothing. If you cannot avoid touching contaminated areas, or you are not sure where the contaminated areas are, wear rubber gloves or put the clothing in the bag using tongs, tool handles, sticks, or similar objects. Anything that touches the contaminated clothing should also be placed in the bag. If you wear contacts, put them in the plastic bag as well.

Seal the bag, and then seal the bag inside another plastic bag. Disposing of your clothing in this way will help protect you and other people from any chemicals that might be on your clothes.

CISAR Response to a Chemical Incident

The first concern must be to recognize a chemical event and protect the CISAR responders. Unless CISAR responders recognize the danger, they will very possibly become casualties in a chemical environment. It may not be possible to determine from the symptoms experienced by affected personnel which chemical agent has been used. Chemical agents may be combined and therefore recognition of agents involved becomes more difficult.

When there is reason to believe that chemical agents may be present, the following should be considered:

- Before approaching the scene, don personnel protective equipment (PPE) specifically designed for use in hazardous chemical environments.

- Approach scene cautiously, from upwind. Resist the urge to rush in and assist victims. Others cannot be helped until the situation has been fully assessed;

- Secure the scene. Without entering the hazard area, isolate the likely or suspected area and assure the safety of people and the environment. Keep people away from the scene and outside the safety perimeter;

- Identify the hazards (evaluate all available information);

- Assess the situation, considering the following:

    o Is there a fire, spill, or leak?

    o What are the weather conditions?

    o What is the terrain like?

    o Who/what is at risk: people, property, or the environment?

    o What actions should be taken? Is an evacuation or shelter in place necessary?

    o What resources (human and equipment) are required and are readily available?

    o What can be done immediately?

- Obtain help (Notify local EMS/911; notify responsible agencies for assistance from qualified personnel);

- Decide on site entry (Any efforts made to rescue persons or to protect property or the environment must be weighed against the possibility that you could become endangered);

- Alert fellow responders to the scene's safest entry route;

- Establish a command post and lines of communication;

- Rescue casualties where possible and evacuate if necessary;

- Maintain control of the site;

- Continually reassess the situation and modify the response accordingly;

- The first duty is to consider the safety of the people in the immediate area, including your own;

- Do not walk into or touch spilled material;

- Avoid inhalation of fumes, smoke and vapors, even if no Weapons of Mass Destruction are known to be involved; and

- Do not assume the gases or vapors are harmless because of a lack of smell – odorless gases or vapors may be harmful.

Contacts

- 911/dispatch to alert police/bomb squad and fire/HazMat;

- Chemtrec, a service of the Chemical Manufacturers Association, can be reached as follows: call CHEMTREC (24 hours) 800-424-9300; 703-527-3997 (For call originating elsewhere; collect calls are accepted);

- Chem-Tel, Inc, an emergency response communication service, can be reached as follows: call Chem-Tel, Inc (24 hours) 800-255-3924; 813-979-0626 (For calls originating elsewhere; collect calls are accepted),

- National Response Center (NRC) – operated by the U.S. Coast Guard, receives reports required when dangerous goods and hazardous substances are spilled. After receiving notification of an incident, the NRC will immediately notify the appropriate Federal On Scene Coordinator and concerned Federal agencies. Call NRC (24 hours) 800-424-8802;

- Military shipments: for assistance in incidents involving materials being shipped by, for, or to DoD, call one of the following 24 hour numbers:

  o U.S. Army Operations Center, for incidents involving explosives and ammunition. 703-697-0218 (call collect); or

  o Defense Logistics Agency, for incidents involving dangerous goods other than explosives and ammunition. 800-851-8061

References

The information in this Section was obtained from the following sources:

- FEMA Are You Ready? Chemical Threats: http://www.ready.gov/chemical-threats.

- Department of Homeland Security. *WMD Response Guidebook (LSU)*, (2006) Version 3.3.

Section 5-3: CBRNE – Biological Incidents

Overview

Delivery Methods

Indicators of Biological Agents

What to Do

CIARS Responder Concerns

References

Overview

A bioterrorism attack is the deliberate release of viruses, bacteria, or other biological agents used to cause illness or death in people, animals, or plants. These agents are typically found in nature, but it is possible that they could be changed to increase their ability to cause disease, make them resistant to current medicines, or to increase their ability to be spread into the environment. Biological agents can be spread through the air, through water, or in food. Terrorists may use biological agents because they can be extremely difficult to detect and do not cause illness for several hours to several days. Some bioterrorism agents, like the smallpox virus, can be spread from person to person and some, like anthrax, cannot.

The three basic groups of biological agents that would likely be used as weapons are bacteria, viruses, and toxins. Most biological agents are difficult to grow and maintain. Many break down quickly when exposed to sunlight and other environmental factors, while others, such as anthrax spores, are long lived.

Bioterrorism agents can be separated into three categories, depending on how easily they can be spread and the severity of illness or death they cause (Table 5-3-1 on the next page). Category A agents are considered the highest risk and Category C agents are those that are considered emerging threats for disease.

Table 5-3-1: Bioterrorism Agent Categories	
Category A	Include organisms or toxins that pose the highest risk to the public and national security because: • They can be easily spread or transmitted from person to person. • They result in high death rates and have the potential for major public health impact. • They might cause public panic and social disruption. • They require special action for public health preparedness.
Category B	These agents are the second highest priority because: • They are moderately easy to spread. • They result in moderate illness rates and low death rates. • They require specific enhancements of CDC's laboratory capacity and enhanced disease monitoring.
Category C	These third highest priority agents include emerging pathogens that could be engineered for mass spread in the future because: • They are easily available. • They are easily produced and spread. • They have potential for high morbidity and mortality rates and major health impact.
CDC Emergency Risk Communication Branch (ERCB), Division of Emergency Operations (DEO), Office of Public Health Preparedness and Response (OPHPR), *Bioterrorism Overview* (Atlanta: CDC, 12 Feb 2007); on the Internet at http://www.bt.cdc.gov/bioterrorism/overview.asp	

Biological weapons in the possession of hostile nations or terrorists pose unique and grave threats to the safety and security of the United States and our allies.

Biological weapons attacks:

- Can cause catastrophic harm by inflicting widespread injury and massive casualties;

- Cause severe economic disruption;

- Can mimic naturally occurring disease, potentially delaying recognition of an attack and creating uncertainty about whether one has even occurred;

- Can be mounted either inside or outside the United States; and

- Because some biological weapons agents are contagious, the effects of an initial attack could spread over a wide geographical region.

Disease outbreaks, whether natural or deliberate, respect no geographic or political borders. Once a biological weapons attack is detected, the speed and coordination of the Federal, State, Tribal, Territorial, and local response will be critical in mitigating the attack's lethal, medical, psychological, and economic consequences. Responses to biological weapons attacks depend on pre-attack planning and preparedness, capabilities to treat casualties, risk communications, physical control measures, medical countermeasures, and decontamination capabilities.

Following a biological weapons attack, all necessary means must be rapidly brought to bear to prevent loss of life, illness, psychological trauma, and to contain the spread of potentially contagious diseases. Provision of timely preventive treatments such as antibiotics or vaccines saves lives, protects scarce medical capabilities, preserves social order, and is cost effective.

Delivery Methods

Disease transmission may occur from:

- Direct contact with an infected individual or animal;

- An environmental reservoir (includes contaminated surface or atmospheric dispersion);

- An insect vector; or

- Contaminated food and water.

Indirect contact transmission may also occur where contaminated inanimate objects (fomites) serve as the vehicle for transmission of the agent.

Figure 5-3-1 below provides specific biological agent delivery methods:

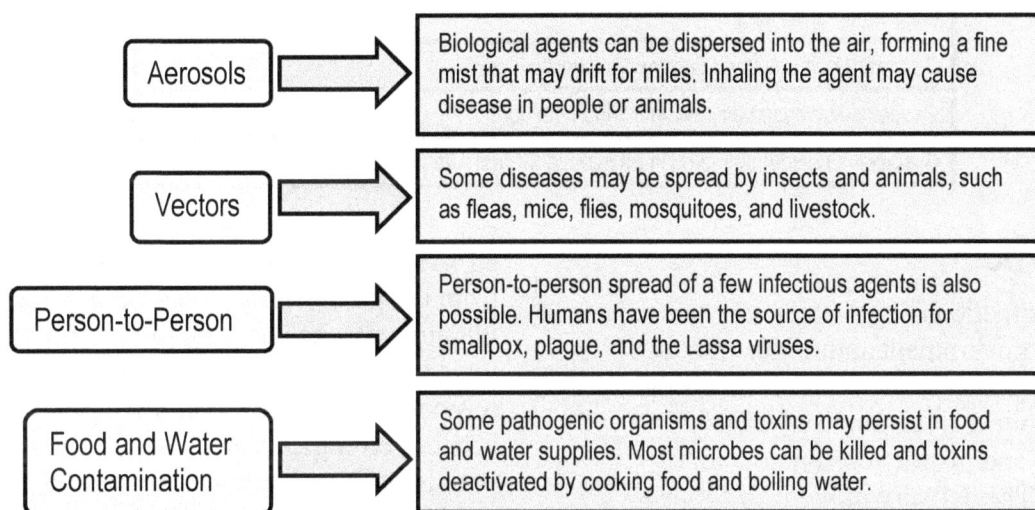

Aerosols	→	Biological agents can be dispersed into the air, forming a fine mist that may drift for miles. Inhaling the agent may cause disease in people or animals.
Vectors	→	Some diseases may be spread by insects and animals, such as fleas, mice, flies, mosquitoes, and livestock.
Person-to-Person	→	Person-to-person spread of a few infectious agents is also possible. Humans have been the source of infection for smallpox, plague, and the Lassa viruses.
Food and Water Contamination	→	Some pathogenic organisms and toxins may persist in food and water supplies. Most microbes can be killed and toxins deactivated by cooking food and boiling water.

Figure 5-3-1: Biological Agent Delivery Methods

Indicators of Biological Agents

Detection of a bioterrorism act against the civilian population may occur in several different ways and involve several different modalities:

- An attack may be surreptitious, in which case the first evidence of dissemination of an agent may be the presentation of disease in humans or animals. This could manifest either in clinical case reports to domestic or international public health authorities or in unusual patterns of symptoms or encounters within domestic or international health surveillance systems.

- A terrorist-induced infectious disease outbreak initially may be indistinguishable from a naturally occurring outbreak; moreover, depending upon the particular agent and associated symptoms, several days could pass before public health and medical authorities even suspect that terrorism may be the cause. In such a case, criminal intent may not be apparent until sometime after illnesses are recognized.

Table 5-3-2 on the next page lists indicators that may signify the presence of biological agents:

Table 5-3-2: Possible Indicators of Biological Agents
Unexplained dead or dying animals (sick or dying animals, people, or fish).
Unexplained casualties (unusual illness for region/area; definite pattern inconsistent with natural disease).
Unusual swarms of insects.
Suspicious bombing incident with little blast or fire damage.
Abandoned spray or dispersion devices.
Laboratory containers.
Biohazard cultures or culture media labels.
Casualty distribution aligned with wind direction.
Suspicious powder delivered in envelopes through the postal system.

What to Do

In some situations (e.g., Anthrax letters sent in 2001), government authorities may be alerted to potential exposure. If this is the case, pay close attention to all official warnings and instructions on how to proceed. The delivery of medical services for a biological event may be handled differently to respond to increased demand. The basic public health procedures and medical protocols for handling exposure to biological agents are the same as for any infectious disease. It is important to pay attention to official instructions via radio, television, and emergency alert systems.

CISAR Responder Concerns

The most practical method of initiating widespread infection using biological agents is through aerosolization, where fine particles are sprayed over or upwind of a target where the particles may be inhaled. An aerosol may be effective for some time after delivery, since it will be deposited on clothing, equipment, and soil. When the clothing is used later, or dust is stirred up, responding personnel may be subject to "secondary" contamination.

CISAR responders need to be protected from the hazard prior to rescuing victims. Planning for the response to bioterrorist acts must include provisioning CISAR responders with the appropriate PPE because biological agents may be able to use portals of entry into the body other than the respiratory tract (NG CBRNE responder protections will include an active immunization program to support victim rescue and management operations). Individuals may be infected by ingestion of contaminated food and water, or even by direct contact with the skin or mucous membranes through abraded or broken skin. Therefore, it is critical for CISAR responders to use appropriate protective clothing or commercially available Level C clothing and to protect the respiratory tract through the use of a mask with biological high-efficiency particulate air (HEPA) filters.

Exposure to biological agents, as noted above, may not be immediately apparent. Casualties may occur minutes, hours, days, or weeks after an exposure has occurred. The time required before signs and symptoms are observed is dependent on the agent used. Symptoms may include: fever, chills, headaches, muscular pain, fatigue, non-productive cough, profuse sweating, chills chest/muscle/joint pain, cramping, abdominal pain, and watery diarrhea (may be somewhat bloody). Even though the above symptoms may become evident, often the first confirmation will come from blood tests or other diagnostic means used by medical personnel.

References

The information in this Section was obtained from the following sources:

- Department of Homeland Security, *Homeland Security Presidential Directive 10: Biodefense for the 21st Century* (April 28, 2004); available online at http://www.nimsonline.com/ presidential_directives/hspd_10.htm.

- FEMA, Biological Incident Annex (August, 2008); available online at http://www.fema.gov/pdf/emergency/nrf /nrf_BiologicalIncidentAnnex.pdf.

- FEMA, "Are You Ready?" *Guidance on Biological Threats*; available online at http://www.ready.gov/biological-threats.

- *Center for Disease Control and Prevention* website, www.bt.cdc.gov.

This page intentionally left blank.

Section 5-4: CBRNE – Radiological Incidents

Introduction

A radiological incident is defined as an event or series of events, deliberate or accidental, leading to the release, or potential release, into the environment of radioactive material in sufficient quantity to warrant consideration of protective actions. Use of a *Radiological Dispersal Device (RDD)* or *Improvised Nuclear Device (IND)* is an act of terror that results in a radiological incident.

Overview: Radiological Dispersion Device (RDD)

An RDD is any device that causes the purposeful dissemination of radioactive material, across an area with the intent to cause harm, without a nuclear detonation.

RDDs pose a threat to public health and safety through the malicious spread of radioactive material by some means of dispersion. An RDD combines a conventional explosive device – such as a bomb – with radioactive material. The explosion adds an immediate threat to human life and property. Other means of dispersal, both passive and active, may be employed.

There is a wide range of possible consequences that may result from an RDD, depending on the type and size of the device and how dispersal is achieved. The consequences of an RDD may range from a small, localized area, such as a single building or city block, to large areas, conceivably several square miles. However, most experts agree that the likelihood of impacting a large area is low. In most plausible scenarios, the radioactive material would not result in acutely harmful radiation doses.

Hazards from fire, smoke, shock (physical, electrical, or thermal), shrapnel (from an explosion), hazardous materials, and other chemical or biological agents may also be present.

Terrorist use of an RDD, often called a "dirty nuke" or "dirty bomb," is considered far more likely than use of a nuclear explosive device. It is designed to scatter dangerous and sub-lethal amounts of radioactive material over a general area.

The primary purpose of an RDD is to cause psychological fear and economic disruption. Some devices could cause fatalities from exposure to radioactive materials. Depending on the speed at which the area of the RDD detonation was evacuated or how successful people were at sheltering-in-place, the number of deaths and injuries from an RDD might not be substantially greater than from a conventional bomb explosion. But the public perception and fear may be considerable.

The size of the affected area and level of destruction caused by an RDD would depend on the sophistication and size of the conventional bomb, type of radioactive material used, quality and quantity of radioactive material, and local meteorological conditions (primarily wind and precipitation). The area affected could be placed off-limits to the public for several months during cleanup efforts.

Overview: Improvised Nuclear Device (IND)

An IND is an illicit nuclear weapon bought, stolen, or otherwise originating from a nuclear State, or a weapon fabricated by a terrorist group from illegally obtained fissile nuclear weapons material that produces a nuclear explosion. The nuclear yield achieved by an IND produces extreme heat, powerful shockwaves, and prompt radiation that would be acutely lethal for a significant distance. It also produces radioactive fallout, which may spread and deposit over very large areas. If a nuclear yield is not achieved, the result would likely resemble an RDD in which fissile weapons material was utilized.

Hazards of Nuclear Devices

The extent, nature, and arrival time of these hazards are difficult to predict. The geographical dispersion of hazard effects will be influenced by the factors in Figure 5-4-1 on the next page.

| Size of the device. | ⟹ | A more powerful bomb will produce more distant effects. |

| Height above the ground when device was detonated. | ⟹ | This will determine the extent of blast effects. |

| Nature of the surface beneath the explosion. | ⟹ | Some materials are more likely to become radioactive and airborne than others. Flat areas are more susceptible to blast effects. |

| Existing meteorological conditions. | ⟹ | Wind speed and direction will affect arrival time of fallout; precipitation may wash fallout from the atmosphere. |

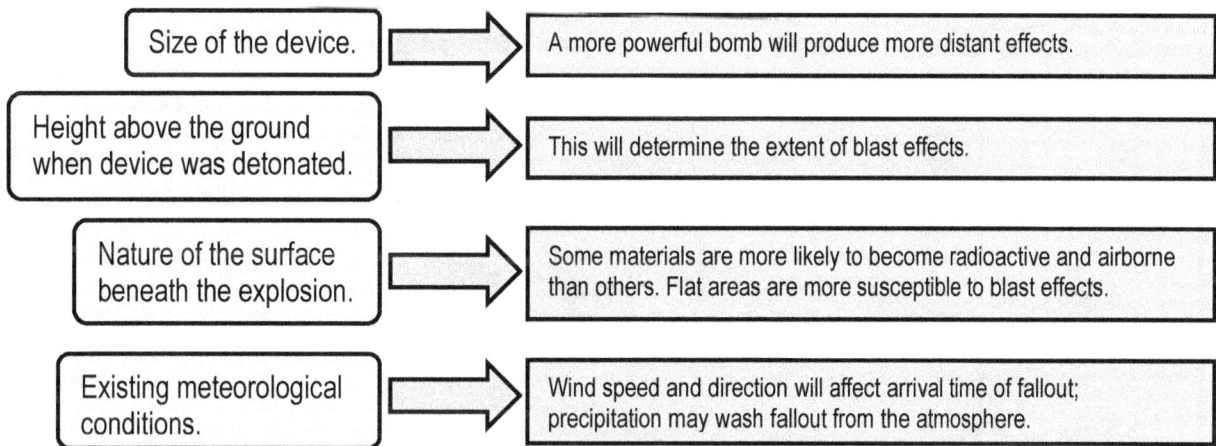

Figure 5-4-1: Factors Influencing Dispersion of Radiological Effects

Phases of Response

Typically, the response to an RDD or IND incident can be divided into three time phases (early, intermediate, and late) that are generally accepted as being common to all radiological incidents.

Although the phases cannot be represented by precise time periods, and may overlap, they provide a useful framework for the considerations involved in emergency response planning.

The phases are provided in Table 5-4-1 below and on the next page.

Table 5-4-1: RDD/IND - Phases of Response	
Early Phase (Emergency Phase)	Period at the beginning of the incident when immediate decisions for effective protective actions are required, and when actual field measurement data generally are not available. Exposure to the radioactive plume, short-term exposure to deposited radioactive materials, and inhalation of radioactive material are generally taken into account when considering protective actions for the early phase. The response during the early phase includes initial emergency response actions to protect public health and welfare in the short term, considering a time period for protective actions of hours to a few days. Priority should be given to lifesaving and first-aid actions. In general, early phase protective actions should be taken very quickly, and the protective action decisions can be modified later as more information becomes available.
	RDD. If an explosive RDD is deployed without warning, however, there may be no time to take protective actions to significantly reduce plume exposure. Also, in the event of a covert dispersal, discovery or detection may not occur for days or weeks, allowing contamination to be dispersed broadly by foot, vehicular traffic, wind, rain, or other forces.
	IND. If an IND explodes, there may only be time to make early phase protective action recommendations (e.g., evacuation, or shelter-in-place) many miles from the explosion to protect areas against exposure to fallout. Areas close to the explosion will be devastated, and communications and access will be extremely limited. Assistance will likely not be forthcoming or even possible for some hours. Self-guided protective actions are likely to be the best recourse for most survivors (e.g., evacuation perpendicular to the plume movement if it can be achieved quickly, or sheltering in a basement or large building for a day or more after the incident). Due to the lack of communication and access, outside guidance and assistance to these areas can be expected to be delayed. Therefore, response planning and public outreach programs are critical measures to meet IND preparedness objectives.
	It is during the Early Phase that the CISAR Responder will conduct most lifesaving operations.

Table 5-4-1 is continued on the next page.

Table 5-4-1: RDD/IND - Phases of Response (continued)	
Intermediate Phase	The intermediate phase of the response; may follow the early phase response within as little as a few hours. Usually assumed to begin after the incident source and releases have been brought under control and protective action decisions can be made based on measurements of exposure and radioactive materials that have been deposited as a result of the incident. Activities in this phase typically overlap with early and late phase activities, and may continue for weeks to many months, until protective actions can be terminated. Decisions must be made on the initial actions needed to recover from the incident, reopen critical infrastructure, and return to a state of relatively normal activity. Local officials must weigh public health and welfare concerns, potential economic effects, and many other factors when making decisions.
Late Phase	The late phase is the period when recovery and cleanup actions designed to reduce radiation levels in the environment to acceptable levels are commenced. This phase ends when all the remediation actions have been completed.

Radioactive Fallout

Even if individuals are not close enough to the nuclear blast to be affected by the direct impacts, they may be affected by radioactive fallout. Any nuclear blast results in some fallout. Blasts that occur near the earth's surface create much greater amounts of fallout than blasts that occur at higher altitudes. This is because the tremendous heat produced from a nuclear blast causes an up-draft of air that forms the familiar mushroom cloud. When a blast occurs near the earth's surface, millions of vaporized dirt particles also are drawn into the cloud. As the heat diminishes, radioactive materials that have vaporized condense on the particles and fall back to Earth. The phenomenon is called radioactive fallout. This fallout material decays over a long period of time, and is the main source of residual nuclear radiation.

Fallout from a nuclear explosion may be carried by wind currents for hundreds of miles if the right conditions exist. Effects from even a small portable device exploded at ground level can be potentially deadly.

Nuclear radiation cannot be seen, smelled, or otherwise detected by normal senses. Radiation can only be detected by radiation monitoring devices. This makes radiological emergencies different from other types of emergencies, such as floods or hurricanes. Monitoring can project the fallout arrival times, which will be announced through official warning channels. However, any increase in surface build-up of gritty dust and dirt should be a warning for taking protective measures.

Emergency Worker Guidelines

The response during the early phase includes initial emergency response actions to protect public health and welfare in the short term. Priority should be given to lifesaving, first-aid, and also include actions such as suppression of fires that could result in further loss of life.

For the purposes of this Guidance, "emergency worker" is defined as any worker who performs an early or intermediate phase work action. Table 5-4-2 on the next page shows the emergency worker guidelines for early phase emergency response actions.

Table 5-4-2: Emergency Worker Guidelines in the Early Phase[a]

Total effective dose equivalent (TEDE)[b] guideline	Activity	Condition
5 *rem*[c] (0.05 *Sv*)	All occupational exposures	All reasonably achievable actions have been taken to minimize dose.
10 *rem* (0.1 *Sv*)	Protecting valuable property necessary for public welfare (*e.g.*, a power plant).	• All appropriate actions and controls have been implemented; however, exceeding 5 *rem* (0.05 *Sv*) is unavoidable. • Responders have been fully informed of the risks of exposures they may experience. • Dose >5 *rem* (0.05 *Sv*) is on a voluntary basis. • Appropriate respiratory protection and other personal protection is provided and used. • Monitoring available to project or measure dose.
25 rem (0.25 Sv)[d]	Lifesaving or protection for large populations. It is highly unlikely that doses would reach this level in an RDD incident; however, worker doses higher than 25 *rem* (0.25 *Sv*) are conceivable in a catastrophic incident such as an IND incident.	• All appropriate actions and controls have been implemented; however, exceeding 5 *rem* (0.05 *Sv*) is unavoidable. • Responders have been fully informed of the risks of exposures they may experience. • Dose >5 rem (0.05 Sv) is on a voluntary basis. • Appropriate respiratory protection and other personal protection is provided and used. • Monitoring available to project or measure dose.

[a] In the intermediate and late phases, standard worker protections, including the annual 5 *rem* occupational dose limit, would normally apply.

[b] The projected sum of the effective dose equivalent from external radiation exposure and committed effective dose equivalent from internal radiation exposure.

[c] In the U.S., radiation doses are measured in units called *rem*. Under the metric system, dose is measured in units called sieverts (*Sv*). One *Sv* is equal to 100 *rem*.

[d] EPA's 1992 Protective Action Guide (PAG) Manual states that "Situations may also rarely occur in which a dose in excess of 25 *rem* for emergency exposure would be unavoidable in order to carry out a lifesaving operation or avoid extensive exposure of large populations." Similarly, the NCRP and ICRP raise the possibility that emergency responders might receive an equivalent dose that approaches or exceeds 50 *rem* (0.5 *Sv*) to a large portion of the body in a short time (Limitation of Exposure to Ionizing Radiation, National Council on Radiation Protection and Measures, NCRP Report 116 (1993a). If lifesaving emergency responder doses approach or exceed 50 rem (0.5 Sv) emergency responders must be made fully aware of both the acute and the chronic (cancer) risks of such exposure.

The emergency worker guidelines were developed for a wide range of possible radiological scenarios, from a small RDD that may impact a single building to an IND that could potentially impact a large geographic region. Therefore, the 5, 10 and 25 *rem* guidelines (Table 2 above) should not be viewed as inflexible limits applicable to the range of early phase emergency actions.

Because of the range of impacts and case-specific information needed, Incident Commands should to establish a "turn-back" dose level for responders.

With proper preparedness measures (training, personnel protective equipment, etc.) many radiological emergencies, even lifesaving operations, may be manageable within the 5 *rem* (0.05 *Sv*) annual occupational limit.

Moreover, ICs should make every effort to employ the "as low as reasonably achievable" (ALARA) principle after an incident.[1]

Still, in some incidents medically significant doses above the annual occupational 5 *rem* (0.05 *Sv*) dose limit may be unavoidable. For instance, in the case of a catastrophic incident, such as an IND, Incident Commanders may need to consider raising the lifesaving and valuable property (i.e., necessary for public welfare) emergency worker guidelines in order to prevent further loss of life and prevent the spread of massive destruction. Ensuring that emergency workers have full knowledge of the associated risks prior to initiating emergency action and medical evaluation of emergency workers after such exposure is essential.

Controlling Occupational Exposures and Doses to Emergency Workers

Appropriate measures should be taken to minimize radiation dose to emergency workers responding to an RDD or IND incident.

Emergency management officials responsible for an incident should take steps to keep all doses to emergency workers ALARA. Protocols for maintaining ALARA should include the following health physics and industrial hygiene practices:

---

[1] ALARA: As low as reasonably achievable; a process to control or manage radiation exposure to individuals and releases of radioactive material to the environment so that doses are as low as social, technical, economic, practical, and public welfare considerations permit.

- Minimize the time spent in the contaminated area (e.g., rotation of emergency responders);

- Maintain distance from sources of radiation;

- Shield radiation source;

- Use hazard controls that are applicable to the work performed;

- Properly select and use respirators and other PPE to minimize exposure to internally deposited radioactive materials (e.g., alpha and beta emitters); and

- Use prophylactic medications, when appropriate, that either blocks the uptake or reduces the retention time of radioactive material to the body.

To minimize the risks from exposure to ionizing radiation, all emergency responders should be trained and instructed to follow emergency response plans and protocols and be advised on how to keep exposures ALARA.

Understanding Radiation Risks

If there is the possibility that emergency workers would receive a radiation dose higher than the 5 *rem* (0.05 *Sv*) guideline, emergency workers should be trained to understand the risk associated with such doses, including a thorough explanation of the latent risks associated with receiving doses greater than 5 *rem* (0.05 *Sv*), and acute risks at higher doses.

Emergency workers should be fully aware of both the projected acute and chronic risks (cancer) they may incur in an emergency response action. Furthermore, emergency workers cannot be forced to perform a rescue action involving radiation doses above regulatory limits, and they should be given reasonable assurance that normal controls cannot be utilized to reduce doses to less than 5 *rem* (0.05 *Sv*). After the event, it is essential that emergency workers be provided with medical follow up.

The estimated risk of fatal cancer for healthy workers who receive a dose of 10 *rem* (0.10 *Sv*) is about 0.46 percent over the worker's lifetime (i.e., .405 fatal cancers per 1000 people, or 0.4-0.5 percent). The risk scales linearly. For workers who receive a dose of 25 *rem* (0.25 *Sv*), the risk is about 1.1 percent. The risk is believed to be greater for those who are younger at the time of exposure. For example, for 20-30 year olds the estimated risk of fatal cancer at 25 *rem* (1.75 percent) is about twice as large as the risk for 40-50 year olds (0.8 percent).

Above 50 *rem* (0.5 *Sv*) acute effects are possible. Where lifesaving actions may result in doses that approach or exceed 50 *rem* (0.50 *Sv*), such as in an IND incident, emergency workers need to have a full understanding of the potential acute effects of the expected radiation exposure, in addition to the risk of chronic effects. The decision to take these lifesaving actions must be based on the estimation that the human health benefits of the action exceed the safety and health risks to the emergency workers.

Symptoms of Radiation Exposure

The IC must strive to limit the radiation exposure of responders when conducting CISAR operations. From extensive studies of persons who suffered radiation exposure, the following information was obtained:

- The more radiation dose a person receives, the greater the chance of developing cancer;

- It is the chance of cancer occurring, not the severity of cancer, that increases as the radiation dose increases;

- Radiation induced cancers do not appear until years after the radiation dose is received; and

- The risk from radiation exposure will vary among individuals.

Acute Exposure

Acute exposure is exposure to a large, single dose of radiation, or a series of moderate doses received during a short period of time. Large acute doses can result from accidental or emergency exposures or from specific medical procedures (radiation therapy).

In most cases, a large acute exposure to radiation causes both immediate and delayed effects. Delayed biological effects can include cataracts, temporary or permanent sterility, cancer, and harmful genetic effects. For humans and other mammals, acute exposure to the whole body, if large enough, can cause rapid development of radiation sickness, evidenced by gastrointestinal disorders, bacterial infections, hemorrhaging, anemia, loss of body fluids, and electrolyte imbalance. An extremely high dose of acute radiation exposure can result in death within a few hours, days, or weeks.

Chronic Exposure

Chronic exposure is continuous or intermittent exposure to low doses of radiation over a long period of time. With chronic exposure, there is a delay between the exposure and the observed health effect. These effects can include cancer and other health outcomes such as benign tumors, cataracts, and potentially harmful genetic effects.

Electromagnetic Pulse (EMP)

In addition to other effects, a nuclear weapon detonated in or above the earth's atmosphere can create an electromagnetic pulse (EMP), a high-density electrical field.

An EMP acts like a stroke of lightning but is stronger, faster, and shorter.

What CISAR responders need to understand is that an EMP can seriously damage electronic devices connected to power sources or antennas. This includes communication systems, computers, electrical appliances, and automobile or aircraft electronic systems. The damage could range from a minor interruption to actual burnout of components. Most electronic equipment within 1,000 miles of a high-altitude nuclear detonation could be affected. Battery-powered radios with short antennas generally would not be affected. Although an EMP is unlikely to harm most people, it could harm those with pacemakers or other implanted electronic devices.

An EMP can severely limit a CISAR responder's ability to coordinate SAR operations, logistics, and the use of communications and medical support equipment. SAR operations should continue, however, they will be more difficult to perform.

Contacts

- 911/dispatch to alert police/bomb squad and fire/HazMat;

- Radiation Emergency Assistance Center/Training Site (REAC/TS) – provides medical care for radiation emergencies. They are equipped to deploy physicians, nurses, EMT paramedics, health physicists, radiobiologists and coordinators with the equipment and supplies needed to treat radiation injury. For info on treatment or training, contact REAC/TS at 865-576-3131 or EMERGENCY 24 hour assistance at 865-576-1005;

- Domestic Nuclear Detection Office Joint Analysis Center (DNDO/JAC) – DNDO improves the nation's capability to detect and report unauthorized attempts to import, possess, store, develop, or transport nuclear or radiological material. The JAC coordinates nuclear detection events and the technical support to Federal, state, and local authorities. The JAC can be contacted 24/7 at 877-363-6522;

- National Response Center (NRC) – operated by the U.S. Coast Guard, receives reports required when dangerous goods and hazardous substances are spilled. After receiving notification of an incident, the NRC will immediately notify the appropriate Federal On Scene Coordinator and concerned Federal agencies; call NRC (24 hours) 1-800-424-8802;

- CDC Emergency Preparedness Branch; 24-hour telephone number: 770-488-7100; and

- Military shipments: for assistance in incidents involving materials being shipped by, for, or to DOD, call one of the following numbers (24 hours);

    o 703-697-0218 (call collect) (U.S. Army Operations Center) for incidents involving explosives and ammunition; and

    o 1-800-851-8061 (toll free) (Defense Logistics Agency) for incidents involving dangerous goods other than explosives and ammunition.

Reference

The information in this Section was obtained from the following source: Department of Homeland Security, *Planning Guidance for Protection and Recovery Following Radiological Dispersal Device (RDD) and Improvised Nuclear Device (IND) Incidents*).

Section 5-5: Radiological Dispersal Device (RDD) Response Actions

CISAR Responder Protective Measures

Protecting the Injured and Exposed after RDD Detonation

Reference

CISAR Responder Protective Measures

As with any radiation, avoid or limit exposure. This is particularly true of inhaling radioactive dust that results from the explosion. If there is visual dust or other contaminants in the air, breathe through the cloth of your shirt or coat to limit your exposure. If you manage to avoid breathing radioactive dust, your proximity to the radioactive particles may still result in some radiation exposure.

In a known RDD environment, CISAR responders must protect themselves. The following guidance is provided for CISAR responders when conducting lifesaving operations after detonation of an RDD:

- Approach the release site with caution. Based on expert advice, position personnel, vehicles, and command post at a safe distance upwind and uphill of the site, if possible. Ensure your own physical safety. Look for fires, exposed high-voltage wires, sharp or falling objects, tripping hazards, or hazardous chemicals. Be alert for changing conditions.

- Wear a mask to reduce the dose from inhalation of radioactive dust. Ideally the mask should be a full face mask with a HEPA filter, but even breathing through a handkerchief or cloth will help. There will be little danger from radioactive gases, so a self-contained breathing mask, while effective, is not necessary unless there are other gasses or toxins present.

- Dust will collect on clothing. After you leave the contaminated environment, remove and discard clothing in a designated area. If you fail to remove clothing you will continue to receive radiation exposure and expose others. Wear loose fitting clothes covering as much of your body as possible. Any removable garment that will prevent the dust from coming into direct contact with your skin will suffice.

- Open wounds or abrasions must be protected from radioactive contamination.

- If running water or showers are available, full body rinsing with lukewarm water is advised. Even a fire hose may remove most contamination not already removed with the outer clothing.

- Do not eat, drink, or smoke while exposed to potentially radioactive dust or smoke. Due to the possibility of heat stress, drinking water may be necessary for people working in high temperatures with bulky protective clothing. If absolutely necessary to drink water, drink from a canteen or other closed container.

- If radiation monitoring instruments are available, wrap them in plastic bags to prevent their contamination. Use them to map the areas leading up to the highest dose rates. Enter the high dose rate areas only when necessary to save a life, make these entries as short as possible, and rotate the personnel who make these entries.

Protecting the Injured and Exposed after RDD Detonation

- Seriously injured people should be removed from the source of radiation, stabilized, and sent to hospitals first.

- After treatment of serious physical injuries, preventing the spread of the radioactive material or unnecessary exposure of other people is paramount. The following immediate response actions should be performed without waiting for any radiation measurements.

  o Establish an exclusion zone around the source. Mark the area with ropes or tapes. Reroute traffic.

  o Limit entry to rescue personnel only. Detain uninjured people who were near the event or who are inside the control zone until they can be checked for radioactive contamination, but do not delay treatment of injured people or transport to a hospital for this purpose.

  o Take action to limit or stop the release of more radioactive material, if possible, but delay cleanup attempts until radiation protection technicians are on the scene.

  o Tell nearby hospitals to expect the arrival of radioactively contaminated and injured people.

- Everyone near the scene should be checked for radioactive contamination. As soon as you can obtain radiation measuring equipment, establish a decontamination area for this purpose. Decontaminate people whose injuries are not life-threatening (broken arms, etc.) before sending them to hospitals. Do not send people without physical injuries to hospitals.

- Record keeping is as important for the long-term health of the victims as it is for the emergency responders. Record contact information for all exposed people so they can be given medical examinations later. The Department of Health and Human Services will request this information later.

Reference

The information in this Section was obtained from the following source: Centers for Disease Control, *Casualty Management After a Deliberate Release of Radioactive Material Fact Sheet* (September 19, 2007).

Section 5-6: CBRNE – Improvised Nuclear Device (IND) Response Actions

CISAR Responder Protective Measures

Protecting the Injured and Exposed after IND Detonation

Reference

CISAR Responder Protective Measures

The following guidance is provided for CISAR responders when conducting lifesaving operations after detonation of an IND:

- Stay away from ground zero. Enter the surrounding area only to save lives. The radiation levels may be very high.

- Ensure your own physical safety. Look for fires, exposed high voltage wires, sharp or falling objects, tripping hazards, or hazardous chemicals. Be alert for changing conditions.

- Wear a mask to reduce the dose from inhalation of radioactive dust. Ideally the mask should be a full face mask with a HEPA filter, but even breathing through a wet handkerchief or cloth will help. There will be little danger from radioactive gases, so a self contained breathing mask, while effective, is not necessary.

- Dust will collect on your clothing. After you leave the contaminated environment, remove and discard clothing in a designated area. If you fail to remove clothing you will continue to receive radiation exposure and expose others. Wear loose fitting clothes covering as much of your body as possible. Any removable garment that will prevent the dust from coming into direct contact with your skin will suffice.

- Open wounds or abrasions must be protected from radioactive contamination.

- If running water or showers are available, full body rinsing with lukewarm water is advised. Even a fire hose may remove most contamination not already removed with the outer clothing.

- Wash vehicles before permitting them to leave the scene, except for emergency vehicles performing life-saving functions.

- Do not eat, drink, or smoke while exposed to potentially radioactive dust or smoke. Due to the possibility of heat stress, drinking water may be necessary for people working in high temperatures with bulky protective clothing. If absolutely necessary to drink water, drink from a canteen or other closed container.

- Use the form attached to this brochure to record contact information for all exposed workers so they can be given medical examinations later. The Department of Health and Human Services will request this information later.

- Wash thoroughly with lukewarm water as soon as possible after leaving the area, even if you decontaminated before leaving the scene.

Protecting the Injured and Exposed after IND Detonation

- Physical injuries are more serious than radioactive contamination. Deal with life-threatening conventional injuries first. When the patients are stable, deal with radioactive contamination. Patients who were treated and are now stable should be evacuated from radiation areas.

- Tell nearby hospitals to expect the arrival of radioactively contaminated and injured people.

- Victims will have radioactive dust on their clothing. If many people are covered with dust, it will not be feasible to conduct a careful survey of each person. Assume all of the dust is radioactive. Set up a facility where each person can remove and discard clothing in a designated area, wash as thoroughly as possible, and don coveralls or wrap in blankets. This facility should be upwind and far enough from ground zero to prevent radiation levels from interfering with monitoring of patients.

- Many people without apparent injuries will leave the scene. Make public service announcements on radio and television advising these people to bag their clothes, place the clothes outdoors, and wash thoroughly. People experiencing nausea, vomiting, reddening of the skin, or unexplained lesions should be advised to report to a hospital immediately and request a checkup for Acute Radiation Syndrome (ARS).

- Table 5-6-1 on the next page can be used to document personal information for future radiological monitoring.

Note: Information on Acute Radiation Syndrome (ARC) can be found at the Center for Disease Control website: http://www.bt.cdc.gov/radiation/ars.asp.

Reference

The information in this Section was obtained from the following source: Centers for Disease Control, *Casualty Management After Detonation of a Nuclear Weapon In an Urban Area Fact Sheet* (May 20, 2005)

Table 5-6-1: Radiological Monitoring Personal Information

(All people exposed to radiation, both emergency responders and members of the public, will require future medical monitoring. Collect the following information from each person who was in the affected area and retain it until requested by the Department of Health and Human Services)

Date _____ **Name** _____

First Responder? ___ **Civilian?** ___

Home address _____
Street or P.O. Box, City, State, Zip Code

Telephone(s) _____

Email address _____

Gender: M F   **Date of birth (or approximate age)** _____

(The following information would be useful for further medical evaluation. Collect the information only if you have time without delaying treatment of the injured.)

Date of exposure _____

Time of exposure _____

Location _____
Describe the location where the person received his or her exposure

What was the exposure? _____

Is the level of exposure known? Y N _____

Device(s) used to determine level of exposure _____

Duration of exposure _____

Did person have any open wounds? Y N

Did person use respiratory protection? Y N   What kind? _____

Did person eat or drink while in the area? Y N

Did you find any external contamination on the person? Y N

<u>Emergency contact information</u>

Name _____

Address _____
(Street or P.O. Box, City, state, Zip Code)

Telephone _____

Employer _____

This page intentionally left blank.

```
┌─────────────────────────────────────┐
│                                      │
│       Appendix A: ESF #9             │
│        (February, 2011)              │
│                                      │
└─────────────────────────────────────┘
```

(Below is ESF #9 as of February, 2011.)

ESF COORDINATOR:

    Department of Homeland Security/Federal Emergency Management Agency

PRIMARY AGENCIES:

    Department of Homeland Security/Federal Emergency Management Agency

    Department of Homeland Security/U.S. Coast Guard

    Department of the Interior/National Park Service

    Department of Defense

SUPPORT AGENCIES:

    Department of Agriculture

    Department of Commerce

    Department of Defense

    Department of Health and Human Services

    Department of Homeland Security

    Department of the Interior

    Department of Justice

    Department of Labor

    Department of Transportation

    National Aeronautics and Space Administration

    U.S. Agency for International Development

INTRODUCTION

Purpose

Emergency Support Function (ESF) #9 – Search and Rescue (SAR) rapidly deploys Federal SAR resources to provide lifesaving assistance to State, Tribal, and local authorities, to include local SAR Coordinators and Mission Coordinators, when there is an actual or anticipated request for federal SAR assistance.

Scope

During incidents or potential incidents requiring a unified SAR response, Federal SAR responsibilities reside with ESF #9 primary agencies that provide timely and specialized SAR capabilities. Support agencies provide specific capabilities or resources that support ESF #9. Federal SAR response operational environments are classified as:

- Structural Collapse (Urban) Search and Rescue (US&R).

- Maritime/Coastal/Waterborne Search and Rescue.

- Land Search and Rescue.

SAR services include distress monitoring, incident communications, locating distressed personnel, coordination, and execution of rescue operations including extrication and/or evacuation, along with providing medical assistance and civilian services through the use of public and private resources, to assist persons and property in potential or actual distress.

Structural Collapse (Urban) Search and Rescue (US&R)

Primary Agency: Department of Homeland Security (DHS)/Federal Emergency Management Agency (FEMA)

Operational Overview: US&R includes operations for natural and manmade disasters and catastrophic incidents, as well as other structural collapse operations that primarily require DHS/FEMA US&R task force operations. The National US&R Response System integrates DHS/FEMA US&R task forces, Incident Support Teams (ISTs), and technical specialists. The Federal US&R response integrates DHS/FEMA task forces in support of unified SAR operations conducted following the U.S. National Search and Rescue Plan (NSP). (The NSP is the policy guidance of the signatory federal departments and agencies for coordinating SAR services to meet domestic needs and international commitments.)

DHS/FEMA develops national US&R policy, provides planning guidance and coordination assistance, standardizes task force procedures, evaluates task force operational readiness, funds special equipment and training within available appropriations, and reimburses, as appropriate, task force costs incurred as a result of ESF #9 deployment.

The National US&R Response System is prepared to deploy and initiate operations immediately in support of ESF #9. The task forces are staffed primarily by emergency services personnel who are trained and experienced in collapsed structure SAR operations and possess specialized expertise and equipment. Upon activation under the *National Response Framework (NRF)*, DHS/FEMA US&R task forces are considered federal assets under the Homeland Security Act of 2002, the Robert T. Stafford Disaster Relief and Emergency Assistance Act, and other applicable authorities.

ISTs provide coordination and logistical support to US&R task forces during emergency operations. They also conduct needs assessments and provide technical advice and assistance to state, tribal, and local government emergency managers.

DHS/FEMA reimburses the parent Sponsoring Agencies for US&R task forces for authorized US&R deployments. DHS/FEMA is authorized to reimburse such activities when there is a Stafford Act declaration or in anticipation of a declaration. For non-Stafford Act US&R deployments, the federal department or agency requesting US&R assistance reimburses DHS/FEMA following provisions contained in the Financial Management Support Annex. DHS/FEMA uses the funding provided by the requesting federal department or agency to reimburse the Sponsoring Agency for the task forces.

Maritime/Coastal/Waterborne Search and Rescue

Primary Agency: DHS/U.S. Coast Guard (USCG)

Operational Overview: Maritime/coastal/waterborne SAR includes operations for natural and manmade disasters that primarily require DHS/USCG air, cutter, boat, and response team operations. The federal maritime/coastal/waterborne SAR response integrates DHS/USCG resources in support of unified SAR operations conducted per the NSP.

DHS/USCG personnel are trained and experienced in maritime/coastal/waterborne SAR operations and possess specialized expertise, facilities, and equipment for conducting an effective response to distress situations. DHS/USCG develops, maintains, and operates rescue facilities for SAR in waters subject to U.S. jurisdiction and is designated the primary agency for maritime/coastal/waterborne SAR under ESF #9. In addition, DHS/USCG staffing at Area, District, and local Sector Command Centers promotes interagency coordination with State, Tribal, and local emergency managers during incidents requiring a unified SAR response in which maritime/coastal/waterborne SAR resources allocation are required.

Land Search and Rescue

Primary Agency: Department of the Interior (DOI)/National Park Service (NPS); Department of Defense (DoD)

Operational Overview: Land SAR includes operations that require aviation and ground forces to meet mission objectives, other than maritime/coastal/waterborne and structural collapse SAR operations as described above. Land SAR primary agencies integrate their efforts to provide an array of diverse capabilities under ESF #9.

DOI/NPS possesses SAR resources that are specially trained to operate in various roles including ground search, small boat operations, swiftwater rescue, helo-aquatic rescue, and other technical rescue disciplines. DOI/NPS maintains preconfigured teams that include personnel and equipment from DOI/NPS, U.S. Fish and Wildlife Service, U.S. Geological Survey, Bureau of Indian Affairs, and other DOI components in planning for ESF #9

When requested, DoD, through U.S. Northern Command (USNORTHCOM) and/or U.S. Pacific Command (USPACOM), coordinates facilities, resources, and special capabilities that conduct and support air, land, and maritime SAR operations according to applicable directives, plans, guidelines, and agreements. Per the NSP, the U.S. Air Force and USPACOM provide resources for the organization and coordination of civil SAR services and operations within their assigned SAR regions and, when requested, to assist Federal, State, Tribal, and local authorities.

DoD's role as a primary agency is based on SAR Coordinator responsibilities stipulated in the NSP and is generally limited to a coordination function.

DoD designation as a primary agency in ESF #9 is not clearly defined in current statutes, authorities, or DoD policies. Under the *NRF*, DoD assists civil authorities by conducting SAR missions on a reimbursable basis pursuant to the Stafford Act or Economy Act, as appropriate.

If DoD SAR capabilities deploy at the direction of the Air Force Rescue Coordination Center in support of the NSP, and subsequently if the Stafford Act is invoked, those capabilities are administered by the *NRF* and ESF #9. As soon as practical, a DHS/FEMA or other department/agency mission assignment are submitted to and approved by DoD for those capabilities' continued support.

Policies

Federal SAR responders assist and support state, tribal, and local SAR capabilities in incidents requiring a coordinated federal response. No provision of this annex is to be construed as an obstruction to prompt and effective action by any agency to assist persons in distress.

ESF #9 SAR operations are conducted following the *NRF* and NSP, and the U.S. National SAR Supplement (NSS), Catastrophic Incident SAR (CISAR) Addendum, and other addenda that define SAR responsibilities and provide guidance to the federal departments and agencies with civil SAR mandates.

If an affected State, Tribal, or local government publishes guidance or a plan for conducting unified SAR operations, that guidance or plan takes precedence.

State-to-State SAR assistance is requested by the affected state through the Emergency Management Assistance Compact (EMAC). Other local SAR resources are requested by the affected locality through mutual aid and assistance agreements. Non-Federal SAR resources are, as appropriate, incorporated into any coordinated SAR operations.

State, Tribal, and local authorities are responsible for SAR within their respective jurisdictions and typically designate a SAR Coordinator to provide integration and coordination of all SAR services.

The following provides primary agency statutory authorities and policy guidance:

- Homeland Security Act of 2002 (as amended); 6 U.S.C. 722: This section codified US&R as a system within FEMA, "There is in the Agency a system known as the Urban Search and Rescue Response System."

- Stafford Act; 42 U.S.C. 5121-5207: This act authorizes the President (assisted by DHS/FEMA) to declare major disasters and emergencies in the United States and provide assistance to State and local governments. The President may use the services of state and local governments for the purposes of the act, which includes addressing immediate threats to life and property (e.g., SAR operations).

- 14 U.S.C. 2: This section requires the U.S. Coast Guard to develop, establish, maintain, and operate rescue facilities for the promotion of safety on, under, and over the high seas and waters subject to the jurisdiction of the United States.

- 16 U.S.C. 1b(1): This section gives DOI/NPS authority to provide emergency rescue, firefighting, and cooperative assistance to public safety agencies for related purposes outside of the National Park System.

- Economy Act; 31 U.S.C. 1535-1536 (2007): This act authorizes Federal Departments and Agencies to provide goods or services, on a reimbursable basis, to other Federal Departments and Agencies.

- 32 U.S.C.: This title authorizes the National Guard to perform DoD-funded activities while remaining under the control of the Governor.

- Post-Katrina Emergency Management Reform Act; P.L. 109-295 (2006): This act expands the scope of ESF #9 from only urban SAR to include all types of SAR activities. Follow on congressional guidance establishes the organizational structure. It codified US&R as a system within FEMA in the Homeland Security Act of 2002 (as amended). It also mandated FEMA to develop a Federal response capability to rapidly and effectively deliver assistance essential to saving lives or protecting property or public health and safety and to carry out the mission of FEMA by conducting emergency operations to save lives and property.

- National Search and Rescue Plan (NSP): The NSP is the policy guidance of the signatory Federal departments and Agencies for coordinating SAR services to meet domestic needs and international commitments.

- National SAR Supplement (NSS): This document provides implementation guidance on the International Aeronautical and Maritime Search and Rescue Manual and the NSP.

- Catastrophic Incident SAR (CISAR) Addendum to the NSS: This document provides a description of the unified SAR response to catastrophic incidents, guides Federal authorities involved in the response, and informs State, Tribal, and local authorities on what to expect of/from Federal SAR responders.

- DoD Support to Civil Search and Rescue (DODD 3003.01): This directive states that DoD shall support domestic civil authorities by providing civil SAR service to the fullest extent practicable on a noninterference basis with primary military duties.

- Military Support to Civil Authorities (DODD 3025.1): This directive identifies the policy and responsibilities by which DoD responds to major disasters or emergencies per the Stafford Act and other authorities.

- Military Assistance to Civil Authorities (DODD 3025.15): This directive states that DoD shall cooperate with and provide military assistance to civil authorities, as directed by and consistent with applicable law, Presidential directives, and Executive orders.

CONCEPT OF OPERATIONS

General

DHS/FEMA activates ESF #9 when an incident is anticipated or occurs that may result in a request for a unified SAR response to an affected area. The ESF #9 response is scalable to meet the specific needs of each incident, based upon the nature and magnitude of the event, the suddenness of onset, and the capability of local SAR resources. Response resources are drawn from ESF #9 primary and support agencies.

As required, the primary agencies are represented at the National Response Coordination Center (NRCC), Joint Field Office (JFO), and state, tribal, and local Emergency Operations Centers (EOCs).

For each incident requiring federal SAR support, DHS/FEMA designates the overall primary agency for that particular ESF #9 SAR response. Designation is dependent upon incident circumstances and the type of response required.

The designated overall primary agency coordinates integration of federal SAR resources, including support agency resources, in support of the requesting federal, state, tribal, or local SAR authority.

All ESF #9 agencies provide support to the designated overall primary agency, as required.

ORGANIZATION

For incidents where DHS/FEMA is the overall primary agency, ESF #9 SAR operations are conducted following the National US&R Response System manuals, directives, NSP, NSS, and CISAR Addendum.

For incidents where DHS/USCG is the overall primary agency, ESF #9 SAR operations are conducted following the SAR response structure as outlined in the NSP, NSS, CISAR Addendum, USCG SAR Addendum, and other USCG directives.

For incidents where DOI/NPS and/or DoD are the overall primary agency, ESF #9 SAR operations are conducted following the SAR response structure as outlined in the NSP, NSS, CISAR Addendum, and other relevant DOI/NPS and DoD SAR procedures, directives, and manuals.

RESPONSIBILITIES

ESF #9 Coordinator: DHS/FEMA

As ESF #9 coordinator, DHS/FEMA:

- Designates the overall primary agency responsible for the coordination of federal SAR operations.

- Coordinates with all other ESFs, as required.

Primary Agencies

For every incident, DHS/FEMA assesses the specific SAR requirements and assigns one of the four primary agencies as the overall primary agency for SAR for that particular incident.

When in the overall primary agency role for a particular incident, that organization conducts the following actions:

- Coordinates planning and operations between primary and support agencies.

- Coordinates resolution of conflicting operational demands for SAR response resources.

Primary Agency: DHS/FEMA

DHS/FEMA serves as the overall primary agency to accomplish the ESF #9 mission during structural collapse SAR operations in incidents requiring a coordinated Federal response.

For incidents in which it is designated the overall primary agency, DHS/FEMA:

- Manages US&R task force and IST deployments in the affected area.

- Coordinates logistical support for US&R assets during field operations.

- Coordinates the provisioning of additional support assets.

- Coordinates with Federal, State, Tribal, and local designated SAR authorities to integrate Federal SAR resources.

- As required, provides representation at the NRCC, JFO, and State, Tribal, and local EOCs.

Provides incident reports, assessments, and situation reports as required.

Primary Agency: DHS/USCG

DHS/USCG serves as the overall primary agency to accomplish the ESF #9 mission during maritime/coastal/waterborne SAR operations in incidents requiring a coordinated federal response.

For incidents in which it is designated the overall primary agency, DHS/USCG:

- Manages USCG SAR resources in the affected area.

- Coordinates the provisioning of additional support assets.

- Coordinates with Federal, State, Tribal, and local designated SAR authorities to integrate Federal SAR resources.

- As required, provides representation at the NRCC, JFO, and State, Tribal, and local EOCs.

- Provides incident reports, assessments, and situation reports, as required.

Primary Agency: DOI/NPS

DOI/NPS and DoD share responsibility as the overall primary agency for a particular incident to accomplish the ESF #9 mission during land SAR operations in incidents requiring a coordinated Federal response.

For incidents in which it is designated the overall primary agency, DOI/NPS:

- Manages DOI/NPS land SAR resources in the affected area.

- Coordinates the provisioning of additional support assets.

- Coordinates with Federal, State, Tribal, and local designated SAR authorities to integrate Federal SAR resources.

- Coordinates logistical support for DOI/NPS resources during field operations.

- As required, provides representation at the NRCC, JFO, and State, Tribal, and local EOCs.

- Provides incident reports, assessments, and situation reports as required.

Primary Agency: DOD

DoD and DOI/NPS share responsibility as the overall primary agency for accomplishing the ESF #9 mission during land SAR operations in incidents requiring a coordinated federal response.

For incidents in which it is designated the overall primary agency, DoD, through USNORTHCOM and USPACOM:

- Manages DoD SAR resources in the affected area.

- Coordinates the provisioning of additional support assets.

- Coordinates with federal, state, tribal, and local designated SAR authorities to integrate Federal SAR resources.

- As required, provides representation at the NRCC, JFO, and State, Tribal, and local EOCs.

- Provides incident reports, assessments, and situation reports as required.

(Table A-1)

Support Agencies

Agency	Functions
Department of Agriculture	**Forest Service** • Develops standby agreements to provide equipment and supplies from the National Interagency Fire Center (NIFC) Cache System at the time of deployment. • Develops contingency plans for use of NIFC contract aircraft during incidents. • If available, provides equipment and supplies from the NIFC Cache System and use of NIFC contract aircraft.
Department of Commerce	**National Oceanic and Atmospheric Administration** • Acquires and disseminates weather data, forecasts, and emergency information. • Provides weather information essential for efficient SAR. • Predicts pollutant movement and dispersion over time (marine and atmospheric). • Assesses areas of greatest hazard following a marine or atmospheric release. • Provides satellite services for detecting and locating persons in potential or actual distress in the wilderness, maritime, and aeronautical environments.

Agency	Functions
Department of Defense	**National Geospatial-Intelligence Agency (NGA)** • Coordinates and manages the timely tasking, acquisition, analysis, and delivery of satellite imagery or imagery-derived products as directed by the primary agency. • Provides expert analysis of imagery to determine damage levels and other elements of essential information, as needed. • Provides technical expertise/analysis from other imagery sources, if such expertise resides within DoD/NGA. • Provides mobile geospatial intelligence including technical experts (imagery analysts and geospatial analysts) and robust communications to support SAR field teams or other DHS/FEMA field teams, as requested by the primary agency. • Provides imagery-derived and geospatial intelligence analysis in preparation for potential disasters or emergencies. • Coordinates the release and dissemination of DoD/NGA products and/or data following applicable security classifications, licensing, copyright agreements, and limited distribution restrictions. **U.S. Army Corps of Engineers (USACE)** • Under the *NRF*, supports the ESF #9 SAR mission by developing, training, and equipping USACE personnel to operate as support to the DHS/FEMA US&R Task Forces. • Through Technical Assistance Structural Engineers (TASEs), supports DHS/FEMA and other agency efforts requiring structural engineering expertise (e.g., evaluate, design, construct, or repair of buildings, bridges, and critical facilities). • Through Structural Safety Assessment Planning and Response Teams (SSA PRTs), provides habitability inspections as required, to support response and recovery efforts for building safety evaluations.
Department of Health and Human Services	**National Disaster Medical System (NDMS)** • Through ESF #8 – Public Health and Medical Services, provides support to ESF #9 primary agencies, including liaisons; medical supplies, equipment, and pharmaceuticals; supporting personnel; and veterinary support. • Provides NDMS personnel to support medical field operations and evacuation. **Indian Health Service:** Maintains specialized response teams to support the medical care of American Indian and Alaska Native people.
Department of Homeland Security	**Customs and Border Protection** • Maintains Border Patrol Search, Trauma, and Rescue (BORSTAR) teams, which are highly specialized units capable of responding to emergency SAR situations anywhere in the United States. • Maintains air and marine assets to support SAR transportation operations.
Department of the Interior	**U.S. Geological Survey:** Provides personnel with appropriate technical disciplines and specialized technology to support geospatial analysis and mapping products in support of ESF #9 primary agencies.
Department of Justice	• Coordinates force protection as required. • Provides assistance with the development and maintenance of tort liability claims coverage for US&R task force and IST personnel engaged in mobilization, deployment, and field operations.

Agency	Functions
Department of Labor	• The Mine Safety and Health Administration provides mine rescue teams, mobile command centers, seismic location systems, TV probe systems, gas sampling analysis, and robot explorers. • The Department of Labor Employment Standards Administration, through its Federal Employees' Compensation Program, provides workers compensation guidance, claims resolution, and coverage for US&R task force and IST personnel while they are engaged in mobilization, deployment, and field operations. • The Occupational Safety and Health Administration implements procedures contained in the Worker Safety and Health Support Annex to provide onsite technical assistance, including the evaluation of SAR team exposure to hazardous substances and the dangers of structural collapse.
Department of Transportation	**Federal Aviation Administration** • Is delegated sole authority to manage the National Airspace System (NAS), which includes operating a safe, secure, and efficient air traffic system; oversight and certification of aircraft and airmen; regulation of airspace; promotion of air commerce; and the support of America's national defense (49 U.S.C.). • Supports activities to protect and recover NAS operations.
National Aeronautics and Space Administration	• Provides personnel in appropriate technical disciplines (e.g., its Disaster Assistance and Rescue Team). • Provides temporary use of facilities for mobilization centers and staging areas for SAR assets.
U.S. Agency for International Development	Manages the support of international SAR teams to a domestic U.S. disaster following a Stafford Act Declaration, under the International Assistance System Concept of Operations (IAS CONOPS), and in support of the NRF's International Coordination Support Annex (ICSA.)
Department of State	If FEMA does not active the IAS and proactive offers of assistance from foreign countries or international/multilateral organizations are received, the State Department may still designate the State Task Force (STF) (or, if the STF has not been established, designate a lead bureau or the Operations Center's Crisis Management Support [CMS] office) as the sole entity within DOS responsible for coordinating formal offers of international assistance, and request all offers be forwarded to the STF (or, as appropriate, the lead bureau or CMS office) for dispensation.

This page intentionally left blank.

Index

Other Selected Books from dbS Productions

Land Search and Rescue Manual

Volume I: Land Search and Rescue Addendum by NSARC
ISBN 978-1-879471-43-6 2012 304p 8.5 x 11 paperback $20.00
e-book   ISBN 978-1-879471-44-3 $9.99

Volume II: Catastrophic Incident Search and Rescue Addendum by NSARC
ISBN 978-1-879471-45-0 2012 336p 8.5 x 11 paperback $25.00
e-book   ISBN 978-1-879471-46-7 $9.99

Volume III: Incident Command System Field Operations Guide for Search and Rescue by Robert J. Koester
ISBN 978-1-879471-47-4 2012 216p 4.25 x 7 wiro $25.00
e-book   ISBN 978-1-879471-48-1 $9.99

Volume IV: Management Field Operations Guide for Search and Rescue
by Robert J. Koester
ISBN 978-1-879471-49-8 2012 xxxp 4.25 x 7 paperback $25.00
e-book   ISBN 978-1-879471-50-4 $9.99

Volume V: Tactical Field Operations Guide for Search and Rescue
by Robert J. Koester
ISBN 978-1-879471-51-1 2012 xxxp 4.25 x 7 paperback $25.00
e-book   ISBN 978-879471-52-8 $9.99

Lost Person Behavior: A Search and Rescue Guide on Where to Look – for Land, Air and Water by Robert J. Koester
ISBN 978-1-879471-39-9 2008   416p 5.5 x 8.5 paperback with semi-concealed wiro $25.00

The Handbook for Aviation Survival Sense by Robert "Skip" Stoffel and Brett C. Stoffel
2009 160p 4.25 x 7 waterproof paper wiro $28.00

Urban Search: Managing Missing Person Searches in the Urban Environment
by Christopher S. Young and John Wehbring
ISBN 978-1-879471-38-2 2007 352p 5.5 x 8.5 paperback $25.00

Foundations for Awareness, Signcutting and Tracking by Robert Speiden
ISBN: 978-0-9817686-0-1 2009 268p 8.5 x 11 paperback $35.00

Incident Commander for Ground Search and Rescue: A Training Course for Incident Commanders by Robert J. Koester
Student Manual   ISBN 978-1-879471-22-1 1997 230p 8.5 x 11 $35.00
Instructor Manual ISBN 978-1-879471-21-4 1999 290p 8.5 x 11 $75.00

Other Selected Books from dbS Productions

Lost Alzheimer's Disease Search Management: A Law Enforcement Guide to Managing the Initial Response and Investigation of the Missing Alzheimer's Disease Subject by Robert J. Koester
ISBN 978-1-879471-34-4   1999 $100.00

Man-Trackers & Dog Handlers in Search & Rescue by Greg Fuller, Ed Johnson and Robert J. Koester
ISBN 978-1-879471-31-1   2000   98p   5 3/8 x 8 1/4   paperback $10.00

Fatigue: Sleep Management During Disasters and Sustained Operations by Robert J. Koester
Instructor Manual with CD   ISBN 978-1-879471-17-7   1997   8.5 x 11 $110.00
Student Manual   ISBN 978-1-879471-18-4 1997   58p   5.5 x 8 $10
e-book   ISBN 978-1-879471-19-1 $10